Unsettling Nature

Under the Sign of Nature: Explorations in Ecocriticism
Serenella Iovino, Kate Rigby, and John Tallmadge, Editors
Michael P. Branch and SueEllen Campbell, Senior Advisory Editors

Unsettling Nature

ECOLOGY, PHENOMENOLOGY, AND THE
SETTLER COLONIAL IMAGINATION

Taylor Eggan

UNIVERSITY OF VIRGINIA PRESS
CHARLOTTESVILLE AND LONDON

University of Virginia Press
© 2022 by the Rector and Visitors of the University of Virginia
All rights reserved
Printed in the United States of America on acid-free paper

First published 2022

9 8 7 6 5 4 3 2 1

Library of Congress Cataloging-in-Publication Data

Names: Eggan, Taylor, author.
Title: Unsettling nature : ecology, phenomenology, and the settler colonial imagination / Taylor Eggan.
Description: Charlottesville ; London : University of Virginia Press, 2021. | Series: Under the Sign of Nature: Explorations in Ecocriticism / Serenella Iovino, Kate Rigby, and John Tallmadge, editors | Includes bibliographical references and index.
Identifiers: LCCN 2021030574 (print) | LCCN 2021030575 (ebook) | ISBN 9780813946832 (Hardcover : acid-free paper) | ISBN 9780813946849 (Paperback : acid-free paper) | ISBN 9780813946856 (ebook)
Subjects: LCSH: Environmentalism—Philosophy. | Ecology—Philosophy. | Phenomenology.
Classification: LCC GE195 .E335 2021 (print) | LCC GE195 (ebook) | DDC 304.2/8—dc23
LC record available at https://lccn.loc.gov/2021030574
LC ebook record available at https://lccn.loc.gov/2021030575

Cover art: *Green River Butte*, photocrom, Detroit Photographic Co., ca. 1898. (Library of Congress, Prints and Photographs Division, LCN 2008678213) for Sweet D

Contents

Acknowledgments   ix

Prologue   1

Introduction. The Trouble with Ecological Homecoming   7

Part I

1. Martin Heidegger and the Coloniality of Nature   49
2. Willa Cather and the Home(l)y Metaphysics of Landscape   77
3. D. H. Lawrence and the Ecological Uncanny   102

   Excursus I. Ecological Realism   131

Part II

4. (Un)settling the Southern African Farm/world   143
5. Allegory, Realism, and Uncanny Ecology on Olive Schreiner's African Farm   169
6. Doris Lessing's Ecological Realism   192

   Excursus II. Exo-Phenomenology   217

Notes   243

Bibliography   265

Index   283

Acknowledgments

I feel astonished and humbled when I reflect on the far-reaching network of generous individuals required for a project like this to see the light of day.

As a first-time book author, it's difficult to resist the temptation to start the accounting early, with those teachers who, like Susan Schierts, affirmed and encouraged my affection for literature, and who, like Amy Hallberg, made learning German such a joy. This project wouldn't have been possible without the early sense of curiosity inspired by these and many others, principal among whom I count my ever-supportive mother, Debbie; my astonishingly generous grandparents, Jack and Eleanor; and the whole Strother clan, which I proudly consider my second family.

I have many gifted pedagogues at Carleton College to thank for their formative influence on my thinking. I'm especially grateful to Arnab Chakladar, Pamela Feldman-Savelsberg, Susan Jaret McKinstry, Jamie Monson (now at Michigan State), Kofi Owusu, and Connie Walker.

I'm thankful for many wonderful people at Princeton. Perhaps no one has helped more to shape my writing (as both practice and product) than Wendy Laura Belcher, whom I'm pleased to count as both a mentor and a friend; I am very much in her debt. I'm also indebted to Maria DiBattista, who embodies a nimble critical vitality that I strive to emulate yet fear I'll only ever imitate. To Simon Gikandi I owe much; it is an honor (and singular pleasure) to have studied with a scholar of such extraordinary breadth and intellectual exactitude. Thanks, too, to Ben Conisbee Baer, for his early contributions to this project, and to Claudia Brodsky for introducing me to the unique pleasures and frustrations of Heidegger.

Harriet Calver and Caitlin Charos, two beloved (and much-missed) friends, have both been important and generous interlocutors over the years. I should also thank Vahid Brown, who first put a copy of David

Abram's *The Spell of the Sensuous* in my hands and set me down the path of eco-phenomenological thought.

For their helpful feedback on the Heidegger material, I want to thank those who participated in a seminar that I co-designed with Caitlin Charos for the American Comparative Literature Association conference in 2015: Pushpa Acharya, Julia Michiko Hori, Kate Kelp-Steppins, Adhira Mangalagiri, Aakash Suchak, Sol Pelaez, and of course Caitlin herself. Thanks to Eric Morel and Erin James for their close scrutiny of the Cather material. Further gratitude goes to Ted Geier, who invited me to speak on D. H. Lawrence for ASLE's panel at the 2015 meeting of the Pacific Ancient and Modern Language Association.

At PNCA I want to thank Shawna Lipton for her ongoing moral support, as well as crucially timed financial assistance via the Critical Studies department, which she chairs with fortitude and poise. I'm also indebted to Qamuuqin Maxwell, along with Randy Meza and Hannah Bakken—three passionate students who first introduced me to the work of Eve Tuck and K. Wayne Yang and helped foster intellectual investment in settler colonial studies at PNCA.

At the University of Virginia Press I'm grateful to Boyd Zenner, who first brought the project in, and to Angie Hogan and Ellen Satrom, who shepherded it the rest of the way through the acquisitions and publication processes, respectively. To the series editors of "Under the Sign of Nature"—Serenella Iovino, Kate Rigby, and John Tallmadge—I'm greatly indebted for their incisive comments on the book proposal. I'm similarly obliged to the two anonymous peer reviewers, whose generous reports have undoubtedly made this a better book. Thanks, too, to Susan Murray, for her deft and perceptive copyediting.

Portions of chapter 2 appeared previously in a 2018 issue of *English Studies* (vol. 99, no. 4). I'm grateful to the publisher, Taylor & Francis, for permission to reprint that material here.

Land acknowledgments may have symbolic importance, but I also recognize that they can't give land back. Even so, I want to express my gratitude to the many Indigenous communities that have long stewarded the lands on which I, as a settler/trespasser, have spent my entire life. In Minnesota, New Jersey, and Pennsylvania I have lived and worked on the traditional lands of Lakota, Sioux, and Lenape peoples. In Oregon I have lived and worked on the traditional lands of several bands of the Chinook, including the Clackamas, Cowlitz, Kalapuya, Kathlamet, Molalla, Multnomah,

Tualatin, Tumwater, Wasco, and Watlala peoples. During the writing of this book I was also fortunate to spend two periods at the gorgeous Caldera Arts Center in Sisters, Oregon, situated on territory traditionally stewarded by the Tana'nma of the Warm Springs bands and Paiute. Though this book focuses on settler imaginaries, I have written it in the ultimate spirit of seeing these and other stolen lands restored to their original caretakers—both here on Turtle Island and elsewhere on this troubled planet.

Finally, my deepest thanks go to my partner, Daniel Addy, who has been a veritable model of love and support throughout this project's long life. I dedicate this book to him.

Unsettling Nature

Prologue

I had never lived in a home with such a meticulously landscaped backyard until my partner and I moved into the house on Terry Street. From the expansive, trapezoidal deck attached to the back of the house you could see the whole yard. The vegetable garden lay to the left, and to the right stood a tall lattice structure densely interwoven with jasmine. Immediately in front of the deck stretched a small patch of grass, bordered by a rock garden. Within the layered contours of the rock arrangement lay nestled a miniature pond with a fountain that burbled in the shade of a Japanese maple. A narrow path threaded between the vegetable garden and the fountain toward the back of the yard, where a flush of flowers and decorative plants grew near an arbor vined with pinot noir grapes. The path ended at the far edge of the property, where there stood a second deck, bordered by a wooden fence. Growing up from a circle cut into the deck was an Oregon white oak, large enough to create a canopy of shade but not yet grown to the full splendor of maturity. Tendrils of ivy had crept through the fence from the neighboring property and begun to climb the oak and cover its bark with lush, ever-green foliage. In the shade of the ivy-covered oak my partner and I set up a table and two chairs, establishing a space of retreat that offered a partially occluded view of the house. From this vantage, with the fountain burbling in the background and jasmine sweetening the warm breeze, the backyard extended the ambit of comfort that radiated out from our new home.

Though we were renters and didn't actually own the property, that didn't stop us from thinking of it as ours and experiencing it as extensions of ourselves. *We made ourselves at home.* Even so, our only explicit outdoor duty was to administer a monthly chemical cocktail that ensured the pond water stayed clear of algae. To prevent the yard from going to seed the homeowners retained the services of a landscaping company. Yet without having to (at)tend to all the corners of this tiny kingdom, without getting into the dirt and seeing for ourselves what was going in the various botanical, fungal, entomological, and microbial communities in the yard, all we could see was the general wash of its beauty, organized into an abstract feeling of loveliness and comfort in which our daily lives played out.

It was not until our second spring in the house when we happened to notice a warp in the visual seamlessness of our backyard landscape. As April rolled in and the city of Portland found itself in a rush of new tree blooms, we noticed that the Oregon white oak lagged behind. Only a few buds appeared, yielding a sparse spray of leaves. And as the weeks passed and no further buds seemed forthcoming, we began to wonder if the tree might be experiencing dieback due to some arboreal disease, depleted soil, or the unusually wet winter. We did some research to determine if other oaks in the region were experiencing anything like this, but came up with nothing. Unable to discern the source of the problem, we decided to wait and see if things got worse.

Life went on. I would walk to a nearby café and read in the morning, then come home, lunch, and spend the afternoon writing in the sunroom on the back of the garage. In the early evening, when my partner finished with his work, we'd relax over a glass of wine in the shade of the oak. One fine summer evening, as we sat on the back deck, a breeze rustled the foliage above our heads, drawing my attention to the sparsely leaved branches. Looking up, I realized that I'd completely forgotten about the sad-looking tree.

It had struck me as odd in the spring that new leaves weren't forming, but since then I had stopped seeing the tree's deficiency. If you didn't look too closely, the ivy that had all but taken over the trunk and branches made the tree look as lush and vibrant as ever. But now, looking up and seeing the flourish of ivy curled around withering branches, it suddenly became clear to me that something was wrong. Very wrong. I don't know how I hadn't noticed it before, but the ivy that had creeped over from the neighboring property was something of a botanical predator, with its long tendrils winding around the tree in a patiently tightening death grip. I thought to myself, *The ivy is killing the oak tree!* What had initially appeared to me as something beautiful, conjuring rustic images of European gardens and ivy-covered ruins, was in fact the very picture of violence. A slow-motion strangulation was transpiring right above our heads, even as we sat there in its shade sipping rosé.

I felt stunned by such a sudden shift in my way of seeing, and a little embarrassed too, both by my previous failure to notice and by the sense of shock this shift in perception had caused. Perhaps the duration of my inattention contributed to the outsized experience of shock. After all, the ivy had been there the whole time, steadily encroaching on the tree and choking out its access to light. And I had stood by, at once ignorant and enjoying the spoils of that ignorance, a banal privilege that I persistently (though unconsciously) sublimated into an experience of beauty, comfort, *home*.

The shock I experienced that evening continued to unfurl as I learned more about the slow violence that had been taking place literally in our own backyard.[1] Upon further investigation, I found that the particular species we were dealing with was English ivy (*Hedera helix*), a plant steeped with tradition. In addition to its classical association with the cult of Dionysus, English ivy also symbolizes binding and connection. The plant plays an important role in Druid weddings, and it famously appears at the end of some versions of the medieval romance of Tristan and Isolde, where ivy grows out of the two lovers' separate graves, joining the dead in a gesture that naturalizes their love in defiance of King Mark. English ivy also connotes civility, collegiality, and wealth. One need only think of the romance of ivy-covered university buildings, and particularly those most associated with elitism and privilege: the universities of the Ivy League.

Yet for all its romantic links to the Old World, English ivy is a much-maligned species in many parts of the New World, including the Pacific Northwest, where it is widely considered invasive. In fact, the State of Oregon, where I live, has legally blacklisted English ivy in an effort to curb the plant's spread, though its colonization is well underway. For example, a walk through the lower sections of Forest Park in Portland will reveal vast swaths of urban forest already succumbing to its tangled, imperialist encroachment.[2]

And *imperial* is the word. Native to Europe as well as Mediterranean Africa, English ivy came to what is now the United States by way of British settlers. No doubt they brought the ivy as an ornamental species intended to make the New World feel more like home. But what these settlers failed to realize was that English ivy, aggressive but held in check in its native environments, would become wildly intrusive wherever it was newly introduced. It isn't difficult to imagine what happened once the ivy entered new territory: "Untended for long enough, the ivy will clamber to the top of a tree and from its high perch transform itself into an evil twin.... In the autumn, the clusters of little white flowers draw bees from afar and then spend the winter turning into black berries. Birds devour them and spread them into the wild. Thus seeded, the ivy begins its patient journey of conquest."[3] English ivy replaces native vegetation with poisonous botanical matter. The sap it produces contains irritating chemical properties, and its berries are toxic to many of the animals that might feel tempted to sample them. And not only does English ivy itself invade and overtake territory, but the rapid expansion of its dense foliage also provides refuge for various insects and rodents, whose presence can further disturb the invaded environments.

This research confirmed what I had already suspected regarding the ivy's durational attack. Yet it also revealed another, more surprising reality. The violence that was transpiring in our yard was not an isolated incident; it was attached to a particular history of conquest, displacement, and occupation. And what's more, this "history" of conquest, displacement, and occupation was still unfolding. The struggle between the ivy and the oak was not a metaphor but rather a real, material (re)enactment of the ongoing, persistently invisibilized violence of settler colonialism, playing out here in the key of ecological imperialism.

But the deeper meaning of my shock did not relate solely to this recognition. For what this incident also taught me was that our backyard was not merely a material battleground but also an ideological battleground in which I, as a settler myself, was fully complicit. The lands that make up what we now call Multnomah County, where I live, are the traditional, unceded territory of the Multnomah, Kathlamet, Clackamas, and Cowlitz bands of the Chinook, as well as the Tualatin Kalapuya, Molalla, and other tribes that lived along the Columbia River. Within this context, my daily enjoyment of "our" beautiful backyard as specifically *home(l)y* constituted an everyday spatial activity of settler reification, an ideological recapitulation of the ongoing subjection of Indigenous peoples through the continued, affective reconstitution of stolen lands as *my* own particular beloved place, *my* "home."[4]

The shock of violence between the ivy and the oak brought to consciousness the forgotten truth of my trespasser status, and in the process deeply unsettled me out of what Mark Rifkin has aptly termed "settler common sense." For Rifkin, settler common sense is a matter of phenomenology. In the opening pages of *Settler Common Sense* he describes his first purchase of a home and the "quotidian affective formations" that attend such a venture into property ownership—formations that "among nonnatives can be understood as normalizing settler presence, privilege, and power, taking up the terms and technologies of settler governance as something like a phenomenological surround that serves as the animating context of nonnatives' engagement with the social environment."[5] Understood in this way, settler common sense organizes a particular way of seeing, a way of experiencing the "phenomenological surround" as a physical extension of one's own body and a metaphysical expansion of one's selfhood. Rifkin uses the term to refer to "the ways the legal and political structures that enable nonnative access to Indigenous territories come to be lived as given, as simply unmarked, generic conditions of possibility for occupancy, association, history, and personhood."[6] Of course, the legal and political structures that govern settler common sense must also be underwritten by ideological structures that determine everyday relations and worldmaking activities.

My own experience with the ivy and the oak intimated a particular ideological formation that also conspires to make nonnative access seem "given" and "unmarked." This ideological formation links the discourses of home and ecology in such a way that sponsors the commonplace notion of the natural world as a primordial dwelling place. Such an ideological formation actively unmarks territory, relegating it to a literal state of nature that erases any preexisting ties or claims to a particular place and enables a more fundamental homecoming to Nature itself. It was precisely this formation that had naturalized my experience of the eco-phenomenological surround as intrinsically home(l)y and made it difficult for me to see the full—that is, historical, material, and ideological—significance of the violence playing out between the ivy and the oak and my deep complicity with it. The book you are reading now stems from and expands on this formative experience of unsettling Nature.

Introduction

The Trouble with Ecological Homecoming

Philosophy is really homesickness, an urge to be at home everywhere. Where, then, are we going? Always to our home.
—Novalis

Ever since Martin Heidegger opened *Fundamental Concepts of Metaphysics* with the now-famous Novalis fragment, it has become commonplace to diagnose the European metaphysical tradition as homesick. Novalis was an eighteenth-century German mystic, poet, and philosopher known for oracular aphorisms and gnostic visions of what he called the "home world," a pseudo-place located at once everywhere and nowhere. The role of philosophy, he believed, lay in its capacity to orient us toward this elusive home world. "Philosophy can bake no bread," as he put it, but it can unveil the most essential truths that make us feel at home in the world. It's in this sense that philosophy is properly understood as *Heimweh*—a primordial homesickness that guides the philosopher's intellectual and spiritual labors.

Of course, for the project of philosophy to survive, the philosopher mustn't hasten homeward; homesickness can only drive the truth-seeker as long as homecoming gets deferred. Something of this deferral reveals itself in Novalis's own pen name. Born Georg Philipp Friedrich Freiherr von Hardenberg, Novalis derived his nom de plume from an ancestral family name, de Novali, which he adapted in a way that signaled a detachment from family history and the establishment of something at once older and newer. Indeed, the sleek mononym draws from the classical tradition to signal a fresh horizon: *novālis* being the old, Latin word for "new land." Judging by his chosen name, to be a philosopher was, for Novalis, to live as a perpetual newcomer settling virgin territory. The philosopher's work allegorizes the originary labor entailed in fashioning a new home on uncultivated land: clearing, plowing, planting, building, and, eventually, dwelling. Yet the philosopher cannot truly *dwell* until all the groundwork has been completed and the foundations firmly laid. And since philosophy is precisely this laying of foundations, its preparatory labors never cease. Perennially tilling the stony fields of prolegomenon, the philosopher cultivates his *Heimweh*.

More than one hundred years on, Heidegger would harness Novalis's prototypical homecoming narrative for his own philosophical project of earthly dwelling. Heidegger had an agonistic relationship to metaphysics, which he felt concealed the meaning of being. He therefore longed to enact what he called a *Destruktion* of metaphysics, which would yield a fundamental ontology and locate the human in that most primordial of all dwelling places: the "house of being" (*Haus des Seins*). Yet as the 2014 publication of his *Black Notebooks* most recently reminded us, Heidegger's philosophical homesickness has chilling links to the politics of his day. Compared, for example, to the Jew's "lack of soil" (*Bodenlosigkeit*), Heidegger's stand-in for human being, called "Dasein," enjoys a rather untroubled relationship to the homeland (*Heimat*). So untroubled is Dasein's relationship to the landscape that it strikes us as an uncanny echo of the autochthonous connection that, for the Nazis, persisted between German "blood" and German "soil" (*Blut und Boden*). Regardless of how direct a link we may draw between Heideggerian *Heimweh* and the Nazi *Heimat*, philosophical homesickness no longer seems as innocent as it once did. After Heidegger, the very notion of *home* would play host to an unhomely specter that haunted it (socially, politically, ideologically) from within.

Perhaps no philosopher has done more to unfurl the uncanny logic of the home than Jacques Derrida, who recalibrated Heidegger's *Destruktion* for his own method of *déconstruction*. Whereas the path Heidegger charted for ontological homecoming required an exit from metaphysics, Derrida understood that there can be no final exit from the metaphysical enclosure. The only way to deal with metaphysics is therefore to move deeper in, and to study how metaphysical logics at once carve the world into oppositional categories and intimately bind those categories together. *Presence* is thus "always already" (*toujours déjà*) shot through with *absence*, just as *the self* inevitably bears the traces of *the other*. Likewise, Derrida insists that the home's comforting interior can only persist through its paradoxical openness to the harsh exterior. Derrida's deconstructive reading of the home draws from Emmanuel Levinas and his phenomenology of alterity. As a Jew who lacked an easy sense of national or ethnic belonging after surviving the horrors of the Second World War, Levinas well understood the deracinating force of otherness, and he elevated this force as a central principle of ethical living. Unlike Heidegger, whose conception of dwelling implies the self-sameness of the German *Volk* residing comfortably in the German *Heimat*, Levinas insisted that "being at home with oneself" (*le chez soi*) fundamentally depended on welcoming the Other inside. A stranger, and a species of estrangement, thus abides at the very heart of the home(l)y.[1] As for Levinas, this paradoxical logic was not, for Derrida, a matter of mere philosophical abstruseness. Derrida was a French citizen of Jewish descent who grew up under the conditions of colonial settlement in Algeria. He understood from lived experience that "home" is a site of unfulfillable desires: a presence shot through with absence, a building founded on an abyss.[2] For Derrida, in even stronger terms than for Freud, the home becomes thoroughly suffused with the unhomely.

Derrida may have undermined the philosophical fiction of home as an originary or primordial state in which human being must once again seek shelter. But he has not, in doing so, eliminated philosophical homesickness, nor has he diffused the environmental valences implicit in Novalis's "new land" or Heidegger's "earthly dwelling." As recent work by the contemporary philosopher Michael Marder has demonstrated, Derrida's attempt to invert the normative topology of home and abstract it from material reality must itself be submitted to deconstructive scrutiny. Marder centers his critique on what he calls Derrida's "allergy" to ecology—that is, his tendency to privilege *economy* at the expense of *ecology*. In fact, Derrida is so allergic to ecology that he refuses to speak its real name, only ever referring to it as a negation of the economic: the "aneconomic." The reason Derrida privileges economy over ecology stems, Marder says, from his erstwhile privileging of *nomos* over *logos*. Etymologically, economy and ecology are linked through their shared prefix eco-, which comes from the Greek *oikos*, meaning "house." But whereas economy, as *oiko-nomia*, spotlights the *nomos*, meaning "law," ecology, as *oiko-logia*, spotlights the *logos*, meaning (loosely) "logic." Derrida's profound resistance to the *logos* and to what he termed logocentrism conditions his rejection of ecology, and his consequent emphasis on the (an)economic fails to account for the intimacy that persists between the *nomos* and the *logos*. Marder, by contrast, argues that we must read these terms as complementary and competing forces that fundamentally unsettle the *oikos*, which houses both. "Ecology," he says, "estranges us from the estrangements of economy," and in doing so "it denies us the entitlement to being at home in the world." Ecology's power to estrange does not literally cast us from our houses, of course. But it does bestow a disturbing new "awareness that we are the ghosts haunting our planetary dwelling." What results is a "nonappropriative" experience of home, where *home* is, unsettlingly, that "to which we have no right whatsoever, legal or otherwise."[3] In restoring ecology to deconstruction, Marder reveals Derridean thought as even more profoundly unsettling than previously thought, as it dissolves any originary claim of belonging to any architectural or environmental space we might call home. By Marder's measure, then, "ecology" doesn't just thwart the desire for a return to some primordial origin; it actively estranges us from that origin.

As the title of this book suggests, the argument I make in these pages has much in common with Marder's assessment of ecology's unsettling force. Yet for me, Marder's method leaves something to be desired. Specifically, I'm skeptical about how thoroughly his style of "eco-deconstruction" dismantles the homesickness enshrined in the likes of Novalis and Heidegger. Marder argues that ecology wields the power to estrange. Yet the insistent nature of his paradoxical wordplay—"the homelessness of the house," the "stable instability" of the home—suggests that a perverse, even ironic homesickness lingers within estrangement. It seems to me that Marder's poststructuralist reliance on etymological games itself paradoxically installs the rhetoric of home even more permanently in philosophy. Furthermore, because ecology emerges in Marder's argument primarily as an etymological object, he cannot fully articulate the social and political stakes of thinking ecology as a "nonappropriative" experience. Consequently, Marder's eco-deconstructive method fails to address the submerged environmental valences that have, since Novalis, implicitly framed philosophical homecoming as a mode of colonial settlement.

To this day, then, and despite our best efforts, the trope of homecoming remains alive in European philosophy. Like the recovering addict who fixates on what they know they can't have, we know in our heads that fetishizing home(l)y discourse can lead to problematic and even dangerous forms of idealism, but we nonetheless find it more than difficult to bid home farewell. And as Marder's example already indicates, this difficulty persists in environmentally inflected thinking as much as it does in Continental philosophy. Even as contemporary ecophilosophy grows increasingly sophisticated in its inquiries and diversified in its methods, one can still find ubiquitous appeals (both subtle and not so subtle) to homecoming. Consider Vicky

Kirby's contribution to the recent volume *Eco-Deconstruction,* the same collection where Marder's essay on Derrida appears. Kirby's essay draws on quantum theory to dispel the normative idea of spacetime as a container for matter. The quantum model insists on a radically emergent reality in which time, space, and matter are all fundamentally entangled and thus create one another in an "intra-active" dynamic that Karen Barad has termed "spacetimemattering."[4] Kirby uses this notion of emergent quantum entanglement to trouble the topological distinction between the insides and outsides of things. Kirby insists that exteriority always inhabits interiority, an "inside" that she readily glosses as "the *oikos* of a subject."[5] She then concludes her essay with an appeal for us to understand this dizzying topology as a home(l)y "place" that quantum theory implicitly authorizes as the site of our most fundamental origin. Kirby asks, "If the alien is not outside, then who are we, and *how might we learn to be at home with différance?*"[6] Kirby's eco-deconstructive take on quantum nature ranks among the most cutting-edge work in contemporary environmental philosophy. But however sophisticated this philosophy may be, the rhetoric of home(coming) remains firmly entrenched in it.

If our most recent and canny thinking hasn't managed to overcome the desire for (ecological) homecoming, then perhaps we need to chart an alternative path that leads us away from the rhetoric of home. In this book I offer one such alternative path. But instead of seeking to supersede the most recent work and hence produce some yet more cutting-edge theory that continues to deploy the same old rhetoric, the path I follow proceeds along a somewhat circuitous route that aims to carry the lessons of eco-deconstruction "back" into the discipline from which it emerges: *eco-phenomenology.*

Just as deconstruction grew from Derrida's close engagement with the work of phenomenologists like Edmund Husserl, Heidegger, and Levinas, so too does eco-deconstruction arise from the important work in eco-phenomenology, a subfield of environmental philosophy that had its heyday in the late 1990s and early 2000s. This dependency becomes particularly clear in the way eco-deconstruction builds on eco-hermeneutics, an intermediary field that followed directly on the heels of eco-phenomenology in the late 2000s and early 2010s.[7] Eco-hermeneutics engaged critically with eco-phenomenology,[8] and eco-deconstruction has since pushed its predecessor's critiques yet further.[9] Without belaboring the matter, my point is simply that however dated the term may seem to us now, eco-phenomenology remains alive today, albeit in new guises. This survival presents an opportunity to revisit eco-phenomenology, not as a historical relic but as a living paradigm, encompassed by its most recent philosophical successors and hence available for ongoing transformation by the lessons of those successors.

Cary Wolfe intimates just such an opportunity in his contribution to *Eco-Deconstruction*. Specifically, Wolfe points to the need for an environmental phenomenology that attunes itself to the complexities of epistemology. Wolfe writes that we must

> realize that "epistemological" isn't the *opposite* of "environmental" but in fact *means* "environmental" in this very specific sense: you can't take embodiment seriously, of *whatever* form of life, without also taking epistemological questions seriously, because if epistemology is precisely the study of how a being knows things, then those modes of knowledge and experience of the world depend directly on the embodied "enaction" (to use Maturana and Varela's phrase), which is a product of the recursive loop between an organism's wetware and how it gets required by external interactions, environmental factors, semiotic systems, cultural inheritances, the use of tools, and much else besides.[10]

Wolfe's description of a recursive loop between perceptual and intellectual capacities yields an embodied epistemology open to new ways of perceiving and knowing. Such an epistemo-phenomenological perspective leads Wolfe to a vision of ecology that avoids the rhetoric of home altogether, instead arriving at something closer to the biologist Jakob von Uexküll's theory of autonomous yet intersecting *Umwelten* (environments; literally "enveloping worlds").[11]

With Wolfe's useful perspective in mind, I return to eco-phenomenology, though I think of this return as a kind of "re-membering," which, as Karen Barad puts it, "is not about going back to what was, but rather about [a] material reconfiguring" that attempts to "produce new openings, new possible histories."[12] Such a return is, as I have already suggested, more topological than geometric. That is, it enables a folding of space and time that diminishes the only apparent distance between eco-phenomenology and more recent elaborations of environmental philosophy.[13] My starting point for thinking through the home(l)y rhetoric of environmental philosophy will therefore take place under the sign of eco-phenomenology, the discursive field in which the narrative of ecological homecoming reached its apex.

ECO-PHENOMENOLOGY AND THE ECOLOGICAL HOMECOMING NARRATIVE

It would be unfair to diagnose eco-phenomenology with as severe a case of homesickness as the European metaphysical tradition at large, but it does suffer from the same basic strain. Though quarantined from infectious idealisms and inoculated against virulent romanticisms, eco-phenomenology maintains a desire to bring us home to our most primordial selves. Such a radical homecoming seems vital in the face of the world's many environmental crises. Crucial as it may be to reject single-use plastics, stand with Indigenous communities against extractivism, and dismantle ecologically hazardous monopolies that sustain and intensify the violence of neoliberal capitalism, none of these measures goes to the root of our contemporary problem. And the problem for eco-phenomenology is not simply that industrial modernity has driven the planet to the brink of collapse, but that it has driven us *out of ourselves,* cleaving mind from body, thought from the flesh of the world. As such, eco-phenomenology advises radical measures meant variously to return us to the natural world and restore us to our embodied sensorium. If only we can reestablish our sense of material enmeshment in nature, we can reverse the degradation that humans, as the drivers of the Anthropocene and the agents of the Sixth Extinction, have wrought. In redeeming ourselves, we can save the earth. And in saving the earth, we can come home in the most essential sense. In the words of Heidegger, a favorite philosopher among eco-phenomenological thinkers, we can once again dwell in the nearness of our own being.

The *Heimweh* that orients eco-phenomenology toward the primordial dwelling has its origins in the pioneering work of twentieth-century phenomenologists such as Edmund Husserl, Maurice Merleau-Ponty, and, of course, Heidegger. It was the earliest of these thinkers, Husserl, who first introduced home(l)y rhetoric within phenomenological discourse. Husserl's rhetoric of home derives from his theory of the "lifeworld" (*Lebenswelt*), a concept he coined in *Crisis of the European Sciences* to refer to the universal framework for all human endeavor and achievement. The lifeworld, as Husserl conceives it, is a pretheoretical attitude toward the milieu in which life actually plays out. As a general structure for how the world at once envelops and grounds conscious life, the lifeworld may be further broken down into two aspects of worldliness that more directly constitute an individual's experience: the "homeworld" (*Heimwelt*) and the "alienworld"

(*Fremdwelt*). As Janet Donohoe explains: "Homeworld is always co-relative with alienworld.... While bodily habit, traditions, and rituals are being established through the constitution of the homeworld, that which is not these habits, that which stands over against these traditions, is also being established."[14] Against that which is alien and unfamiliar, *Heimwelt* designates those embodied habits of perceiving and acting that establish a unified field of meaning and constitute our most basic sphere of familiarity—what Husserl elsewhere termed our "primitive home-place."[15]

Whereas *Heimwelt* designates just one concept of worldliness among several in Husserl's thinking, home(l)y discourse occupies a position of central importance in the work of Martin Heidegger. As I've already indicated, Heidegger's chief aim throughout the *longue durée* of his career was to restore proper concern for what he called "the question of being," which he claimed the tradition of metaphysics had reduced to an apparent absurdity. He therefore sought to rescue ontology from metaphysics by separating the wheat of essence (*Wesen*) from the chaff of mere presence (*Anwesen*), thereby restoring human being to its originary fullness in the wake of its having "fallen out of Being."[16] Heidegger's concept of "dwelling" (*Wohnen*) plays a crucial role in this restoration, and his overall project represents a sustained homecoming narrative that aims to reinstall Dasein in its most essential dwelling place: the "house of being."

Unlike Husserl and Heidegger, Maurice Merleau-Ponty does not make explicit use of the rhetoric of home, sublimating it instead into a discourse of primordiality.[17] As he puts it on the first page of his magnum opus, phenomenology provides a method for "re-achieving *a direct and primitive contact with the world, and endowing that contact with a philosophical status.*"[18] Such a restoration of "primitive contact" suggests a homecoming by way of the human sensorium—not so much a return to the fullness of being as a return to the perceiving self and its material enmeshment in the sensual world. It is precisely through more attentive habits of embodiment that an individual can come home via what Merleau-Ponty calls "inhabitation." Patricia Moya summarizes: "Merleau-Ponty frequently expresses the close relationship between body and world with the term 'inhabit,' as referring to that which is known by the body and which translates into a knowledge of what to do with an object without any reflexion coming in between."[19] Like Husserl's *Heimwelt*, the sense of inhabitation afforded by cumulative habits of embodiment comport the individual in their perceptual world, guiding action and experience in ways that center the perceiving body as the individual's most primitive dwelling place.

One need not search far to see that phenomenology plays a centrally important role in the various contemporary discourses of the home, whether philosophical,[20] literary,[21] sociological,[22] or even medical.[23] But the chief concern here is the influence of home(l)y rhetoric on eco-phenomenology. Although a certain strain of Heideggerian eco-phenomenology directly picks up on the rhetoric of home (see chapter 1), what appears more frequently is a sublimated version of home(l)y discourse that resurrects scientific naturalism, the roots of which lie not in strict Enlightenment rationality but in embodied experience. As Brown and Toadvine claim in the introduction to their edited volume on the subject, eco-phenomenology is revolutionary because it enables the "rediscovery of a natural world that is inherently and primordially meaningful and worthy of respect."[24] Restoring primordial meaning(fulness)—one that scientific methodologies have forgotten and in turn lead us all to forget—thus represents a key aim of eco-phenomenological thought.

The ultimate horizon for such a return is our earthly home on this planet. Brown and Toadvine indicate as much in the subtitle to their edited volume: *Back to the Earth Itself*. This phrase is, of course, a pun on the famous phenomenological call to arms, of which two versions exist. Whereas Husserl initially urged philosophers to "go back to the things [*Sachen*] themselves,"[25] Heidegger later issued a slight corrective by advocating that we "[stay] with the things [*Dinge*] themselves."[26] The difference between *Sachen* and *Dinge* is key. Husserl famously brackets, or "suspends," the material universe in order to focus on phenomenal appearances. Hence, when he made his original call, the things (*Sachen*) to which he referred were precisely *not* material entities but rather sensible intuitions (*Anshauungen*) given to us in experience. Heidegger, however, replaces *Sachen* with *Dinge*, implying that the return to "things themselves" means staying with actual entities that appear to us in the world, and not reducing them to mere objects for consciousness. Staying with the things themselves becomes the basis for a Heideggerian ethics of care that aims at saving the earth. I take Brown and Toadvine as suggesting a similar goal when they claim that eco-phenomenology sponsors a return to "the earth itself." Such a return also provides the basis for what I call *the ecological homecoming narrative*.

Paradigmatically, and as if to assert the naturalness of an earthly homecoming, eco-phenomenological thinkers often gesture to the etymology of "ecology."[27] Referencing its Greek roots, such thinkers define ecology as the *logos* of the *oikos*—that is, the discourse (or science, or logic) of home. This etymology appears innocuous enough. After all, the science of ecology is all about how organisms interact with each other and with the physical environments in which they live. Yet it would be a mistake to equate *environment* with *home*. All living entities may require a sustaining environment, but *home* is an undeniably human construct—not simply a physical living space, but a culturally defined and ontologically charged dwelling place. Home names that "felicitous space" which, in Gaston Bachelard's phenomenology, "concentrates being within limits that protect."[28] History has shown that home is also a surprisingly chameleonic concept, one that can be used to refer to a wide range of ideological spaces. It is at once a domestic, economic, and geopolitical signifier; it consolidates identity and status at the same time as it concentrates being. Home is a multivalent metaphysical conceit; it is anything but natural.

Aside from the issues attending the conceptual extension of *home*, the etymology itself presents additional problems. Consider *oikos*. In the ancient Greek context, the word gathered three closely related concepts: house, property, and family.[29] Taken together, physical houses, their associated wealth and property (*oikoi* were often attached to large farms), and patrilineal lines of inheritance served as the building blocks of Greek city-states. *Oikos* is thus a social and political unit as much as a domestic one, which explains its etymological connection to "economy," a term that originally referred to the management of household resources and affairs. The link to economics is certainly not lost on ecologists[30] or on ecocritics.[31] Even so, the metaphor of nature as an economy cannot itself naturalize the *oikos*. *Logos* proves even trickier given its disputed status in philosophical and theological discourse. From Heraclitus to Aristotle, and from the Stoics to the Gnostics, *logos* has referred to a pseudo-spiritualized form of reason. Some thinkers articulate logos as the hidden yet rational logic of the material universe; others figure it more esoterically, as the soul of the cosmos. Regardless of the specific articulation, an obsession with logos has dominated European thought since its advent.[32] For Derrida, "logocentrism" names the characteristic way that texts, representational modes, and signification systems produce a yearning for unmediated access to *presence*, which becomes a shorthand for other metaphysical species like *meaning, essence,* and *being.*

The logos stands at the center of what Derrida calls the "metaphysics of presence"; it is thus also implicated in the search for a "transcendental signified" that would, so to speak, bring us home to the origin of all things. But Derrida's point is that the object of logocentrism is fundamentally illusory, as is the desire it generates for direct access to primordial presence. The logos, as origin, remains forever inaccessible, nonlocatable. If ecology really means the *logos* of the *oikos*, then what we are talking about is not a scientific notion of how organisms interact with their environments or even a softer notion of the discourse (or "poetics") of home. It is, rather, a profoundly human desire for direct access to the very essence of our being and to the transcendental presence in which it is housed: capital-N Nature. The term *ecology* houses within itself a persistent desire for a homecoming that would restore the lost purity of Nature as an earthly dwelling place.[33]

Such a desire for ecological homecoming emerges partly as a corrective to a widespread European narrative that posits humans as existing in a state of alienation from the natural world. Much like the Christian paradigm of exile from the Garden of Eden, this narrative sees European civilization as having alienated itself from its natural origins through the development of various cultural, political, religious, and intellectual institutions.[34] Such institutions include, principally, Enlightenment rationalism, the technological drive of the Industrial Revolution, and (speaking of logocentrism) the perverse modern tendency to privilege writing over speech.[35] More radical perspectives locate the origins of human alienation as far back as the advent of agriculture during the Neolithic Revolution, circa 10,000 BCE;[36] the intensification of big-game hunting during the period of Europe's last glaciation starting some thirty-five thousand years ago;[37] or even with the evolution of the earliest-known hominids, *Ardipithecus ramidus*, nearly six million years ago.[38] Raymond Williams writes of an analogous perspective in the British pastoral literary tradition, which has long mourned the passing of rural life and yet never seems able to pinpoint the time of death, always pushing it further back into time immemorial.[39] Those who lament ecological alienation likewise seem unable to identify the moment of our collective exile from Nature. And the further back in time we situate our alleged exile, the more urgent this need for home seems to become. Much like Odysseus's final *nostos*, the power of which depends on his having suffered through not one but two epics, ecological homecoming seems sweeter the longer we imagine ourselves to have been exiled from Nature.

What, exactly, is at stake in relentlessly recapitulating narratives that project a triumphant return to Nature? As this book demonstrates at various points throughout its six chapters, the reiterative force of ecological homecoming institutes a self-fulfilling mode of thinking and perceiving that often functions to reify a normative understanding of the human. Such a reifying function appears particularly in eco-phenomenological writing that advocates for a return to the senses. Consider a touchstone example focused on somatic practice: Mitchell Thomashow's *Bringing the Biosphere Home*. In this work Thomashow develops a place-based "perceptual ecology" through which he hopes to teach others how to perceive changes in the biosphere. He advocates cultivating a curiosity-based practice that enhances the senses, but which also entails deeper emotional work: "Awareness of biodiversity and megaextinctions transcends ecological and political considerations and opens you up to all kinds of existential dilemmas," such as "love and loss, life and death, creation and extinction."[40] Thomashow writes compellingly about how to refine a "deliberate gaze" that "combines wonder, intent, and consideration," and that enacts an "improvisational concentration across the senses."[41] Yet the kind of perceptual practice Thomashow elaborates also entails a circular logic. The refinement of the senses aims to foster an environmental ethic, but it strikes me that such an ethic would already need to be in place for an individual to orient their sensorium toward the biosphere in the first place. Can an ecologically inflected phenomenological practice—and the environmental affects to which such a practice would attend—really induce ecophilia?[42] Or does such a practice primarily foster self-development, refining an individual's own cognitive and perceptual abilities?

Such appears to be the view of David Abram, whose work offers lyrical, moving, and often brilliant meditations on the embodied experience of the natural world. These meditations are meant to inspire readers to develop their own personal practices that, through mindfulness, will open them to their local environments and to the "natural magic of perception."[43] Whereas Abram's first book, *The Spell of the Sensuous*, spends more time examining how the shift from speech to writing has progressively drawn humans (and particularly Westerners) away from our natural origins, his second book, *Becoming Animal*, sets out into the dense thicket of phenomenal experiences that he implies may help to redeem both our earthly selves and the earth itself. Yet Abram's method leads to a problem: the aim of his phenomenological approach is, as he says explicitly, not primarily to connect us to the sensuous web of natural phenomena but to enable us to become "fully human."[44] As for the lone hiker who treks into the wilderness to make himself whole again—*himself* because this desire is also historically gendered—for the eco-phenomenologist, environmental perception becomes a tool for self-actualization, and for a surprisingly solitary homecoming in the midst of what Abram terms "the more-than-human world."

As a dancer and performance maker with a long-standing somatic practice, I understand how much embodied exploration has to offer the individual practitioner. However, the attempts by Thomashow, Abram, and others to elevate such practices to what Brown and Toadvine call "an ecological philosophy" also strike me as dangerous with regard to the way they reify a notion of humanness—*the* human—as a fundamentally free and able-bodied subject(ivity), located, as we are led to presume by the typical lack of overt positionality awareness, in places of relative privilege and environments relatively less afflicted by extractivism, pollution, and climate change. Only recently has such reification of the human come under scrutiny in environmental discourse, for example in Sarah Jacquette Ray's work on the trope of disgust. Ray investigates how disgust functions to separate "natural" bodies from "unnatural" ones, and she claims that the "whole" body (i.e., mentally and physically fit, nondisabled, straight, cis-gendered, and white) is the most environmentally "pure" body; hence, paradigmatic ecological others include those who are mentally ill, obese, disabled, queer, transgendered, and/or racially marginalized. The rhetoric of wholeness in environmental discourse provides the foundation for what she considers "the uncritical turn to the body as an ideal way to connect with the environment."[45] In addition to Ray's explicit critique of eco-phenomenology, it is also important to mention other interventions in phenomenological discourse that draw our attention to the ways such discourse silently reiterates normative, privileged ways of perceiving and moving within the world. Sara Ahmed, for instance, argues that whiteness functions as a normative orientation that, in institutionally white spaces, repeatedly *disorients* nonwhite bodies.[46] These crucial interventions remind us that there is no such thing as *the* body, *the* human. Eco-phenomenology would thus be well advised to resist the unreflective desire for a return to the body and the biosphere as concentrically encompassing and holistic *homes*.

HOMECOMING NARRATIVES AND THE
SETTLER COLONIAL IMAGINATION

To understand one of the dangers eco-phenomenology courts with its home(l)y rhetoric, consider the recent social movement known as "rewilding." Though the term originates in conservation biology, where it refers to the restoration of land to an uncultivated state, in the cultural sphere *rewilding* has been rescripted as "[the restoration of] ancestral ways of living that create greater health and well-being for humans and the ecosystems that we belong to. . . . It means returning to our senses, returning to ourselves, and *coming home to the world* we never stopped belonging to."[47] The rewilding movement, and the narrative of ecological homecoming on which it is founded, harnesses seemingly innocuous language that, like eco-phenomenology, seeks to return to the earth through a radical return to "our senses" and "ourselves." Yet the language of the above quote also performs a deft sleight of hand that at once appeals to "ancestral ways of living" and immediately lays claim to those ways of living, universalizing them as an ideal for all humans. The language renders this slippage almost invisible by making the solution seem simple: all "we" need to do is cast back into the past and reel into the present certain ways of living that have long been out of practice. The obvious critique here relates to the pat celebration of "ancestral ways of living," which recapitulates the long-lived Romantic stereotype of Indigenous peoples as being inherently close(r) to Nature—what in the U.S. context Shepard Krech refers to as the myth of the "ecological Indian."[48] Rewilding thus depends on a logical tactic that disappears Indigenous peoples into Nature and casts living Indigenous practices as things of the past.[49] This past is then conceived as a kind of archaeological free-for-all in which all those who belong to the "we" of the present—supposedly *all* humans, but by implication just *"us"* alienated Euro-Americans—may unearth and appropriate our way to a greener, more holistic future. Such a strategy lays claim to a dream of indigeneity to ensure non-Indigenous survival, all the while placing Indigenous peoples under erasure.

Nor is the contemporary rewilding movement alone in appealing to sublimated Indigenous ways of being in service of a new (and ideally universalizable) mode of living and perceiving. Similar gestures appear nearly everywhere phenomenological methods have been adopted. Bachelard, for instance, invokes a generalized *primalness* as an impetus for accessing the imagination: "Primal images, simple engravings are but so many invitations to start imagining again. They give us back areas of being, houses in which the human being's certainty of being is concentrated, . . . a life that would be our own, that would belong to us in our very depths."⁵⁰ Yet nowhere is this tendency more common than in certain corners of eco-phenomenology. For instance, David Abram frequently references the animisms, shamanisms, and oral traditions of various Indigenous communities, threatening variously to essentialize and tokenize along the way. Mitchell Thomashow makes a yet more troubling invocation when he describes a visit to a monarch butterfly sanctuary near Mexico City. Adopting the phrase "aboriginal abyss of radical amazement" from the Jewish theologian Abraham Joshua Heschel, Thomashow writes: "In those speechless moments, when you're surrounded by such grandeur and fragility, you feel as if you are bearing witness to the magnificence of creation. You gaze through *the aboriginal abyss*. The living landscape swallows you whole. . . . It's the place from which you originate. There is no need for explanation."⁵¹ Thomashow leans into a spiritualist language that, redirected from its original context, recalls the logical maneuver witnessed in the rewilding example. The passage begins by establishing an intimacy between the indigenous status of "the aboriginal" and the vitality of "the living landscape." It then makes a universalizing gesture that lays claim to that intimacy: "It's the place from which *you* [i.e., any 'you'] originate." Everyone, it would seem, and particularly those eco-phenomenologists-in-training Thomashow describes as "pilgrims" and "explorers," gets to be native.⁵² And what's more: *"There is no need for explanation."* Whether made explicit, as in Bachelard's houses for being, or implicit, as in Thomashow's sublimated discourse of origins, the images deployed in these eco-phenomenological accounts foster a rhetoric of ecological homecoming.

Homecoming narratives like those examined above bear an uncanny resemblance to narratives that underwrite settler colonialism. Crucially, the resemblance is *structural*, which requires a distinction between colonial and settler colonial narratives. Colonial narratives are circular in form: like Odysseus's journey in the *Odyssey*, they include both an outward voyage and a final *nostos*. By contrast, settler colonial narratives have a linear form: like Aeneas's journey in the *Aeneid*, there is a movement from point *A* to point *B*, with no ultimate return. Yet this form is linear in a strange way. Colonial narratives emphasize discovery and adventure—what Lorenzo Veracini refers to as "a multiplicity of 'middle passages'" between the voyage out and the journey home.[53] In settler colonial narratives, however, "settlers do not discover: they carry their sovereignty and lifestyles with them. . . . [T]hey transform the land into their image, they settle another place without really moving . . . [and they] construe their very movement forward as a 'return' to something that was irretrievably lost: a return to the land, but also a return to an Edenic condition, . . . to a Golden Age of unsurrendered freedoms."[54] The settler colonial narrative is thus strangely tautological: arrival and homecoming fuse into a simultaneity. A similar tautology appears in eco-phenomenological homecoming narratives, where we are always, as in the words of the rewilding movement, "coming home to the world *we never stopped belonging to.*" The paradox of returning to a place we have never really left structurally mirrors the logic of settler colonial homecoming, in which home is a *terra nullius* anticipating settler arrival—blank, yet also proleptically inscribed as the settler's home in waiting.

Importantly, the settler's paradoxical return/arrival is not a discrete event; it is an ongoing process guided by the imagination. In order to inhabit new territory as home, the settler has to refocus their perceptual schema in a way that rescripts the alien as familiar. Thus, the settler must imaginatively transform space into place. Terence Ranger, a British historian of southern Africa, describes such a process in his account of British settlers' imaginative claims on the Matopos hills in colonial Rhodesia (modern Zimbabwe):

> The Matopos country had an unrivalled capacity to touch "civilised man," in his Rhodesian settler manifestation. The hills became a place of meditation and communion with Nature; more than anywhere else they symbolised the white Rhodesian's special relationship with the landscape. . . . But this sort of intimacy with landscape is not, of course, just given. It has to be worked for. When whites first saw the Matopos in the nineteenth century, they saw a confusing jumble of rocks rather than scenery. It took a long time for whites to turn the Matopos into landscape and even longer for them to turn them into their own particular beloved place.[55]

If it took white settlers a long time to imaginatively transform the visual jumble of the Matopos into "their own particular beloved place," it was because the traditional categories of British landscape aesthetics got in the way. Ranger reports that the Matopos "struck the first artist travellers as *almost* picturesque."⁵⁶ The hills weren't quite picturesque, because they lacked any sign of human presence, like a castle or hermitage, and "without such human signs the Matopos were hard to see."⁵⁷ But with time the British learned to see anew, and this perceptual shift reconfigured the hillscape into the settlers' "own particular beloved place." According to Ranger, the imaginative process of learning to see the Matopos as a home(l)y place underwrote the material processes by which European settlers dispossessed local Ndebele and Shona communities.

The process of imaginative transformation Ranger outlines in his work represents just one of many strategies settler communities have used in diverse times and places to legitimize the appropriation and occupation of their (new) homelands. Scholars of settler colonialism have elaborated a host of other tactics. Lorenzo Veracini, for instance, has itemized a list of twenty-four "transferist approaches" by which settler communities enact their founding fantasies to "'cleans[e]' the settler body politic of its (indigenous and exogenous) alterities."⁵⁸ Veracini explains that some of these approaches "operate discursively," whereas others "operate at the level of practice." He also indicates that various approaches may operate simultaneously or in succession, and that "different transfers are antithetical and mutually exclusive, but they can also overlap and blur into each other." Regardless, "all these strategies aim to manipulate the population economy by discursively or practically emptying the indigenous sector of the population system."⁵⁹ In a similar vein, Eve Tuck (Unangax̂) and K. Wayne Yang have outlined a series of "settler moves to innocence" by which contemporary settler communities reframe decolonization as a metaphor, thereby legitimizing settler presence at the expense of Indigenous sovereignty.⁶⁰

The collective range of these and similar strategies constitutes what I call *the settler colonial imagination*. The settler colonial imagination refers broadly to ways of thinking, being, and perceiving that continuously naturalize settler occupancy by erasing Indigenous pasts and invisibilizing Indigenous presence. Settler modes of thinking, being, and perceiving also disable Indigenous futurity by legitimizing settler governance and hence furthering the forever-unfinished project that is the settler nation. In these senses, the settler colonial imagination is most generally characterized by its recapitulation, in various forms, of what Patrick Wolfe has deemed the logic of elimination. According to Wolfe, settler colonialism is always premised on a genocidal logic that requires the elimination of "the native." Although the logic of elimination includes "the summary liquidation of Indigenous people," it also exceeds the narrow definition of genocide and includes "both negative and positive dimensions":

> Negatively, it strives for the dissolution of native societies. Positively, it erects a new colonial society on the expropriated land base. . . . In its positive aspect, elimination is an organizing principal of settler-colonial society rather than a one-off (and superseded) occurrence. The positive outcomes of the logic of elimination can include officially encouraged miscegenation, the breaking-down of native title into alienable individual freeholds, native citizenship, child abduction, religious conversion, resocialization in total institutions such as missions or boarding schools, and a whole range of cognate biocultural assimilations. All these strategies, including frontier homicide, are characteristic of settler colonialism.[61]

The positive and negative dimensions of the logic of elimination, paired with the various other cognitive, perceptual, and narrative strategies indicated above, collectively mark the settler colonial imagination and condition the ongoingness of settler colonial projects in the present. As Wolfe puts the matter, "invasion is a structure not an event."[62] If decolonization is ever to happen in a *material* form and not just metaphorically, it will depend on dismantling settler colonialism as a living structure. The ecological homecoming narrative represents one component of that living structure; it is thus a thread we must pull to instigate a larger undoing.

To demonstrate how narratives of ecological homecoming often unwittingly ground the settler colonial imagination, I turn to a landmark event of environmental humanities scholarship that productively unsettled normative ways of thinking about Nature. In the edited volume *Uncommon Ground: Rethinking the Human Place in Nature*, the environmental historian William Cronon and his distinguished contributors set out to elaborate on the apparently straightforward claim that "'nature' is a human idea, with a long and complicated cultural history."[63] That this claim was only apparently straightforward becomes clear in Cronon's foreword to the paperback edition, where he reflects on the backlash the book received from environmental thinkers who took issue with the project's revisionist ideas. Noting that the book had been published at a time when a Republican-controlled legislature was launching an attack on environmental(ist) ideals, these scholars felt concerned that exposing Nature's fraught ideological underpinnings implicitly supported environmental destruction. Yet Cronon insists that reckoning with the unnaturalness of Nature is an increasingly necessary aspect of living among a diverse array of human cultures: "once we recognize that not all human groups and cultures view nature in the same way, it becomes at least more complicated to assert that one group's ideas of nature should take precedence over another's. . . . [T]his can seem to make the work of protecting nature more difficult," precisely because the very nature of Nature is contested.[64]

Cronon leads the charge of "rethinking the human place in nature" with his own contribution to the volume, titled "The Trouble with Wilderness; or, Getting Back to the Wrong Nature," where he argues that the concept of *wilderness* underwent a significant ideological transformation in the United States over the course of the eighteenth and nineteenth centuries. Wilderness had initially been associated with the concept of *wasteland* but became redefined as something beautiful, sacred, and hence worthy of protection. The redefinition depended on two interrelated concepts: *the sublime* and *the frontier*. These concepts, one aesthetic and the other geopolitical, "converged to remake the wilderness in their own image, freighting it with moral values and cultural symbols that it carries to this day."[65] Consider the creation of the first national parks in the nineteenth century. Cronon asserts that early national park policy stemmed from a double belief: first, that wilderness spaces should be understood as sublime landscapes infused with the supernatural; and second, that such wilderness spaces were quickly disappearing due to westward expansion. And indeed, those wilderness landscapes marked out for preservation at the closing of the frontier all embodied the transcendentalist aesthetics of the sublime. As Cronon puts it, if the presence of the supernatural could be found "on the mountaintop, in the chasm, [and] in the waterfall," then Americans were most likely to discover it in the first sites chosen for public preservation: Yosemite, Yellowstone, the Grand Canyon, Mount Rainier, and Zion.[66] Yet if wilderness ideology arose from the convergence of the sublime and the frontier, it also articulated with the Romantic tradition of primitivism, which, going back to Rousseau, fostered a "belief that the best antidote to the ills of an overly refined and civilized modern world was a return to simpler, more primitive living."[67] National parks, as zones of preservation as well as retreat, enabled the concept of wilderness to become fully idealized as "the natural antithesis of an unnatural civilization that had lost its soul." Wilderness thus came to figure "the ultimate landscape of authenticity"—that is, "a place of freedom in which we can recover the true selves we have lost to the corrupting influences of our artificial lives. . . . [I]t is the place where we can see the world as is really is, and so know ourselves as we really are—or ought to be."[68]

For Cronon, the wilderness ideal that arose in the nineteenth-century United States marks an ideological site of homecoming that has proven detrimental to environmental discourse. If we are meant to "get back" to Nature, wilderness is the *"wrong* Nature," and it is wrong because, as an ideal, it sets up an unlivable standard: "By imagining that our true home is in the wilderness, we forgive ourselves the homes we actually inhabit."[69] The sacralizing ideology of wilderness sections off areas of land to be preserved against human interference. The only responsible way to treat wilderness becomes not to interfere with it at all. Cronon explains: "The wilderness dualism tends to cast any use as *ab*-use, and thereby denies us a middle ground in which responsible use and non-use might attain some kind of balanced, sustainable relationship. . . . The middle ground is where we actually live. It is where we—all of us, in our different places and ways—make our homes."[70] This middle ground where we actually "make our homes" is not an ideal(ized) place for Cronon, and it is one that necessarily remains open to mixed use. It is neither saved nor fallen, pristine nor ruined, natural nor unnatural. Because Cronon's concept of home occupies this middle ground, it cannot be located in wilderness, in Nature. A home must actively be *made,* not simply found: we *"make* our living there," and hence home "is the place for which we take responsibility, the place we try to sustain so we can pass on what is best in it (and in ourselves) to our children."[71]

Cronon articulates a powerful alternative to conventional ecological homecoming. Yet his essay only obliquely links the ecological homecoming narrative to the broader paradigm of settler colonialism. As the author of *Changes in the Land,* Cronon of course understands the close ties between land-use ideology and the legacy of settler colonial violence. And indeed, he does refer to the historical irony whereby the myth of wilderness as uninhabited land enabled the mass displacement of Indigenous communities and animated the subsequent vilification of Indigenous land-use practices. "To this day," Cronon notes, "the Blackfeet continue to be accused of 'poaching' on the lands of Glacier National Park that originally belonged to them and that were ceded by treaty only with the proviso that they be permitted to hunt there."[72] However, to stop with the barest recognition of Indigenous dispossession here misses something crucial about the idealization of

Nature in U.S. (environmental) history: namely, that the transformation of the wilderness concept—along with its increasingly important ideological links to a notion of *home*—was at once driven by and helped legitimize settler colonial expansion. In her contribution to *Uncommon Ground*, Carolyn Merchant comes closer to making this connection explicit through her reading of U.S. expansion as a "recovery narrative" that figured the nation as a lost Paradise waiting to be found. Merchant notes that "the idea of recovery functioned as ideology and legitimation for settlement of the New World."[73] But this claim arises only in passing, effectively subordinated to the recovery narrative itself.

Whereas Cronon and Merchant merely recognize the link between the idealization of Nature and the colonial settlement of the United States, this book aims to push that recognition further and assert a foundational link between the ecological homecoming narrative and the settler colonial imagination.[74] In making this claim I echo a rich body of existing work by Indigenous scholars who have powerfully documented the ways settler colonial occupation is always also a form of violence against the environment as well as human relations to it.[75] The Indigenous philosopher Kyle Whyte (Potawatomi) puts the matter in the bluntest, most urgent terms possible: "Settler colonialism, as an ecological form of domination, is environmental violence."[76]

THE COLONIALITY OF NATURE

While drawing on scholars and activists who link settler colonialism to environmental injustice, this book also situates that link within a lineage of decolonial thinking in order to argue that the (settler) narrative of ecological homecoming must be understood as fundamentally bound up with *the coloniality of Nature*. "Coloniality" names a predominant analytic within decolonial thought, a field that emerges from the observation that modernity and coloniality were born together in the fifteenth century alongside the invention of race as a mechanism of domination. One of the central suppositions of decolonial thinking is therefore that the link between coloniality and modernity is constitutive: "There is no modernity without coloniality, thus the compound expression: *modernity/coloniality*."[77]

The Peruvian scholar Aníbal Quijano initially formalized the analytic of coloniality by rerouting the language of globalization. Quijano demonstrates that "globalization" and its processes for homogenizing the planet's spatial and temporal structures are not simply products of neoliberal capital but "the culmination of a process that began with the constitution of America and colonial/modern Eurocentered capitalism as a new global power."[78] Coloniality describes a model of power that began with the Spanish arrival in the Americas and that has, since that historical moment of rupture, constituted "our" modernity. If the world order prior to the fifteenth century was polycentric and noncapitalist, the voyages to the New World that kicked off the West's global expansion in the latter half of that century instigated a new world order centered on European power. This power expropriated territory and accumulated wealth through processes of resource extraction that depended, in turn, on dispossession, subjection, and enslavement. The reconfiguration of power that took place via the West's expansion required what Sylvia Wynter has termed "a new order of cognition."[79] This new order of cognition instituted a global racial imaginary that systematically subordinated Indigenous peoples and peoples of African descent to positions of subrationality and subhumanness in order to legitimize expropriation and enslavement.[80] As the work of Quijano, Wynter, and other decolonial thinkers powerfully demonstrates, the interrelated logics of domination and racialization may have "a colonial origin and character," but over the centuries that have passed since their instigation they have "proven more durable and stable than the colonialism in whose matrix [they were] established."[81] What this means for decolonial thinkers is that the model of global power that remains hegemonic in the present is in fact continuous with that of the fifteenth century. Contra Michel Foucault and his emphasis on the epistemic breaks that reconfigure the order of knowledge, decolonial thought underscores a powerful continuity that alters the *terms* of knowledge while retaining the same basic forms that *structure* it.[82] The result is that, as Walter Mignolo cogently puts it, "Coloniality is not over; it is all over."[83]

If coloniality is all over, then no one and no thing escapes its reach. Indeed, coloniality applies as much to concepts as to people. Expanding on Quijano's original analysis of the colonial matrix of power, decolonial thinkers have developed specific analyses of a range of key concepts: the *coloniality of power*,[84] the *coloniality of knowledge*,[85] the *coloniality of being*,[86] and the *coloniality of gender*.[87] In each case, the work has not focused on the *colonization of* power, knowledge, being, or gender but rather on an investigation of the coloniality intrinsic to each concept. Power, for instance, should not be understood as a general force that preexists coloniality but rather as a particular, *internal* "instance of the colonial matrix in which all of us, human beings, are being ruled."[88] Likewise, as Maria Lugones argues, European expansion gave rise to "the modern/colonial gender system" that, once forced on communities virtually everywhere on the planet, did not simply transform local meanings of gender but also actively (re)gendered the relations between men and women—especially men and women of color. *Power* and *gender* therefore embody and enact their own coloniality. These and other manifestations of coloniality inform the coloniality of knowledge and its effects on the form, content, and accessibility of various epistemologies.

Among the growing list of terms subject to the analytic of coloniality, this book contributes to the as-yet nascent body of work devoted to the *coloniality of Nature*. To the best of my knowledge the first use of the phrase *coloniality of nature* appears in Walter Mignolo's introduction to a 2007 special issue of *Cultural Studies*.[89] However, he does not define the term. A fuller elaboration of the coloniality of Nature would have to wait until the publication of Héctor Alimoda's essay "The Coloniality of Nature: An Approach to Latin American Political Economy," which first appeared in Spanish in 2011. Alimoda adopts a decolonial framework in order to link political ecology and environmental history in a way that highlights "the persistent coloniality that affects the Latin American nature (as in 'environment')."[90] He specifically analyzes the way Latin America—constituted by both its "bio-physical reality" and its "territorial configuration"—appears to regional and global hegemonic elites as "a subaltern space, which can be exploited, levelled, and reconfigured according to the necessities of the prevailing regimes of accumulation."[91] In other words, Alimoda examines how both local and foreign power organizes Latin American environments as spaces for resource extraction. The reduction of the environment to a resource reserve for the wealthy emblematizes the coloniality inherent in Latin American political economy.

Though decolonial scholars have not yet adopted the term *coloniality of Nature*, its basic premise, as initially laid out by Héctor Alimoda, plays a central role in recent work linking decolonial methods and the study of social ecology. Macarena Gómez-Barris's book *The Extractive Zone: Social Ecologies and Decolonial Perspectives* represents an important example of such work. More explicitly than Alimoda, Gómez-Barris notes that what has become known as "extractivism" plays a key role in Aníbal Quijano's concept of the colonial matrix of power. She develops Quijano's decolonial understanding of extractivism by explaining that, "in its longue durée, extractivism references colonial capitalism and its afterlives, extending from its sixteenth-century emergence until the present day," when it has given rise to what Gómez-Barris terms *the extractive zone*—that is, a particular manifestation of coloniality that "marks out regions of 'high biodiversity' in order to reduce life to capitalist resource conversion."[92]

As for Alimoda, for Gómez-Barris it is not "Nature" in itself that becomes subject to the analytic of coloniality but rather the process of extractivism that indicates a particular way in which global power reduces the environment to a demarcated zone of exploitability. On the surface this analysis does not seem altogether different from other well-known critiques of modernity that lament the instrumentalization of the environment along with the damage this instrumentalization has wrought. A similar critique animates Heidegger's late essays, as well as the Heideggerian strain of eco-phenomenology (see chapter 1). This similarity appears even stronger in light of Gómez-Barris's emphasis on phenomenology, and on how "decolonial thinkers put into motion a range of methods and epistemologies that give primacy to renewed perception."[93] What's different is that, unlike Heideggerian (eco-)phenomenology, decolonial strategies for renewing perception do not depend on home(l)y rhetoric. Decolonial phenomenologies, like the "Andean phenomenology" Gómez-Barris describes, do not seek a restoration of domestic comfort or a radical return to some stable or primordial way of being in the world. Instead, they unsettle the status quo instituted by coloniality in order to unearth those perspectives that have been "submerged" by the centuries-long imposition of a colonial matrix. Though submerged, such perspectives can still "pierce through the entanglements of power to differently organize the meanings of social and political life," and so "the possibility of decolonization moves within the landscape of multiplicity that is submerged perspectives."[94]

To demonstrate the potential for decolonial phenomenology to unsettle the coloniality of Nature, Gómez-Barris narrates an experience she had while walking alone in the darkness one evening in Peru's Sacred Valley: "I had no flashlight and only the night sounds as my guide. The sky hung in a shadowy thick black, pricked only by a dim arc of stars, making the path between danger and safety a narrow one, both physically and existentially."[95] The recognition of being alone in an unfamiliar world took hold, and "a sensation of utter powerlessness and intense solitude came over me." Gómez-Barris reflects that the perception of her aloneness and the negative sensations that accompanied it were linked to a particular concept of the discrete individual fostered by a lineage of European philosophers and integrated into Western modes of living. She also notes how this way of perceiving stands in contrast to Andean phenomenology, which "locates the subject in multirelational terms and blurs the binary distinctions between the human and biomatter into porous interactivity."[96] Without claiming to have full access to Andean modes of perception, and also without pretending simply to abandon ingrained Western ways of perceiving, Gómez-Barris leverages her experience of navigating two contrary phenomenologies to reflect on how embodiment is itself subject to coloniality:

> While [the] unboundedness of being [in Andean phenomenology] was not completely unfamiliar to me, because of my queer femininity and perhaps also because of my propensity to reach beyond the limits imposed by masculine, Western, and reason-oriented subjectivity, *the reorientation of my encounter with existential fear made me realize the extent to which such an affective state is also a colonial and consumer capitalist imposition*. It was only through the realization of an externally imposed relation to fear that I could usher in a different sensibility, one that extended into the unseen and unknown without the interruptions of the logics of separateness.[97]

What makes Gómez-Barris's observation here so significant is the way it points out how perception is always subject to the imposition of ideology, and in her case subject to the global, capitalist ideology marked out by modernity/coloniality. What she terms "masculine, Western, and reason-oriented subjectivity" encourages the subject to see itself as divorced from the environment. On its own this observation merely echoes a vast literature that has long bemoaned "the logics of separateness" that define modernity's ideological separation of Nature and culture. But Gómez-Barris further emphasizes how coloniality comes to bear equally on perceptions of the environment and the self. Even though she does not use the particular phrase, her narrative nonetheless implies that the ideological mechanism that enforces an individual's perceived alienation from the environment is itself an expression of the *coloniality of Nature*. This observation allows an analysis of the coloniality of Nature to go beyond the critiques of extractivism, green-grabbing, and other unjust practices that, as Romain Francis puts it, "demonstrate the inextricable link between oppression of the subaltern and nature."[98] Though such critiques are essential for contemporary political ecology, Gómez-Barris's decolonial perspective also indicates a need to situate the coloniality of Nature within the historical context of modernity/coloniality.

It is only with Walter Mignolo's recent work that the coloniality of Nature comes fully into view within decolonial thought. Drawing on and extending the work of Sylvia Wynter, Mignolo shows that the fifteenth-century invention of "Man" as the racially unmarked (i.e., white, European), universal, and innately superior genre of *the human* both coincided with the invention of Nature and depended on it. Mignolo writes that, in addition to inventing themselves, "Western imperial subjects . . . invented also the idea of *nature* to separate their bodies from all living (and the very life-energy of the biosphere) organisms on the planet."[99] According to Wynter's account, this separation enabled two interconnected developments: first, the post–Copernican Revolution development of the physical sciences, which depended on the notion of an inert Nature whose physical properties could be objectively determined; and second, the global process of racialization by which non-European subjects were ejected from the category of Man and reinscribed as lower animals that, enslaved to irrationality, maintained a primitive link to Nature.[100] Mignolo echoes Wynter when he writes, "The classification and invention of 'Indians'; the classification and invention of 'Blacks' to homogenize the African population; the identification of the New World with 'nature' and with the wealth of 'natural resources' after the Industrial Revolution—all of these are epistemic inventions of ontological natural and cultural entities."[101] Born together in the crucible of coloniality, the two key fictions "Man" and "Nature" collaborated in the expropriation and subjection of untold numbers of "subhumans" and other invented "ontological natural and cultural entities."

Decolonial thinkers provide compelling reasons to abandon the narrative of ecological homecoming and its logic of exile and return. If "Man" is exiled from "Nature," it is not due to the advent of agriculture, big-game hunting, or the Industrial Revolution but rather to the foundational violence that constituted and continues to uphold the modernity/coloniality matrix.[102] As such, there is no going "home" to Nature, because there is no "Nature" to recuperate in the first place. Echoing the work of many scholars who have labored to denaturalize the concept, Mignolo reminds us that Nature exists only as an ontological fiction, and that what really exists, instead, is "the relentless generation and the regeneration of life in the solar system from which processes emerged a species of living/languaging organisms," only some of whom "were able to define themselves as human and impose their self-referential description as standard for all living organisms of the same species."[103]

Attending to the coloniality of Nature demonstrates that, instead of coming home to Nature, *we must decolonize it*. Such a claim will no doubt unsettle many, as indeed it should—and deeply. Nor should the notion of decolonizing Nature be understood as merely metaphorical. This book argues that if the ecological homecoming narrative plays a foundational role in the settler colonial imagination, then a critique of the coloniality of Nature must ultimately agitate for the social and political goal of decolonization, which "involves the repatriation of land[;] . . . that is, *all* of the land, and not just symbolically."[104] This makes matters even more unsettling. Indeed, as Eve Tuck (Unangax̂) and K. Wayne Yang argue, "settler colonialism and its decolonization implicates and unsettles everyone," and for the precise reason that the pathway forward is unknown: "the details are not fixed or agreed upon."[105] This book proceeds in solidarity with Tuck and Yang's implication that unsettling the coloniality of Nature will itself prove deeply unsettling, and that a practice of becoming *and remaining* unsettled constitutes important decolonial work.

LANDSCAPE, ECOLOGY, AND THE UNCANNY

Crucial to the project of unsettling (the coloniality of) Nature will be the parallel project of unsettling landscape. In this book, *landscape* names a modality of representation by which Nature can be framed and given (aesthetic) form in a way that naturalizes the perceived distance that separates the viewer from their environment.[106] As the first two chapters of this book elaborate, landscape also denotes a perceptual and descriptive practice that actively gathers otherwise unfamiliar space into a safe and comforting place; it ushers alienating remoteness into pleasurable closeness. Not unlike the paradox of needing to return to the world we never actually left, landscape is founded on both distance and proximity, separation and identification.

Underscoring this paradox, the French philosopher Jean-Luc Nancy claims that landscape, understood as a perceptual and representational mode, only arises following an experience of alienation. Nancy links alienation to industrial and urbanizing processes that transform the country (*le pays*) into a landscape (*un paysage*). Such processes drive out "the divine . . . presence" and its attendant meaningfulness, such that "peasants find themselves unsettled, straying and lost. It is *thus* that we encounter the question of landscape, that is, of the representation of the country and the peasant, but perhaps also of estrangement and uncanniness."[107] Landscape may appear as a result of alienation, but for Nancy landscape perception enacts an additional logic of phenomenological (self-)absorption:

> [Landscape] is a representation of the land as the possibility of a taking place of sense, a localization or a locality of sense, which makes sense only by being occupied with itself, making itself "itself" as this corner, this angle opened onto an area opposite or onto a spectacle already laid out; but it is an angle opened onto itself, creating an opening and thus a view, not as the perspective of a gaze upon an object (or as vision) but as a springing up or a surging forth, the opening and presentation of a sense that refers to nothing but this presentation.[108]

According to Nancy, a landscape does not afford a vantage of an external reality but rather frames a view of the internal self, perversely externalized. More mirror than environmental expanse, landscape leads the gaze inward, "an angle opened onto itself." Consequently, landscape provides a site for psychic preoccupation rather than physical occupation: it "makes sense only by being [pre]occupied with itself." The unsettling in(tro)version by which landscape occupies and is preoccupied by itself reveals the fundamental illogic that projects an affect of belonging onto landscape. Landscape belongs to itself. The alienation that produces landscape in the first place cannot be redressed through physical or psychic occupation. To occupy landscape physically is to exist perpetually within the psychic experience of exile that produced it. Landscape, the product of alienation, alienates.

John Wylie echoes Nancy's unsettling implication and makes explicit the charge that landscape cannot be a home. Writing from the disciplinary vantage of critical geography, Wylie addresses the "habitual" and "surreptitious" infiltration of landscape studies by a species of "homeland thinking." This type of thinking "asserts an original and indefinitely sustained link between, and perhaps even a fusion of, a site and its inhabitants."[109] According to Wylie, scholars in landscape studies frequently make the ontopological assumption that landscape grounds identity, that it homes in on something essential about an individual person or community. Wylie questions this received wisdom by pointing out the exilic origins of the homeland concept: "Homeland shimmers into existence as something already lost, remote and absent. Exile and displacement are, in a strong sense, the very *preconditions* of any thought of homeland."[110] If homeland has its ground in exile, then so does the assumption of landscape *as* homeland: "A landscape cannot be a homeland—cannot aspire to be such—because there are no such homelands for us to inhabit."[111] Wylie therefore asserts that any future work on landscape must concern itself with "displacing as much as dwelling" and labor actively "to unsettle" the very notion of landscape as the container for "unified communities, regions, nations and worlds."[112]

That said, Wylie's suggestion that the concept of landscape might be unsettled yet retained ignores his own argument that, when conceived phenomenologically, "landscape almost inevitably becomes homeland."[113] As a representational and perceptual mode, landscape plays a crucial role in ecological homecoming narratives. Such narratives typically route themselves through the figure of landscape, which they invest with home(l)y desire. For this reason, working against the ecological homecoming narrative and its ties to both the settler colonial imagination and, more broadly, the coloniality of Nature, would necessarily entail a dismantling of landscape, understood as that most familiar modality of a home(l)y Nature. Dismantling landscape would further require unsettling the entire project of settlement, understood as a homemaking practice aimed at securing one's sense of being in and belonging to a particular beloved place. Contra Wylie, what is left after landscape is unsettled is not landscape at all, but that which landscape represses.

As the terms *unsettling, home(l)y,* and *repression* already suggest, attending rigorously to the discourse of home opens eco-phenomenology to a Freudian logic. In his 1919 essay "The Uncanny," Freud proposed an intimacy between the homely and the unhomely: "*heimlich* is a word the meaning of which develops in the direction of ambivalence, until it finally coincides with its opposite, *unheimlich*."[114] Usually translated as "the uncanny," *das Unheimliche* names a peculiar affective experience that "arouses dread and horror" in response to phenomena that aren't simply strange, but strangely familiar. For Freud, the strange familiarity of the uncanny generates an unsettling muddle of attraction and repulsion, and the experience itself can arise from numerous psychological origins. Freud locates one source of uncanny experience in substitutive relations, such as the Oedipal association between the eye and the phallus. The uncanny also emerges in instances of doubling, such as the appearance of a doppelgänger that obscurely reflects the (super)ego. What these and other sources have in common, however, is *repression*. According to Freud, "The essence of repression lies simply in turning something away and keeping it at a distance, from the conscious."[115] As the horror genre teaches us, keeping unwanted knowledge hidden from view is a dangerous game: we don't recognize it when it inevitably resurfaces, and terror ensues. Hence why repression and its psychic dynamics of denial and repetition generate the perverse obsession with one's own death known as morbid anxiety. However, what distinguishes the uncanny from other causes of morbid anxiety is that it emerges from the repression and recurrence of something more personal, closer to home: "[The] uncanny is in reality nothing new or alien."[116]

If repression generates the uncanny, then what, exactly, is being repressed? According to Freud, it is, in some sense, the sources of the self. He notes that the uncanny can erupt from the involuntary return to a primitive past, "a regression to a time when the ego had not yet marked itself off sharply from the external world and from other people."[117] Elsewhere in the same essay he writes: "Nowadays we no longer believe in [primitive superstitions], . . . but we do not feel quite sure of our new beliefs, and the old ones still exist within us ready to seize upon any confirmation. As soon as something actually happens in our lives which seems to support the old, discarded beliefs we get a feeling of the uncanny."[118] In addition to wrenching us back to our primordial selves, uncanny experience also returns us to a primordial home of sorts. Freud notes that many of his male patients are filled with uncanny sensation at the sight of female genitalia: "This unheimlich place, however, is the entrance to the former *Heim* [home] of all human beings, to the place where each one of us lived once upon a time and in the beginning."[119]

But what does all of this have to do with eco-phenomenology and the ecological homecoming narrative? The links may at first seem remote, given that the home(l)y discourse pervading eco-phenomenology rarely accounts for uncanny experience. But one possible reason for this rarity has already emerged: ecological homecoming narratives function according to the logic of alienation. The alienation paradigm generates an unreflective longing for return and restoration and thus promotes comforting rather than unsettling imagery. As already demonstrated, however, the preoccupation with alienation is an ideological one, and every ideology comes with its own strategies for repression. Indeed, alienation is itself a repressive strategy.

Precisely *what* the alienation paradigm represses cannot easily be summarized, and the complexities of that repression will unfold over the course of the book. In its briefest formulation, however, what alienation represses is the profoundly contingent nature of all being. The ecological homecoming narrative is all about the longing to become whole again, to reinsert the human more fully into being and hence to find a radical sense of belonging in the world. This longing inflates our understanding of what it means to be human, and in doing so it represses the strange way in which to be human is itself already to be other than human. As long as the word "human" designates a particular kind of entity that is produced by genetics as well as by various social, political, and scientific institutions, then we at once over- and underestimate the human. We overestimate the human for its apparent exceptionalism in the natural world—for our superior consciousness, our ingenuity, our innate desire for progress, and so on. Yet we also underestimate all that's required to produce that unique sense of being human. The more we learn about biology, the more we understand that organic life greatly depends on inorganic matter and that an individual human body is comprised of more "foreign" cells than "native" ones.[120] Likewise, the more we learn about consciousness, the more we understand just how selective and limited our perceptual capacities need to be for us to perceive anything at all.[121] Like all other living entities, human beings are radically impure and incomplete, and I suspect that we humans tend to repress what is strange and estranging about Nature because we also repress what is strange and estranging about ourselves. Furthermore, the age of the Anthropocene has licensed far too many environmental humanities scholars to repress the fact that the "we" we refer to tends to be quite limited in terms of race, class, gender, sexuality, and geopolitical status. This unreflective "we" points to an uncritical representation of the human as some kind of universal ontological category to which all biological humans have social, political, and juridical access. Hence, even as "we" extol the virtues of the human, "we" also neglect to see its incompleteness and insufficiency as a category.

The simultaneous over- and underestimation of the human reflects that category's status as a boundary object that at once straddles the social and the ecological and demonstrates a deep incompatibility between them. Race provides another useful example of such a boundary object. We know that from the perspective of genetics, race doesn't exist as a biological actuality. Yet race absolutely exists as a social reality, one that greatly influences all lives through the globally distributed hierarchies of power and privilege that constitute modernity/coloniality. One cannot dismantle the fiction of race simply by pointing to the absence of significant difference in genetic code. The social reality of racial difference polices who does and does not have access to the category *human* as a social reality. Even so, this social policing, along with the various forms of degradation that attend it, cannot change the fundamental, biological fact of humanness.[122] As such, one's *ecological* humanness remains untouchable, even if one's *social* access to the category remains troubled. Like race, "the human" functions as an uncanny boundary object that exposes the otherwise repressed, twofold nature of the contingency of being: at once social and ecological.

Reckoning with either the social or the ecological aspect of contingency can be unsettling enough, but the task is ultimately to reckon with both. This is why other explorations of uncanny ecology prove insufficient for the project at hand. Consider the recent emergence of the critical discourse known as "ecogothic," which examines uncanny and even horrific manifestations of Nature for what they reveal about issues related to various forms of oppression and alterity. Like the present book, ecogothic scholarship pursues otherwise concealed links between the social and the ecological. As Sharae Deckard explains: "[The] ecogothic turns around the uncanny manifestation of the 'environmental unconscious,' particularly those forms of environmental violence or crisis that have been occulted."[123] Even more pertinent to the present project, Kerstin Oloff argues that postcolonial examples of the ecogothic, such as those from the Caribbean, have long indicated "the need to reconceptualize the relation between humans and their environment as central to the project of decolonization."[124] Although ecogothic scholarship spans the social and the ecological, it does so primarily by reading the environment allegorically. Whenever horrific manifestations of Nature appear in stories with strong gothic elements, those *environmental* manifestations always allegorize *social* forms of violence. As an analytical mode, then, the ecogothic does not fully reckon with the twofold contingency of ecological and social being. Instead, it reads the one for the other in a way that the later chapters of this book will complicate, both by pursuing a literary theory of ecological realism (excursus 1) and by putting that theory into practice through readings of fiction that would not generally be characterized as "gothic" (chapters 4–6).

Unlike scholars of the ecogothic, propagators of the ecological homecoming narrative tend not to find Nature *unheimlich* at all, and if this is the case it's because they are not paying attention. Freud suggests as much near the end of his essay on the uncanny, when he turns from psychological forms of the uncanny to literary ones. Literature provides especially fertile ground for thinking about the uncanny because, as Freud admits, it doesn't require the same kind of reality test: the effectiveness of the literary uncanny is "[not] a question of the material reality of the phenomena."[125] This lack of restriction is one of the things that makes fictional worlds so absorbing for readers, who often pay closer attention to fictional realities than they do to their own environments.[126] In literature, authorial manipulation makes it easier for uncanny phenomena to appear: "By means of the moods he can put us into, [the storyteller] is able to guide the current of our emotions, to dam it up in one direction and make it flow in another."[127] Furthermore, the multiple levels of mediation involved in fiction make it possible for the uncanny to manifest in various guises. It is perhaps most common for characters in the storyworld to have uncanny experiences, and readers may or may not share the sensation. It is also possible for readers to pick up on uncanny resonances not experienced by the characters. The narratological distinction between story and plot enables these divergences. There is, however, a third possibility, one that Freud forecloses in his distinction between real and imagined worlds and the kinds of attention we afford each. Fictional environments may be more capable of provoking uncanny experience because they are constructed and manipulated. But in making this distinction Freud also implies that if "our physical environment" seems less uncanny, *this is only because we aren't paying attention*. As Samuel Weber has theorized, the uncanny can only appear when the perceiver is keenly aware of his or her surroundings and has the "desire to penetrate, discover and ultimately to conserve the integrity of perception: perceiver and perceived, a wholeness of the body, the power of vision."[128] Only in such cases does denial—the root of repression—take an active form: "a [form of] denial that in turn involves a certain structure of narration, in which this denial repeats and articulates itself."[129] If, as the eco-phenomenologists tell us, we need to attune our senses to the more-than-human world, then we are as likely to find an *unheimlich* Nature as a *heimlich* one.

Because this is an issue of different levels of attention, the ecological uncanny can creep up on you. Unlike the sublime, the uncanny is not an aesthetic category; one does not seek it out, nor is it a way to frame, package, or consume Nature. Special attention is required, however, to distinguish the affective quality of the ecological uncanny from that of the sublime.

Of the latter Edmund Burke writes: "Whatever is fitted in any sort to excite the ideas of pain, and danger, that is to say, whatever is in any sort terrible, or is conversant about terrible objects, or operates in a manner analogous to terror, is a source of the *sublime.*"[130] Yet as Immanuel Kant makes explicit, sources of the sublime can only be found in Nature, and particularly in natural things that are apparently purposeless.[131] Animals are thus not sublime, but mountains and oceans are. And as nineteenth-century paintings such as Caspar David Friedrich's *Wanderer above the Sea of Fog* (1818) suggest, proper distance and careful framing are usually required for the sublime aesthetic to register fully.

By contrast, the ecological uncanny operates on much smaller scales, and it works through proximity rather than distance. It can also involve inanimate as well as animate entities. An example from Olive Schreiner's *Thoughts on South Africa* illustrates the matter. The first chapter of this work provides a historical and geographical survey of South Africa, and Schreiner indicates that her home country's "colossal plenitude" is primed for the sublime: "If Nature here wishes to make a mountain, she runs a range for five hundred miles; if a plain, she levels eighty; if a rock, she tilts five thousand feet of strata on end; our skies are higher and more intensely blue; our waves larger than others; our rivers fiercer."[132] As if to assist her readers in experiencing this colossal plenitude, Schreiner includes vignettes that read like traveler's guides for framing an optimal landscape experience. In one instance, describing the coastal lowland bush, she writes:

> To see this land typically one should outspan one's wagon on the top of a height on a hot summer's day, *when not a creature is stirring,* and the sun pours down its rays on the flaccid, dust-covered leaves of the bushes, if you stand up on the front chest of the wagon, and look out, as far as your eye can reach, you will see over hills and dales, the bush stretching, silent, motionless, and hot. *Not a sound is to be heard;* your hand blisters on the tent of the wagon; *suddenly* a cicada from a clump of bush at your right sets up its keen, shrill cry; it is glorying in the heat and the solitude of the bush. You listen to it in unbroken silence, till you and it seem to be alone in the world.[133]

Schreiner begins by locating a spot devoid of all inhabitants, as if to suggest that a viewer can only access a profound aesthetic experience of landscape when guaranteed absolute solitude. Yet despite the traveler's efforts, and precisely at the moment when she stands atop the wagon like a conqueror to gaze upon the intense purity of South African Nature, a screeching cicada interrupts her solitude; the insect's sudden irruption troubles the traveler's framing efforts and undermines the sublime aesthetic that is this outing's raison d'être.

The significance of this passage resides not only in the way it empties the landscape only to reveal one of its (presumably many) hidden inhabitants but also in the subtle way the passage represses this revelation, seeking to preserve a sense of Nature as a backdrop for the solitary subject. Despite the suddenness of the nearby cicada's cry, the traveler doesn't flinch. The insect's shrillness quickly becomes recast as "glorying," and regardless of its forceful and alien presence, the traveler absorbs it into her experience to remain paradoxically *alone* in the world *together* with the cicada. This reading might seem too forced if the passage did not also employ second-person direct address. The guide-like nature of the vignette has an instructional quality, and its strategy for repressing the cicada has a quietly inculcating effect. The narrator tells us to keep our focus outward. Only in this process of imaginatively clearing space to *"seem to be* alone in the world" can we fully experience Nature's ennobling beauty and thus consolidate our sense of self. Particularly telling is how Schreiner wants to purify this vignette of all anxiety: "It is not fear one feels, with that clear, blue sky above one; that which creeps over one is not dread."[134] In resisting the uncanny cicada that disturbs her landscape vision, Schreiner also resists the potential defamiliarization of the "I" that might result from it.

The reason the aesthetics of the South African landscape win out over the ecological uncanny has something to do with Kant's understanding of the sublime. For Kant, the power of the sublime doesn't reside in the ability of natural objects to elicit existential dread. Instead, sublime experiences are ultimately about human superiority; they are characterized by the feeling that human reason is more powerful than Nature. This is particularly true for what Kant calls "dynamically sublime" experience, which occurs when natural phenomena inspire fear even though we know we are safe. The dynamically sublime produces a kind of fearful delight in which "the irresistibility of [nature's] power certainly makes us, considered as natural beings, recognize our physical powerlessness; but at the same time it reveals a capacity for judging ourselves as independent of nature and a superiority over nature . . . whereby the humanity in our person remains undemeaned even though the human being must submit to that dominion."[135] The sublime therefore affirms human exceptionalism, placing the species both exterior and superior to Nature. Hence why Friedrich's wanderer appears *above* the sea of fog.

When Schreiner tells her prospective traveler to bask in the great vastness of the landscape view, she is insisting on landscape's sublime power to bolster our own self-conception. And when the cicada irrupts into the narrative, Schreiner adjusts the frame to return it safely to Nature, where it belongs; the story can proceed apace as if nothing has happened. But what would happen if we readers stayed with the moment of irruption and read it for the uncanniness it betokens—that is, as a crack in the aesthetico-ideological edifice that otherwise conceals landscape's foundation in alienation and repression? Not only would the cicada's disruptive presence be allowed to unsettle the ideological framing of Nature that Schreiner's text works so hard to maintain. It would also be allowed to unsettle the narrative device by which Schreiner, herself a product of settler colonial culture, sets up a landscape vision of "her" South Africa. Such represents the twofold project of *Unsettling Nature*.

UNSETTLING NATURE

This book's argument unfolds over the course of two parts, each of which features three chapters and concludes with a theoretical "excursus" that weaves the threads of the previous chapters into something new. Although the full reach of the argument will emerge most clearly through a chronological reading, each chapter may also be productively read on its own.

Part 1's broadest function is to introduce the book's three guiding tropes: the coloniality of Nature (chapter 1), the home(l)y metaphysics of landscape (chapter 2), and the ecological uncanny (chapter 3).

Chapter 1 opens the book by turning to eco-phenomenology's favorite philosopher: Martin Heidegger. The chapter begins with a recognition that Heidegger's hermeneutic phenomenology doubles as a metaphysics of home, grounded in the concept of dwelling. Drawing on decolonial and Black studies scholars who have elaborated connections between the Heideggerian figure of Dasein and the coloniality of being, the chapter extends the analytic of coloniality to understand how Heidegger's complex and frequently strained topology of being outlines a peculiar concept of landscape that recapitulates *the coloniality of Nature*.

The concern with landscape continues in chapter 2, which considers the novelistic theory and practice of Willa Cather. Specifically, chapter 2 explores *the home(l)y metaphysics of landscape* as it manifests in Cather's 1925 novel *The Professor's House*, which uses architecture as a thematic and narratological tool that focalizes landscape into a primordial dwelling place. On a thematic level, the novella at the novel's center, "Tom Outland's Story," narrates how Tom Outland's settlement and excavation of a Native American archaeological site on New Mexico's Blue Mesa imaginatively reconsecrates the Mesa as Tom's own sacred ground—a meaningful landscape that coalesces around him and brings him home to himself, all the while sublimating Indigenous history into Tom's Native American inheritance. On the level of narrative architecture, the novella serves as a kind of narratological picture window that offers the novel's erstwhile protagonist, St. Peter, a magisterial view of a landscape and a mode of existence that resonates deeply with his own desire for a revitalized sense of being and belonging.

Chapter 3 begins to trouble the conceptual underpinnings of landscape by examining D. H. Lawrence's unsettling landscapes of the U.S. Southwest. Emphasizing the "weird anima" of landscape, Lawrence's New Mexican writings—and particularly his 1925 novella *St. Mawr*—reconceive home as an intrinsically unhomely place of being without belonging. The chapter pays special attention to Lawrence's use of spectral rhetoric to signal the vital and frequently disturbing complexity underlying his notion of the "spirit" of place. As such, the chapter turns from the home(l)y metaphysics of landscape to *the ecological uncanny*. The final section of the chapter addresses Lawrence's representations of Native Americans to trouble the claim that his writing recapitulates the settler colonial logic of Indigenous elimination. Without celebrating Lawrence as a model of political correctness, the chapter concludes by suggesting that Lawrence's unique depiction of New Mexico's uncanny spirit of place may nonetheless help unsettle fantasies of settler colonial homemaking.

Part 1 ends with a bridging text—called an "excursus"—that prepares the way for part 2 by introducing a (literary) theory of *ecological realism*. Excursus 1 turns to literary realism in order to think through the challenges that interpretive practices based in symptomatic reading—broadly speaking, reading for allegory—pose for a specifically ecological mode of analysis. As D. H. Lawrence's engagement with the crisis of ecological representation indicates, the uncanny reality that comes into view when cracks appear in the beautiful, mirror-like surface of Nature signals the need for an unsettling ecological realism. Excursus 1 makes it possible to turn to the material of part 2, much of which engages with the possibility of working against certain allegories of reading to reveal uncanny ecological realisms that, in turn, draw attention to the contingent complexities of settler colonial occupation.

Chapter 4 serves broadly as an introduction to part 2, which makes a geographic shift from the U.S. Southwest (the scene of writing for both Cather and Lawrence) to southern Africa (the original site of the [post]colonial farm novel). Chapter 4 begins by defining and historicizing the "farmworld" ideology at the heart of the Afrikaans *plaasroman* (farm novel) tradition, wherein the farm becomes enshrined as a self-enclosed and hence self-justifying space—a world in itself that remains metaphysically separate from its surrounding environs. The chapter investigates how this ideology plays out through a reading of C. M. van den Heever's Afrikaans-language novel *Somer* (1935). After outlining the basic ideological structure of the farmworld, the argument returns to a discussion of uncanny ecology and demonstrates why ecological realism is both philosophical and literary, at once suggesting a new frame for thinking and introducing a new way of reading.

The theoretical expansion of ecological realism prepares the way for the final two chapters, which explore how two English-language farm novels—Olive Schreiner's *The Story of an African Farm* (chapter 5) and Doris Lessing's *The Grass Is Singing* (chapter 6)—progressively dismantle the Afrikaner farmworld ideology. Each of these chapters examines quasi-apocalyptic visions that occur in Schreiner's and Lessing's novels when, in moments of mental breakdown, white characters imagine the destruction of the farmworld by way of ecological uprising. These world-ending visions do not simply allegorize a racial insurgency that will eventually restore native lands to Black African hands, as many readings have it. Rather, these novels disrupt (post)colonial allegory by developing uncanny forms of ecological realism, opening these novels to speculative modes of reading that reorient our attention from the world of human concern to a wider universe of entities. Thus, although these novels attend closely to the social, political, and historical inequalities of settler colonial society in southern Africa, they also encompass the violence that human settlers have perpetrated against nonhumans.

Unsettling Nature concludes with a second excursus, which addresses the impasse that arises in the chapters of part 2 between the ecological and the social. Trading the chiefly social concept of *complicity* for the more fecund socioecological concept of *contingency*, excursus 2 draws the book to a close by turning away from eco-phenomenology and outlining a speculative conception of *exo-phenomenology*—that is, a phenomenological mode that engages deeply with the alterity of others, and with the self as its own Other. Excursus 2 finds a curious model for exo-phenomenology in Magda, the protagonist of J. M. Coetzee's experimental farm novel of 1977, *In the Heart of the Country*. Following a reading of Magda's exo-phenomenological thought and practice, the excursus concludes with a suggestion that alienation may in fact be "natural" to the human condition and hence something worth embracing instead of repressing. Rather than enabling further narratives of ecological homecoming, such a reckoning with the inheritance of alienation may serve the urgent and profoundly unsettling project of decolonization.

PART I

CHAPTER 1

Martin Heidegger and the Coloniality of Nature

> Through our questioning, we are entering a landscape; to be in this landscape is the fundamental prerequisite for restoring rootedness to historical Dasein.
> —Martin Heidegger, *Introduction to Metaphysics*

Martin Heidegger may well be the twentieth century's greatest thinker of being, but he is also one of that century's foremost philosophers of home. The rhetoric of home stands at the very center of Heidegger's work, from his early analysis of Dasein in *Being and Time* to his exploration of dwelling in the late essays. Indeed, the philosopher's lifelong project sought to wrest humanity out of its exile in the postindustrial age of technology, and to manifest a radical homecoming to being. Such a homecoming did not, for Heidegger, entail the construction of a new philosophical edifice so much as the "destruction" (*Destruktion*) of the European ontometaphysical tradition, which understood being strictly in terms of beings. Metaphysics—whose main analytical instruments included schematization, calculation, and a claim to objectivity ensured by rational logic—had long ago fallen under the spell of discourse that understood being as the constant presence of physically manifest entities. Metaphysics had thus also long since forgotten to ask the crucial question, "How does it stand with being?"[1] For Heidegger, relearning how properly to pose the question of being goes beyond issues of grammar and syntax. Such relearning requires a new mode of existing; it requires us to inhabit an essential mode of "dwelling" (*Wohnen*) that the ontometaphysical tradition had rendered obscure. Only in rediscovering how to dwell—that is, in rediscovering how to be at home in the world—can we return to ourselves and restore our primordial residence in the "house of being" (*Haus des Seins*).

This chapter begins and ends with the question of home, of dwelling. Heidegger's thinking harnesses phenomenology in order to lead us home to our most essential dwelling in being.[2] But where, exactly, does such dwelling take place? The question of the "location" of dwelling is a difficult one, and one that always haunted Heidegger. He felt particularly deflated by persistent misunderstanding of *Dasein*, the term he used to refer to human being in its particular capacity as a being (*Seiende*) that thinks about being (*Sein*). Dasein is the ordinary German word for "existence," but its literal meaning, "being-there" (or "being-here"), has fueled much confusion. Where exactly does Dasein, as a being-*there*, dwell? Heidegger frequently guards against all literally spatial understandings of the *Da* of Dasein, as when he complains about its rendering in French as *être-là*, a literal translation of "being-there" that fails to signify Dasein's colloquial meaning as existence (*existence* in French). With this rendering, Heidegger complains: "Everything that was gained as a new position in *Being and Time* is lost. Are human beings there like a chair? . . . Dasein does not mean being there and being here [*Dort- und Hiersein*]."[3] Gaston Bachelard falls prey to this mistake when he asks: "Where is the main stress . . . in *being there* (être-là): on *being*, or on *there*? In *there*—which would be better to call *here*—shall I first look for my being? Or am I going to find, in my being, above all, certainty of my fixation in a there?"[4] But for Heidegger there can be no fixed "hereness" or "thereness" of being. Dasein dwells in a more abstract location: not a specific place but rather an "openness where beings can be present for the human being, and the human being also for himself."[5]

Aside from the topological challenge of locating Dasein's proper dwelling place, there is also the matter of the dark side implicit in what Theodor Adorno once termed Heidegger's "homey murmurings."[6] The recognition of Heidegger's entanglement with National Socialism has led some critics to draw damning connections between his concept of dwelling—itself frequently paired in his writing with the notion of homeland (*Heimat*)—and the Nazi ideology of *Blut und Boden*, which invokes an exclusive definition of German identity based on the pairing of blood descent (*Blut*, blood) and loyalty to the homeland (*Boden*, soil).[7] For scholars such as David Harvey and Neil Leach, the concept of dwelling marks an implicitly fascist strain in Heidegger's thinking.[8] Others, such as Hannah Arendt, Philippe Lacoue-Labarthe, and Herbert Marcuse, denounce such criticism for its ad hominem thrust, its failure to engage rigorously with Heidegger's writing, or else its attempt to "discredit ideas explicit in the later thinking largely on the basis of the political engagement apparently present in the earlier."⁹

There are thus two major issues with Heidegger's home(l)y articulation of dwelling: the topological difficulty related to "where" Dasein dwells, and the ideological difficulty related to dwelling's disturbing political resonances. Neither issue has been satisfactorily resolved, and they continue to cause trouble for those scholars who draw inspiration from his philosophical works. Rather than seeking to resolve these difficulties, however, this chapter puts further pressure on them. Instead of exonerating or condemning Heidegger, my purpose here is to pry his work open further and demonstrate how the link between Dasein and dwelling in his thinking functions as both crux and crisis point.

Prying Heidegger open in this way requires echoing and amplifying ongoing debates within the environmental humanities. Crucially, it also requires going beyond these debates, and indeed, beyond the pale of the explicitly "ecological." Thus, after offering a brief evaluation of the philosopher's foundational influence on the disciplines of ecophilosophy and ecocriticism in the next section, the following section turns to recent work in decolonial theory and Black studies. This recent work situates Heidegger within a historical lineage of modernity/coloniality and shows how his analysis of Dasein recapitulates the coloniality of being. Whereas Heidegger takes Dasein to be a universal figure for the human, these scholars convincingly argue that "the colonized *is not* this ordinary Dasein."¹⁰ The final three sections of this chapter expand the frame to consider how the coloniality of being, as embodied by Heidegger's articulation of Dasein in *Being and Time* and other early writings, relates to the philosopher's shifting elaboration of the topology of dwelling across the *longue durée* of his career. In particular, I examine Heidegger's anomalous use of the word *landscape (Landschaft)* and read this term—one of many in the philosopher's proliferating topological lexicon—as emblematic of the crisis in Heidegger's ongoing struggle to articulate the proper place of human being, the *Da* of Dasein. Landscape represents Dasein's proper dwelling place, in the sense that it is the place where Dasein *is* most fully in being. Yet as the privileged domain of Dasein, which is an already exclusive category of human being, landscape also proves improper—a doubly exclusive dwelling that recapitulates the coloniality of Nature.

As a whole this chapter reevaluates the utility of Heidegger's work for present and future ecological thinking. It also develops a heuristic that later chapters will draw on and elaborate further. This heuristic not only affords a strategy for reading landscape for its *ideological function*. It also specifically locates this ideological function vis-à-vis the colonial matrix of power that grounds the coloniality constitutive of modernity.[11] At its most ambitious, then, this chapter situates the Heideggerian understanding of dwelling-in-landscape within the long history of modernity that has, from the sixteenth century to the present, animated and given voice to one of the sustaining narratives of the (settler) colonial imagination.

A HEIDEGGERIAN ECOLOGY?

It has become a commonplace among scholars in the environmental humanities to cite Heidegger as a foundational antecedent to the ecophilosophical and ecocritical vanguards of the 1970s and 1980s. Energized by the momentum of the environmental movement, which had grown steadily since the publication of Rachel Carson's *Silent Spring* in 1962, these vanguards answered Carson's call for an ideological challenge to Western culture's anthropocentrism. For many of those who responded to Carson's call, Heidegger's hermeneutic phenomenology seemed indispensable for the task. Among philosophers, Heidegger proved central to the founding of deep ecology in the 1970s,[12] and he remains a significant figure for ecophenomenology.[13] Among literary critics, Heidegger's influence has proven equally significant. Heidegger has most obviously left his mark on Jonathan Bate's seminal study *The Song of the Earth*, as well as on work that responds to Bate.[14] Heidegger has also left a more oblique but no less profound mark on other crucial works of first-wave ecocriticism, such as Robert Pogue Harrison's *Forests*.[15] Heidegger's name also appears consistently across numerous essays reprinted in important ecocriticism anthologies,[16] as well as in key appraisals of the growing discipline,[17] which at the very least demonstrates his ongoing relevance to ecocritical discourse.

What has made Heidegger indispensable to the environmental humanities? Perhaps most important has been his critique of anthropocentrism, as well as his call for an ethics of "letting-be," in which humans would cultivate care for and preserve fellow beings. These principles stand at the very heart of deep ecology, the central precept of which emphasizes the inherent worth of all living entities. Michael Zimmerman, the leading authority on Heidegger and ecology, cites as further reasons for the philosopher's appeal "his meditation on the possibility of an authentic mode of 'dwelling' on the earth, his complaint that industrial technology is laying waste to the earth, [and] his emphasis on the local place and 'homeland.'"[18] For ecocritics, there is also the matter of Heidegger's use of language and his later emphasis on poetry. Heidegger was an enthusiastic reader of Hölderlin and Rilke, and his writing on poetry and poetics has enabled a notion of Heideggerian "ecopoiesis" as itself a mode of saving the earth.[19]

Yet despite the celebration of Heidegger threaded throughout ecophilosophy and ecocriticism, dissenting voices have emerged from within both. One of the first to question the philosopher's eco-cred was Michael Zimmerman. Zimmerman initially expressed his concerns in a 1993 essay in *Environmental Ethics*,[20] and he later revisited these concerns in his contribution to Brown and Toadvine's anthology *Eco-Phenomenology*.[21] In the earlier essay he emphasizes two points that weaken Heidegger's compatibility with ecological thinking: first, his relationship to National Socialism, and second, his failure to see humans as animals. In the later essay Zimmerman explores the even more problematic possibility that "Heidegger's own thought—despite his own personal or political preferences—is consistent with modernity's project of the technological domination of nature."[22] Zimmerman's questioning ultimately leads to ambivalence rather than outright dismissal, leaving Heidegger's utility for ecological thinking up for debate.

The debate Zimmerman started recently resurfaced in the pages of the premier journal of ecocriticism, *Interdisciplinary Studies in Literature and the Environment* (*ISLE*), when Greg Garrard published a controversial article titled "Heidegger Nazism Ecocriticism." Though Garrard echoes Zimmerman's concerns about Heidegger's connection to National Socialism, he also makes the more forceful claim that, regardless of the philosopher's flirtation with Nazism, Heidegger's project is doomed from the get-go, since "there is no question of Being" in the first place, and further: "Once the mood music of 'saving the earth' is set aside, we can see that Heidegger's 'question of Being' is actually so contemptuous of mere beings, and so suspicious of science as a mode of enframing, that ecological crisis could not but fall into the realm of the 'inessential.'"[23] Garrard's appraisal of Heidegger's ongoing value for ecophilosophy and ecocriticism is obviously bleak. Two years after Garrard's essay, *ISLE* published a response by John Claborn, who dismisses Garrard's major claims as "problematic" and urges ecocritics to recuperate Heidegger, using his work to develop a (re)new(ed) "eco-ontology."[24] The exchange between Garrard and Claborn is important at the very least for its having taken place at the very heart of contemporary ecocritical discourse. The high visibility of such an exchange suggests that now more than ever, Heidegger's foundational influence on the contemporary environmental humanities is primed for reevaluation.

As my own reevaluation should indicate, I seek to develop Garrard's critique further. Yet unlike Garrard, who I think discards Heidegger before registering the full extent of his centrality to eco-phenomenology, I see the need to carry the critique of Heidegger into environmental philosophy at large. In doing so I follow Michael Marder, whose recent collection of essays on Heidegger seeks to understand how the philosopher's politics function in concert with his ecological thinking. Like Marder, I am interested in how the political and ecological converge in Heidegger's topology of being. However, my own reading of Heidegger differs quite drastically from Marder's, both in terms of the texts I choose to emphasize, my methods for reading them, and my fundamentally critical conclusions.[25]

HEIDEGGER AND THE COLONIALITY OF BEING

Beyond the realms of ecophilosophy and ecocriticism, a reevaluation of another sort has been taking place in two seemingly remote outposts of academia: decolonial theory and Black studies. As outlined in the introduction to this book, decolonial theory stands at the forefront of the analysis of the global paradigm of modernity/coloniality. And though not often articulated explicitly within a decolonial framework, the work of many scholars in contemporary Black studies has felicitous affinities with decolonial thinking. This section briefly outlines the particular affinities that relate to the critique of Heidegger's philosophical project. Put briefly, just as the ontometaphysical tradition had been plagued by the forgetfulness of being, as Heidegger claimed, decolonial and Black studies scholars argue that the reception of Heidegger, not to mention Western philosophy more broadly, has been plagued by the "forgetfulness of coloniality."[26]

Decolonial thinking begins from the observation that modernity and coloniality were born together in the sixteenth century and have shaped world systems from that point to the present. Developing from the foundational work of the Peruvian sociologist Aníbal Quijano, who originated the concept and analysis of coloniality, a contingent of scholars based throughout South America and the United States have extended Quijano's work to consider specific modalities through which coloniality is expressed. The extension of Quijano's work has thus given rise, within decolonial thought, to specific analyses of the *coloniality of power*, the *coloniality of knowledge*, and, most important here, the *coloniality of being*. Whereas the coloniality of power and the coloniality of knowledge refer, respectively, to modern forms of domination and their impact on the production and circulation of knowledge, the coloniality of being "thematize[s] the question of the effects of coloniality in lived experience."[27]

Although Walter Mignolo coined the phrase, the foremost thinker of the coloniality of being since its origination in 2003 has been the Puerto Rican scholar Nelson Maldonado-Torres. According to Maldonado-Torres, the ongoing significance of Heidegger's ontology makes his work "an inescapable reference point" for the discussion of the coloniality of being.[28] Maldonado-Torres begins his discussion of Heidegger's entanglement with the coloniality of being by recalling Mignolo's decolonial refrain, "I am where I think,"[29] to situate the "where" of Heidegger's thinking. Specifically, he situates Heidegger within a historical context that elucidates the ideological mechanism on which he grounded his philosophy: that is, his search for the very roots of thinking in antiquity. As Maldonado-Torres notes, this search for roots was itself rooted in a myth that had originally gained popularity around the turn of the nineteenth century, and that claimed a unique affinity linking Germany to ancient Greece. The myth of Greco-German affinity was grounded in two ideas. The first was the idea of a shared linguistic inheritance. The second was the idea that Greeks and Germans each enjoyed an autochthonous relationship to their homeland. The Greek political myth of belonging derived from the story of Erichthonius of Athens, who sprung forth from a spot of soil where a drop of Hephaestus's semen landed. Likewise, the German political myth of origin depended on the nationalist figure of the *Heimat,* or homeland: the very wellspring of Germanness itself.

According to Maldonado-Torres, Heidegger, as a twentieth-century inheritor of the Greco-German myth, sought to extend the affinity between Greece and Germany from the matters of language and territory to the matter of thinking: "While Erichthonius remains as the model for the political myth of roots in the land, Heidegger posits Pre-Socratic thinking, 'sprung up from the arche [i.e., origin] of being itself,' as the authentic root of thinking—a way of thinking that would contrast sharply with Western metaphysics and epistemology."[30] Heidegger posits himself as the inheritor of pre-Socratic philosophy, in which case the "arche of being" that sprung forth in Greek thinking implicitly serves to legitimize German philosophy (and Germany itself) as "the new arche of Europe."[31] For Maldonado-Torres, then, the claim Heidegger makes to the lineage extending from ancient Greece to modern Germany helps pinpoint the "where" of his thinking and reveal the geopolitics of knowledge within which the philosopher orients himself—a geopolitics that grounds the idea of an autochthonous relation between the German *Volk* and the German *Heimat* simultaneously on "a politics of the earth and a politics of exclusion."[32] Heidegger's geopolitics of knowledge recapitulates the foundational logic of modernity, which posits European exceptionalism through the exclusion (sometimes explicit but more often implicit) of Europe's others. And if Heidegger's geopolitics of knowledge repeats the logic of modernity, that's just another way of saying that it repeats the logic of coloniality—since coloniality, as Mignolo repeatedly intones, is "constitutive, not derivative, of modernity."[33]

How does Heidegger's embeddedness in the logic of modernity/coloniality affect his analysis of Dasein? As the privileged being that asks the question of the meaning of its own being, Dasein has central importance for Heidegger. Dasein serves as the point of departure for his ontological analysis in *Being and Time*, and it remains the protagonist as he explores the various temporal and spatial manifestations of Dasein's experience in and understanding of the world. Significantly, Heidegger's work depends on Dasein not only as an analytic but also as a universal. Despite his attempt to de-struct metaphysics, Heidegger must still posit a universal notion of the human as the ground for his analysis. Without Dasein there is no question of the meaning of being, and so no need to recover that question in the first place.

For Maldonado-Torres, however, the presumed universality of Dasein represents a crisis point in Heidegger's analysis. To isolate and elaborate this crisis, Maldonado-Torres points to the second division of *Being and Time*, where Heidegger identifies "being-toward-death" as the primary factor that individuates humans, enabling each "to determine her ownmost possibilities, and to resolutely define her own project of ek-sistence."[34] The singularity of death situates Dasein most fully in being, and Dasein must therefore be allowed to encounter its own death on its own terms. By contrast, the conditions of living for colonized and enslaved peoples, as well as marginalized and oppressed peoples who live in the wake of colonialism and enslavement, do not often allow for such a singular encounter with death. Indeed, the modern world is, for the colonized and racialized subject, a world of perennial war: "The lived experience of racialized people is deeply touched . . . by the constant encounter with violence and death . . . [and] shaped by the understanding of the world as a battle field in which they are permanently vanquished."³⁵ The implications of this observation are clear: "The encounter with death is no extra-ordinary affair, but a constitutive feature of the reality of colonized and racialized subjects. *The colonized is not this ordinary Dasein.*"³⁶ What this means is that, just as Heidegger orients his thinking within a geopolitics of exclusion (i.e., an exclusive idea of German autochthony), he also grounds his philosophical work on a fundamental ontological exclusion. It is precisely this ontological exclusion of the figure that Frantz Fanon termed *les damnés de la terre* (the wretched of the earth) that Maldonado-Torres identifies as "the very radical starting point to think about the coloniality of Being."³⁷

Maldonado-Torres's analysis of Dasein as an emblem of the coloniality of being has important affinities with recent scholarship in Black studies, and particularly with the critical framework known as Afro-pessimism. Scholars in Black studies have recently echoed the claims of decolonial scholars when they assert that anti-Black racism is constitutive of modernity.³⁸ Extending this basic assertion, scholars associated with Afro-pessimism think and write in response to the ongoing effects of global anti-Black racism and the continued oppression of people of color "in the wake" of colonialism and slavery.³⁹ When the Congolese writer Sony Labou Tansi originated the term *Afro-pessimism* in 1990, he was referring to the way economic conditions had led to the dehumanization of Africans: "Afro-pessimism, a terrible word used to conceal the greatest mess of all time . . . that dooms us to construct and build garbage economies in the depths of the most cruel, unbearable, and inhuman form of indignity that humans can swallow."⁴⁰ Afro-pessimist thinking has maintained Tansi's emphasis on dehumanization, and it has also placed further pressure on the assumption that *the human* offers a paradigmatic figure of universal relevance. Perhaps most significant in this vein of thinking is Sylvia Wynter, a Jamaican theorist whose work presents a tour de force analysis of the complex historical conditions that defrauded Blacks of access to the normative "genre" of humanity, "Man"—that is, "our present ethnoclass (i.e., Western bourgeois) conception of the human, . . . which overrepresents itself as if it were the human itself."⁴¹ Wynter's critique of the overrepresentation of Man has proven central to Afro-pessimism and its interlocutors, and particularly to those who have placed special emphasis on the figure of the human in relation to the question of Black being.⁴²

Although Heidegger's name appears sporadically in the fast-growing archive of Black studies,[43] no work in this field has subjected Heidegger's thinking to as thoroughgoing a critique as Calvin Warren's recent book *Ontological Terror.* Like Maldonado-Torres, Warren observes that the lineage of postmetaphysics to which Heidegger gave birth has a foundational problem, in that it posits being as universal and "proceed[s] as if the question of being has been settled and that we no longer need to return to it."[44] The failure to unsettle being's universality requires Warren to pose the question of Black being: "Is the black, in fact, a human being?"[45] Warren responds unflinchingly in the negative. To demonstrate why, he reiterates the basic aim of Heidegger's postmetaphysical project. Heidegger wants to de-struct metaphysics, leaving behind the one thing that metaphysics cannot eliminate: the primordial relationship between Dasein and being. But as Warren argues, this primordial relationship can exist only because the question of being itself is irreducibly dual. After all, the question of being is just another way of asking, "Why are there beings at all instead of nothing?"[46] The very structure of this question places "beings" at odds with "nothing," requiring the negation of the latter to enable the former: "Dasein flees the anxiety nothing stimulates and projects it as terror onto blacks."[47] Anti-Black violence recapitulates the primordial logic of being, because "black being incarnates metaphysical nothing, the terror of metaphysics, in an antiblack world."[48] Warren's analysis here is worth quoting in detail: "Black being's function within metaphysics is to inhabit the void of relationality—relationality between it and Being and relationality between it and human-being-ness and the world itself. Thus, we must reconceptualize black being ontometaphysically as *pure function* and not relation. . . . In a word, black being helps the human being re-member its relation to Being through its lack of relationality. The essence of black being, like the essence of technology, is to open up an understanding of Dasein, it is always being for another."[49] Blackness in metaphysics therefore has a function, but no access to being. Its function, as the incarnation of nothing(ness) that must constantly be negated, is to enable and sustain the primordial relation between Dasein and being. Read in this way, Heidegger's philosophy becomes, shockingly, "an allegory of antiblackness and black suffering," and hence a recapitulation of "the metaphysical violence of the transatlantic slave trade."[50]

DASEIN AND THE PROPER PLACE OF DWELLING

If Heidegger's geopolitics of knowledge situates his work within the colonial matrix of power, and if his analysis of Dasein is based on the false premise of *the human*'s universality and hence alienates *les damnés de la terre* from being, then what does this mean for ongoing debates in the environmental humanities regarding Heidegger's value for ecological thinking? More specifically, since ecological thinking tends to privilege the spatial more than the temporal aspects of Heidegger's ontology, we must ask what the critiques from decolonial and Black studies scholars means for the understanding of Heidegger's topology of being. If the coloniality of being marks the figure of Dasein, then to what extent does it infiltrate Dasein's proper place of dwelling? Answering this question will occupy the remainder of this chapter, and the process of doing so will require two analytical steps. This section starts by defining the "where" of Dasein's proper dwelling place. The final two sections go further, showing how Heidegger's topology comes into crisis and what this means for a Heideggerian ecology.

Critical consensus among Heidegger scholars posits a *Kehre*, or "turning," midway through the philosopher's career as he moves away from the "meaning of being" (the preoccupation of *Being and Time*) to the "truth of being" (the preoccupation of his writing from *Contributions to Philosophy* onward). Heidegger's *Kehre* also apparently entails a conceptual shift in emphasis from "World" (*Welt*) to "Earth" (*Erde*).[51] This apparent shift inspires Heidegger's eco-interpreters to talk up the significance of his organic philosophical vocabulary and emphasize the preeminence of Earth in the late essays. According to these scholars, as the Earth takes center stage, it also becomes the site of Dasein's most primordial dwelling place.

Unlike these interpreters, however, I remain skeptical of any significant turning in Heidegger's work. Although a clear stylistic transformation occurs, Heidegger's emergent earthiness does not constitute a major advancement in his thinking as such. On the contrary, from *Being and Time* to the late essays there exists profound and even monotonous emphasis on the constant, dynamic oscillation between appearance and withdrawal. Within Heidegger's broader project, Earth emerges simply as one polarity of this enduring oscillation: the polarity of withdrawal, or *concealment*. And as part of a twofold dynamic, Earth cannot be thought independently from its companion, World, which names the polarity of appearance, or *unconcealment*. These polarities exist in a mutually sustaining relationship of tension. However, despite the essential belonging-together of Earth and World, this section aims to demonstrate that for Heidegger, who is primarily a philosopher of appearances, Earth is in fact a secondary concept that serves dialectally to define World as the open region within which Dasein most essentially dwells. The point I shall argue, then, is that whereas Heidegger's eco-interpreters claim that Dasein dwells most primordially on the Earth, Heidegger himself situates the proper place of Dasein's dwelling in the World.

Heidegger first investigates the relation between Earth and World in his essay "The Origin of the Work of Art." Heidegger claims that a work of art, as a "work," possesses the power to gather two opposing forces that he designates Earth and World. He introduces these terms in his famous analysis of a Van Gogh painting, but it is not until the later passage on the Greek temple that Heidegger attempts to clarify his meaning. He begins by establishing these concepts' cryptic interrelation through the example of a building and its surroundings:

> Standing there, the building rests on the rocky ground. This resting of the [temple] draws up out of the rock the obscurity of that rock's bulky yet spontaneous support. Standing there, the building holds its ground against the storm raging above it and so first [*erst*] makes the storm itself manifest in its violence. . . . The temple's firm towering makes visible the invisible space of the air. The steadfastness of the work contrasts with the surge of the surf, and its own repose brings out the raging of the sea. Tree and grass, eagle and bull, snake and cricket first [*erst*] enter into their distinctive shapes and thus come to appear as what they are.[52]

Heidegger describes an ontogenetic relationship between the temple and its surroundings. The temple stands between ground and sky. The very fact of its standing there reveals something about the sturdiness of the rock on which it is built, and through this sturdy grounding the temple brings the aerial chaos of the storm into appearance. Ground, sky, and temple each come into appearance "first" through their interrelations.

Yet Heidegger's dialectical narration proves trickier than it initially seems. After all, Heidegger is no traditional dialectician, which has made his point here easy to misapprehend. Mark Wigley, for instance, takes the temple as an entity that actively creates its surroundings rather than simply making them apparent: "It does not stand on a ground that preceded it and on which it depends for its structural integrity. . . . *The building's structure makes the ground possible.*"⁵³ But Heidegger does not mean that the temple produces its own metaphysical foundation in the philosophical sense of *Grund*. Rather, he says that the temple only emerges in and through its relation to its surroundings. What is primary is the relation.

If the relation between Earth and World is primary, then what does that mean for the understanding of Earth and World as distinct polarities? For Heidegger, the term *Earth* does not correlate in any simple way with the surroundings (e.g., the rocks, the sky, the sea), nor does "World" simply signify the temple. Heidegger's language poses a challenge due to its simultaneous concreteness and abstractness. Many interpreters—both ecologically inclined and not—have stressed the concreteness of Earth. Joseph Fell, for instance, argues that "Earth is not a category, nor is it advanced by Heidegger as a speculative ground," but is, rather, "intended concretely."⁵⁴ Fell is correct insofar as Heidegger uses concrete examples to elaborate his concepts. In the "Origin" essay, for instance, Heidegger discusses a stone to clarify the meaning of Earth. Although Heidegger does talk about an actual stone, he is not claiming that the stone's physicality constitutes its earthy character. Instead, the stone "shows itself only when it remains undisclosed and unexplained. Earth thus shatters every attempt to penetrate it." Hence: "The earth appears openly cleared as itself only when it is perceived and preserved as that which is essentially undisclosable."⁵⁵ Although the stone in the example really is a stone, the earthy character Heidegger identifies does not simply lie in the stone's concreteness as a dense assemblage of minerals. Earth signifies precisely that which cannot be disclosed about the stone, that which *withdraws* and hence cannot be directly equated with the stone as a mere mass of matter (*Stoffmasse*). The apparent concreteness of Earth recedes into the realm of the undisclosable.

If Earth represents a mode of concealment, then its dialectical companion, World, manifests what Heidegger calls "unconcealment." Just as Earth indicates something other than mere matter, World refers to something other than a realm stuffed with beings. World is not a container. Instead, it is better understood as that which allows phenomena to appear in the first place. As Heidegger explains: "The world worlds [*Welt weltet*], and is more fully in being [*seiender*] than the tangible and perceptible realm in which we believe ourselves to be at home [*worin wir uns heimisch glauben*]." For this reason, "World is never an object that stands before us and can be seen. World is the ever-nonobjective to which we are subject as long as the paths of birth and death, blessing and curse keep us transported into Being."⁵⁶

Rather than designating the realm of the tangible or perceptible as such, World names the essential (and apparently self-opening ["world worlds"]) openness that allows beings to become perceptible to other beings in the first place: the clearing within which beings "come into appearance."

Crucially, World is the clearing within which beings come into appearance *for Dasein*. As Heidegger argues in *Being and Time*, World belongs exclusively to Dasein, which he describes as most primordially a "being-in-the-world." Being occurs *for Dasein* in and through the exclusivity of World.[57] Likewise, if World belongs exclusively to Dasein, then Dasein belongs exclusively to World. As Heidegger indicates in his discussion of equipment in *Being and Time*, the two are mutually constitutive. At its most basic level, Heidegger's analysis of "equipmental being" demonstrates that Dasein typically deals with things not by observing them as present-to-hand (*vorhanden*) in consciousness, but by quietly relying on them as ready-to-hand (*zuhanden*). Useful things like hammers or spectacles are only present to us when they break, at which point they draw attention to themselves and the functions for which we unwittingly relied on them. Before becoming present in this way, useful things withdraw into obscurity, setting forth their usefulness without ever coming into appearance. In this way, Heidegger's discussion of the *Vorhandenheit* and *Zuhandenheit* of equipmental being inaugurates the conceptual dynamic of appearance and withdrawal that he will later formalize as the strife between World and Earth. Importantly, just as the terms World and Earth do not refer to specific types of beings, neither do *Vorhandenheit* and *Zuhandenheit* signal particular types of entities. As Graham Harman points out: "Ready-to-hand and present-to-hand do not give us a taxonomy of different *kinds* of objects.... Instead, tool and broken tool make up the whole of Heidegger's universe. He recognizes these two basic modes of being, and *only* these two: entities withdraw into a silent underground while also exposing themselves to presence."[58] Because of this dynamic, Heidegger insists that "there 'is' no such thing as *a* useful thing."[59] That is, because useful things withdraw into inaccessibility, we cannot say that their usefulness or their use fully constitutes their being. By the same token, however, insofar as useful things come into appearance, they do so within a larger structure of relations that are manifest in their use. Hence, the hammer is useful *for building*, just as spectacles are useful *for seeing*. Because we tend to notice only phenomena that are already involved with our concerns and therefore always already have meaning for us, useful things can come into appearance as useful only within a context that is already organized into a structure of meaningful relations.

Heidegger clarifies this last point through the example of a room full of objects, which in his analysis becomes symbolic of World:

> In accordance with their character of utility, useful things always are *in terms of* their belonging to other useful things: writing utensils, pen, ink, paper, desk blotter, table, lamp, furniture, windows, doors, room. These "things" never show themselves initially by themselves, in order then to fill out a room as a sum of real things. What we encounter as closest to us, although we do not grasp it thematically, is the room, not as what is "between the four walls" in a geometrical, spatial sense, but rather as something useful for living. On the basis of this an "organization" shows itself, and in this organization any "individual" useful thing shows itself. A totality of useful things is always already discovered *before* the individual useful thing.[60]

As a system of equipment that allows individual useful things to come into appearance, the room becomes an emblem for World. World is thus a "contextual" phenomenon in that it refers foremost to "the context of the self-referral of Dasein."[61] Furthermore, as a context fashioned out of "referential relations," World specifically designates a sphere of meaning, where meaningfulness (*Bedeutsamkeit*) itself constitutes the "primordial totality" of such referential relations.[62] In short, meaningfulness "is what constitutes the structure of the world, of that in which Dasein as such always already is. . . . In that it *is*, Dasein has always already referred itself to an encounter with a 'world.'"[63] The place in which Dasein dwells therefore designates a total context of reference, a network of meaningful relations that must exist prior to Dasein's concerns. Dasein is thus inextricably enmeshed in a sphere of meaning, and as such must be understood as most primordially at home in the World. Or, as Heidegger puts it elsewhere in *Being and Time*: as Dasein, "historical man *grounds his dwelling in the world.*"[64]

As discussed at the beginning of this section, one reason why Heidegger's eco-interpreters focus on Earth instead of World has to do with a supposed turning in the philosopher's thought: as World recedes from view, Earth advances into prominence. However, understood as the name for concealment within the dialectic of strife, Earth constitutes less of a significant shift in Heidegger's thinking than it appears. More important is the primordial dynamic between concealment and unconcealment, which animates his work throughout his career, from the elegant tool analysis in *Being and Time* all the way to the relatively more complex concept of the "the fourfold" (*das Geviert*) that he develops in his later essays. It may seem reductive to dismiss the changes in Heidegger's language and style as ornamental variations; after all, rhetorical nuances often make all the difference. Yet as Alfred North Whitehead claims, in the context of philosophy "each phraseology leads to a crop of misunderstandings."[65] And as the next section goes on to show, in Heidegger's case the shifts in emphasis and rhetorical style have indeed generated more misunderstanding than clarity.

LANDSCHAFT AND/AT THE LIMITS OF HEIDEGGER'S TOPOLOGY OF BEING

As the name for the openness in which Dasein dwells and *is* most primordially "at home," World plays an essential and primary role in Heidegger's topology of being. But for all of the emphasis Heidegger's early writings put on the conceptual primacy of World, it remains unclear how something like a sense of place becomes manifest for Dasein within World. In other words, if World designates the broader sphere of meaning into which Dasein is always already "thrown" (*geworfen*), what enables Dasein to situate itself as an individual being within this broader sphere of meaning? Heidegger sought to answer this question throughout his later essays. In the process of doing so, however, he conjured a dizzying lexicon that, in its very proliferation, unwittingly gestured to the limits of his topology of being. At these limits, it becomes radically undecidable whether the matters of place and placemaking are primarily physical or metaphysical. Here it is worth recalling Heidegger's claim, in the face of persistent misunderstanding, that "Dasein does not mean being there and being here [*Dort- und Hiersein*]."[66] But then where does *Da*-sein, as a being-*there*, actually dwell? The foregoing analysis of World goes some distance to answering this question. World is phenomenal, a hermeneutic architecture built of appearances as well as Dasein's concern for those entities that come into appearance for it. Such a description of World obviously cannot be spatial in any straightforward, literal way. But then what does that mean for how we understand Dasein's physical environment? What space remains in Heidegger's topology for the *materiality* of dwelling?

This issue represents a major crux in ecologically inclined interpretations of Heidegger, which effectively need Heidegger to address the material actualities of the environment in order for his work to have value. And Heidegger does frequently appear to make room for the materiality of other entities. In his essay "Building Dwelling Thinking," for example, he claims that Dasein dwells in a realm in which even entities not immediately within reach remain close, such that "we are always staying with the things themselves [*bei den Dingen selbst*]."[67] Here Heidegger offers a corrective to Edmund Husserl's phenomenological call to arms: "We must go back to the things themselves [*die 'Sachen selbst'*]."[68] Husserlian phenomenology famously brackets, or "suspends," the universe of physically present entities in order to focus on phenomenal appearances. Hence, when Husserl penned these words, the things (*Sachen*) to which he referred were precisely *not* external objects but rather sensible intuitions (*Anshauungen*) given to us immediately in experience. Heidegger, however, replaces *Sachen* with *Dinge*, implying that the return to "things themselves" means staying with actual entities in the World and not reducing them to mere objects for consciousness. Examples like this validate Heidegger's work for ecophilosophers and ecocritics, making it possible to appropriate a Heideggerian notion of dwelling for an analysis of material places and spaces.[69]

Yet it takes only a slight lexical shift before physical topology becomes metaphysical again. Consider, for instance, Bruce Foltz's attempt to develop from Heidegger's work a "genuine understanding of the environment itself" that would reveal "the earth as homeland."[70] According to Foltz, Heidegger points toward such a genuine understanding when, in *Being and Time*, he rejects the use of the term *Umwelt* (environment; that is, a "world that surrounds") in the biological sciences. What the biological sciences (ecology included) understand as *environment* actually refers to *environs*, or mere surroundings (*Umgebung*). *Umgebung* is the province of animals, plants, and inorganic entities like stones, which, according to Heidegger, do not have a consolidated perceptual World like Dasein. Instead, nonhuman entities languish in states of relative worldlessness (*Weltlosigkeit*).[71] Humans are thus the only beings with an environment in the "genuine" sense of *Umwelt*. Hence, as Foltz asserts in his discussion of "the human environment," "the 'surrounding' character of the environment must be understood with regard to the phenomenon of '*world*,' that which in every case is 'already previously uncovered and from which we return to the entities with which we have to do and among which we dwell.'"[72] What this means is that the environment, understood as *Umwelt*, cannot possibly enable an understanding of "the *earth* as homeland," as Foltz claims. As an avatar of

Welt, Umwelt has nothing to do with either material earth or Heideggerian Earth. Materiality reverts to the immaterial, and Foltz's "genuine understanding of the environment itself" simply returns us to the phenomenal World of appearances.

Is it possible to resolve the undecidability that appears inherent to Heidegger's topology of being? Jeff Malpas, who stands at the forefront of investigating the topological nuances of Heidegger's thinking, has made a Herculean effort toward just such a resolution. In the two books that he has devoted to the subject to date, Malpas deftly navigates the confusing topological lexicon Heidegger developed throughout his career.[73] Beginning with "The Origin of the Work of Art" and continuing with later essays such as "Building Dwelling Thinking" and "The Thing," Heidegger cultivates a growing spatial vocabulary meant to explain not only how the World comes into appearance but also how places of inhabitation are set up *within* the World. The proliferation of terminology is especially confusing in later essays like "Building Dwelling Thinking," where Heidegger distinguishes between up to six different words for "place." Although rigorous definitions of Heidegger's terms remain elusive, Malpas distinguishes between two main clusters of spatial terms: *Ort, Ortschaft,* and *Stätte* on the one hand, and *Platz, Stelle,* and *Statt* on the other. The first cluster of terms "refer[s] to place in the ontologically significant sense . . . [i.e.,] place as the open region in which things are gathered and disclosed," whereas the second cluster "invariably refer[s] to place merely in the sense of location or position—usually the location or position of some already identified and determined entity."[74]

However, in attempting to corral Heidegger's topological lexicon and reduce it to a pair of basic categories, Malpas implies a degree of clarity in Heidegger's later work that is, in fact, absent. Instead of insisting on a usable typology, it may be more instructive to demonstrate how Heidegger's topological language persistently fails to elucidate the mechanisms by which placemaking works for Dasein. I attribute this failure in part to Heidegger's argumentative style, which characteristically unfolds through a series of concepts that emerge from one another like links in a chain, often giving rise to philosophical wheel spinning. This is especially the case when Heidegger tries to move beyond the appearance–withdrawal polarity in order to explain what opens up *between* these two poles. Let us return, for instance, to the temple example from the "Origin" essay. After describing how World and Earth relate through a dynamic tension of "strife" (*Streit*),

Heidegger suggests that this strife in turn opens up an "open region" (*das Offene*) between World and Earth, which he subsequently calls a "rift" (*Riss*). Such a rift does not represent an unbroachable gap so much as it manifests a kind of bridging. Indulging in his characteristic wordplay, Heidegger writes: "The rift [*Riss*] is the drawing together, into a unity, of sketch [*Aufriss*] and basic design [*Grundriss*], breach and outline [*Umriss*]."[75] The philosophical punning on *Riss* brings the notions of the rift as an abyss and the rift as a unifying design or ground plan into an intimacy; the rift becomes, paradoxically, a grounding abyss. In this way, the primordial strife between World and Earth opens up a rift, and it is within this rift—a *breach* in the double sense of an opening and a bridging—that Dasein dwells.

Yet Heidegger goes on to claim that the "open region" of the rift does not, in itself, establish a dwelling place for Dasein. Instead, it marks a space that still needs to be gathered into a meaningful place. How, then, does this rift help constitute the place of human dwelling? Heidegger answers through another round of obscure wordplay, this time featuring variations on a theme of the verb *stellen* (to put, set, place): "The strife that is brought into the rift and thus set back [*zurückgestellt*] into the earth and thus fixed in place [*festgestellt*] is the figure [*Gestalt*]. . . . What is here called figure is always to be thought in terms of the particular placing [*Stellen*] and enframing [*Ge-stell*] as which the *work* occurs [*west*] when it sets itself up and sets itself forth [*auf- und herstellt*]."[76] This passage turns on the verb *feststellen*, meaning "to fix in place." At first, Heidegger characterizes this fixing as a "figure" but then immediately recasts it as a mode of "enframing." *Enframing* is a common translation of Heidegger's neologism *Ge-stell*, which conceptually gathers into itself all of the ways in which *stellen* occurs, much in the same way the term *Gebirge* (mountain range) semantically gathers together all of the *Berge* (mountains) in a chain. As Joan Stambaugh defines it, *Ge-stell* represents "a unity (but *not* a unity in the sense of a general whole subsuming all particulars under it) of all the activities in which the verb '*stellen*' (place, put, set) figures: *vor-stellen* (represent, think), *stellen* (challenge), *ent-stellen* (disfigure), *nach-stellen* (to be after someone, pursue him stealthily), *sicher-stellen* (to make certain of something)."[77] This list could include other examples, such as those appearing in the passage above: *fest-stellen*, *zurück-stellen* (set back), *auf-stellen* (set up), and *her-stellen* (set forth). What is at issue here, however, is not so much a precise definition of *Ge-stell*, which always remains elusive, but rather the way in which the term functions as a gathering together of various, complex modes of placemaking. It is in this sense that Heidegger says the temple occurs essentially (*wesen*) *as* a mode of enframing—that is, as a setting-up, a setting-forth, a radical *placing* that establishes an open region for human being and "first affords the possibility of a somewhere [*Irgendwo*] and of sites [*Stätten*] filled by present beings."[78]

Despite Heidegger's proliferative punning, his language games remain but a prologue to the point he is actually trying to make. Maddeningly, *Ge-stell* still merely "affords the possibility" of placemaking rather than elucidating this process directly. Heidegger therefore returns to the problem of place in "Building Dwelling Thinking," where he deploys an even more dizzying array of spatial terms. In a series of especially dense passages, Heidegger plots a confusing and abstract map of how "the" space (*"der" Raum*) emerges from mere space (*Raum*), and how a locale (*Ort*) or site (*Stätte*) opens up within a place (*Platz*) or location (*Stelle*). There is little point in taking this analysis further. As the above examples already suffice to indicate, Heidegger's incessant play threatens to become what he elsewhere called "mere toying with words."[79] Gregory Fried and Richard Polt insist that Heidegger's use of language depends on "shifting patterns of meaning" that arise from his dense wordplay.[80] They also imply that such language play further reflects the very "happening" that is the event of being: "Being is thus essentially verbal and temporal."[81] Yet in this case, even if the "shifting patterns of meanings" do conjure something of being's unfolding, they also indicate the persistent way the object of Heidegger's analysis recedes into the distance. Heidegger's terminology obfuscates more than it clarifies. And even more than the insufficient clarity of the terminology, it is the searching quality of the writing itself that most plainly indicates the limits of Heidegger's topology.

As suggested earlier in this section, the reason Heidegger cannot find the proper language for the topology of dwelling is that he wants it two ways simultaneously: Dasein's proper dwelling place must be both material and phenomenal, both physical and metaphysical. Yet in his desperation to prevent the most common misunderstanding of Dasein as primarily a physically present being-*there*, Heidegger overemphasizes the phenomenal and the metaphysical. In other words, what Heidegger lacks is a term that would bridge the distinction Malpas makes between Cartesian space and ontologically significant place. And yet there is, in fact, a term Heidegger uses that accomplishes that precise task. Although generally ignored by his interpreters, no word in Heidegger's topological lexicon signifies the (meta)physical duality of dwelling more clearly than "landscape" (*Landschaft*).

What is the status of *landscape* in Heidegger's lexicon? This is an important question, because he uses the word infrequently and nowhere defines it as a rigorous philosophical concept. Even so, everywhere the word appears it serves as a kind of intermediary between appearance and withdrawal. For example, as early as *Being and Time* Heidegger uses the term *landscape* to refer to the mysterious way in which "nature" could be said to hover between World and Earth: "As the 'surrounding world' is discovered, 'nature' thus discovered is encountered along with it. We can abstract from nature's kind of being as handiness; we can discover and define it in its mere objective presence. But in this kind of discovery of nature, nature as what 'stirs and strives,' what overcomes us, entrances us as landscape [*Landschaft*], remains hidden."[82] Here *landscape* appears to refer to something more primordial (and hence entrancing) than the way the material environment appears ontically within the World. This is not quite the case, however. For Heidegger, it is not worldly knowledge of "nature" that entrances but rather the understanding that worldly knowledge does not exhaust the being of natural beings. To recognize that something else remains "hidden" behind appearances in the World is not the same thing as knowing what that something else actually is. Thus, in *Being and Time, landscape* names a mystery; it signals the gap between worldly perception and earthly essence, but only as this gap is perceived from within the World.

Landscape returns in the essays of Heidegger's later career. For instance, he uses the word in his 1953 lecture "The Question Concerning Technology."[83] However, a similar and more instructive use of the term appears in the 1951 essay "Building Dwelling Thinking." Heidegger employs the word in a passage describing how a bridge that crosses a stream does more than "just connect banks that are already there":

> The banks emerge as banks only as the bridge crosses the stream. The bridge expressly causes them to lie across from each other. One side is set off against the other by the bridge. Nor do the banks stretch along the stream as indifferent border strips of the dry land. With the banks, the bridge brings to the stream the one and the other expanse of the landscape [*Landschaft*] lying behind them. It brings stream and bank and

land into each other's neighborhood. The bridge *gathers* the earth as landscape around the stream [*Die Brücke versammelt die Erde als Landschaft um den Strom*].[84]

Landscape appears twice in this passage, with two different meanings. The first use of the word refers collectively to the various individual elements of the scene, which exist without any apparent relation to one another. In this sense, *landscape* seems aligned with what Bruce Foltz referred to as *Umgebung*—that is, the merely extant, physically present surroundings, which include other-than-human animals as well as plants and inorganic entities like stones. In the second instance, however, *landscape* expresses a different meaning. It names the unity into which the bridge actively "gathers" these surroundings. Here, by denoting what merely extant entities become once the bridge gathers them into a meaningful unity, *landscape* seems aligned with the ontologically significant meaning Foltz attributes to *Umwelt*—that is, the world that surrounds Dasein, which is to say the World. In "Building Dwelling Thinking," then, *Landschaft* expresses a double meaning that encompasses both individual entities and the network of relations into which they become unified for Dasein. How is it possible for *landscape* to unify the meanings of *Umgebung* and *Umwelt*? Instead of simply hovering between these two poles or marking the gap between them, *landscape* names the mechanism by which World comes into appearance from out of Earth's radical withdrawal. *Landscape is* this emergence. As such, it manifests the process of "gathering" (*Versammlung*) as opposed to the dynamic that Heidegger, following Heraclitus, elsewhere called *polemos*, or "strife" (*Streit*).

If *landscape* designates a process, then a full understanding of the concept depends on narrative. All of Heidegger's landscapes come into focus through descriptive passages that gather individual entities that otherwise remain concealed from Dasein into a unified environment that brings them into appearance for it. Recall, once again, the passage on the Greek temple in the "Origin" essay where Heidegger employs dialectical language to describe how the structure draws together a continuous environment around itself. As with the bridge scene in the later essay, the landscape is not simply a backdrop but rather an emergent phenomenon that appears *in and through the relation* between the architectural structure and its setting. Jeff Malpas notes that, for Heidegger, "the temple brings into view a 'sacred' landscape, which is also a meaningful landscape, and it does so through the ways in which it works in relation to the landscape in which it is situated."[85] Just as Heidegger himself does in "Building Dwelling Thinking," Malpas deploys *landscape* in two distinct senses that seem respectively linked to the concepts of Earth and World: merely extant surroundings on the one hand, and a relational totality of signification on the other. This slippage occurs for both Heidegger and Malpas because each foregrounds the dialectical process that precipitates the coming-into-appearance of landscape as a unified totality. In emphasizing process, both writers silently rely on narrative to describe how one sense of *landscape* (as concealment, Earth) transmutes into another (as unconcealment, World).

However, as Heidegger memorably puts it, description has a tendency of "get[ting] stuck in beings."[86] Thus, as a specifically narratological act that deals with phenomena as they initially appear to Dasein, the gathering of surroundings into landscape must therefore be understood as a descriptive and hence "ontic" process that takes the raw materials of the World and organizes them into a second-order, World-like manifestation. This process plays out along a temporal as well as a spatial axis. Although in its ontologically significant sense World does not emerge from Earth in time (it has always already "occurred" for Dasein), landscape must be understood as an innerworldly phenomenon that is organized by an individual human consciousness and implicates time. In Heidegger's landscape descriptions, the philosopher himself stands as the narrating subject, gathering mere space into meaningful place. If Dasein is ontologically "at home in the World," then landscape occurs as an ontic mechanism that brings into focus precisely what is home(l)y about the World in which Dasein dwells. Landscape brings what is familiar closer to Dasein; it secures Dasein's sense of being in and belonging to the World in which it always already dwells. In other words, if the World represents the Earth brought into appearance for Dasein, then landscape represents the World brought into a more *personal* proximity—wrapped around the individual human being like a blanket.

In this sense, landscape encompasses the concept of "nearness" (*die Nähe*), which Heidegger elaborates in another late essay. "Nearness," Heidegger writes, "cannot be encountered directly. We succeed in reaching it rather by attending to what is near. Near to us are what we usually call things."[87] Dasein's potential for understanding nearness depends on its ability to encounter things *as* things. The experience of nearness entails holding things close in attention and maintaining a level of care and respect so as not to reduce them to mere utility. Like landscape, nearness emerges through a process of "bringing into salience"[88]—that is, a process that gathers innerworldly objects into a meaningful proximity. Therefore, and also like landscape, the experience of nearness depends on the World as a relational totality of signification, within which thinking can derive "salience" for Dasein without also instrumentalizing other entities that appear for Dasein in the World. Crucially, this is what it means for Dasein to dwell. So long as Dasein exists (i.e., barring death), it does so within a preexisting sphere of meaning (World) that is further gathered into a particular, meaningful nearness (landscape). Dasein inhabits the nearness of landscape like a house. As Heidegger puts it, the locale manifested by *die Nähe* offers "a shelter [*Hut*] for [dwelling] or, by the same token, a house [*ein Huis, ein Haus*]. Things such as locales shelter or house [*behausen*] men's lives. Things of this sort are housings [*Behausungen*], though not necessarily dwelling-houses [*Wohnungen*] in the narrower sense."[89] In other words, dwelling means quite simply to *be* at *home* in the World as it is drawn into personal nearness as landscape.

HEIDEGGER AND THE COLONIALITY OF NATURE

The term *Landschaft* wields a powerful albeit largely unacknowledged force in Heidegger's philosophy. On my reading, *landscape* designates a worlding figure, one that gathers mere surroundings into a personal nearness, and hence actively manifests a dwelling place for Dasein within the World. This gathering takes place—and indeed *makes place*—within the World. In this sense, *landscape* refers to an ontic or "innerworldly" manifestation of World: *Dasein's own particular beloved place.*

This final section aims to draw the various threads of this chapter together and indicate how the connections between Dasein, dwelling, and landscape help elucidate a relationship between Heidegger's thinking and what in the introduction to this book I termed "the coloniality of Nature." In the context of a discussion on Heidegger, it is important to remember that the coloniality of Nature does not simply concern the instrumentalization of beings. Instead, it references a wider range of forces constitutive of the global paradigm of modernity/coloniality. These forces have contributed to an idealized and reductionist notion of "Nature" that has, in turn, underwritten a diverse collection of narratives of existential homecoming. As amply documented by decolonial scholars, such narratives have played key historical and ideological roles in naturalizing the West's self-image as well as its imperialistic forays into the rest of the world. Despite Heidegger's vociferations against technology and the violence of instrumentalization, his sustained project of philosophical homecoming reveals a fundamental embeddedness in the colonial matrix of power.

Recall, for instance, that Dasein is profoundly marked by coloniality. As the decolonial and Black studies scholars introduced in this chapter's third section show, Dasein poses a problem because it cannot apply universally. Dasein is not simply equivalent to human beings, to which Heidegger otherwise refers using the word *Menschen*. Instead, Dasein refers to a particular *condition* into which human beings enter, a condition in which being itself is at issue. For thinkers like Nelson Maldonado-Torres and Calvin Warren, it is the particular *condition* referred to as Dasein that is not accessible to all humans. Not only does this inaccessibility reflect what Maldonado-Torres calls the coloniality of being, but it also explains why he needs to introduce the Fanonian figure of *les damnés de la terre* as a counterpart to Dasein. For his part, Warren reads the limited accessibility of the condition of Dasein as a symptom of the internal logic of metaphysics itself, which depends on the negation of nothing in order for there to be "beings at all." Black being incarnates this metaphysical nothing that must be negated. Hence, "The essence of black being, like the essence of technology, is to open up an understanding of Dasein, it is always being for another."[90]

Taking this same logic a step further, if *les damnés* lack the access to being that is proper to Dasein, then they also lack access to dwelling in its most proper sense. To understand why this is the case, we have to demonstrate the dangers inherent in the way Heidegger's rhetoric of home infiltrates his topology of being, and how Heidegger's peculiar use of the term *Landschaft* makes these dangers manifest.

As discussed at the beginning of this chapter, home(l)y rhetoric permeates nearly every aspect of Heidegger's writing, and it comes into particular focus through the concept of dwelling. Many commentators have noted that this rhetoric cannot be separated from the *Blut-und-Boden* politics of National Socialism. But as Maldonado-Torres suggested, the force of Heidegger's rhetoric of home does not come from its Nazi resonances so much as from the myth of Greco-German affinity that allowed German intellectuals from the early nineteenth century on to see Germany through the lens of its purported historical link to Greece: the supposed point of origin (*arche*) of Western civilization, and hence of thinking itself. The myth of Greco-German affinity made it possible for German thinkers—Heidegger included—to transfer the Greeks' claim to an autochthonous origin to their own context, thereby justifying another myth: that of an originary relationship between the German *Volk* and the German *Heimat*. With these historical points in mind, it becomes clear that Heidegger's home(l)y rhetoric has its deepest roots in a metaphysics that underwrites the myth of an originary and hence *authentic* tie between a people and their land—or, to return to Heidegger's philosophical register, between Dasein and its dwelling place. In Heidegger's topological lexicon, there exists no better term than *landscape* for capturing the nuances of how an originary, authentic relation between Dasein and its surroundings enables Dasein to gather particular elements within a broader sphere of meaning into a personal nearness that manifests its own particular beloved place. Within the context of Heidegger's topology of being, then, we could understand landscape as a physical place framed by a metaphysics of home. This dwelling-in-landscape marks the place where Dasein *is* most fully in being.

Landscape thus depends on a metaphysics of home in which authentic Dasein relates in an authentic way to place. But what does such a claim to an authentic relationship between people and land mean in the context of modernity/coloniality, whose global paradigm of bio- and necropolitical power has subjected and continues to subject vast proportions of the world's population to perennial conditions of war and dispossession? In this paradigm of violence and displacement, no territorial claim is unproblematic. This is true even for the Indigenous peoples of the world, whose very "indigeneity" is a product of colonialism, and who thus can never again embody the mythic "authenticity" of old. And speaking of mythic authenticity, can there ever be a truly embodied autochthonous status that ensures authentic dwelling, or can such an idea only ever be established retrospectively, through the epistemic violence of historical narrative? Consider the example of Erichthonius, whose mythic birth from Athenian soil establishes the authenticity of his belonging. Yet Erichthonius's relationship to Athens is also preceded and made possible by a founding act of violence. That is, he is only born as a result of one god (Hephaestus) attempting to rape another (Athena) and spilling his seed on the ground. This example of foundational violence repeats the logic of negation that, as Calvin Warren reminds us, lies at the core of the metaphysical tradition: the negation of nothing that makes being possible. Just as Warren argues that Black being is the (negated) incarnation of nothing and hence lacks access to Dasein, he also asserts that the negation of Black being results in a denial of access to dwelling: "black ~~being~~ lacks precisely [the] historical place (there-ness) that situates the human being in the world. Black ~~being~~, then, lacks not only physical space in the world (i.e., *a home*) but also an existential place in an antiblack world."[91] If Black ~~being~~ lacks access to Dasein and to an authentic relationship to its surroundings, then it necessarily lacks access to dwelling and the metaphysics of home that grounds it. What this means is that the Heideggerian figure of landscape is a process available only to particular individuals who already possess the privileges of being in and belonging to the World. In fact, as the process by which place is gathered *for Dasein* (and by implication *not for others*), landscape is the very mechanism that does the excluding.

With this observation, it at last becomes possible to disclose how Heidegger's philosophy recapitulates the coloniality of Nature. The concept of dwelling is profoundly marked by coloniality. If this is the case, then ecophilosophers and ecocritics who underscore the importance of Heideggerian dwelling to ecological thinking and environmental ethics need seriously to reconsider their basic position that dwelling offers a paradigm that situates human being firmly within the material actualities of the other-than-human environment. As the previous two sections demonstrated, no matter how much Heidegger talks about material actualities (i.e., things as *Dinge*), he always ends up back within the frame of Dasein's consciousness (i.e., things as *Sachen*). What we might call "the environment" is, for Heidegger, always already gathered into personal nearness *for Dasein*.

I argued above that the concept of landscape helps us see the exclusionary mechanism that opens a place of dwelling for Dasein while simultaneously closing this possibility for those without access to the condition of Dasein (i.e., *les damnés de la terre*). But landscape also helps us see an additional exclusionary mechanism, one that brings Dasein home to its ownmost dwelling in the World while simultaneously condemning the vast multiplicity of other-than-human worlds and worldings to the void of worldlessness (*Weltlosigkeit*). Thus the mechanism that excludes the wretched of the earth from access to Dasein and dwelling necessarily also excludes all nonhuman entities from access to a sphere of meaning.

Framed in this way, Heidegger never offers a philosophical ecology, nor does he provide what Bruce Foltz calls "a genuine understanding of the environment." Heidegger may announce a desire to "save the earth" through an ethics of "letting-be," but he also remains committed to an exclusionary metaphysics of home to which only Dasein has access. He cannot, alas, have it both ways. Furthermore, the overarching aim of his philosophical project to manifest Dasein's radical homecoming to being indicates that Heidegger may in fact have *betrayed* the earth. In this sense, my reading echoes Michael Zimmerman's suggestion that Heidegger's thinking is, despite itself, "consistent with modernity's project of the technological domination of nature."[92] Not only does Dasein's primordial dwelling in the World, gathered near as landscape, fail to enable ecological thinking; it also depends on a violent act of exclusion that denies the wretched of the earth a home in the world and reduces to mere background the material environment and the multiplicity of entities that constitute it. Such is the very definition of the coloniality of Nature.

This chapter has laid out the book's first guiding trope. The following two chapters, which comprise the remainder of part 1, introduce the book's other two guiding tropes: *the home(l)y metaphysics of landscape* (chapter 2) and *the ecological uncanny* (chapter 3). In moving into these later chapters, which respectively focus on the work of Willa Cather and D. H. Lawrence, I turn from philosophy to literature. Although these chapters leave explicit phenomenological discourse behind, they nonetheless aim to explore socioperceptual apparatuses that guide two rather different types of comportment toward the other-than-human environment. Furthermore, by turning from the German *Heimat* to the contested settler territories of the U.S. Southwest, chapters 2 and 3 at once develop and complicate the relationship between *Landschaft* and the coloniality of Nature outlined in chapter 1.

CHAPTER 2

Willa Cather and the Home(l)y Metaphysics of Landscape

[Windows] establish the relation between the inside of the house and the landscape, making the latter what, as seen from a room, it logically ought to be: a part of the wall-decoration.
—Edith Wharton and Ogden Codman, *The Decoration of Houses*

In 1891, during her freshman year at the University of Nebraska, Willa Cather penned her first noteworthy publication: a brief essay entitled "Concerning Thomas Carlyle." More of a poetic sketch of the English writer than a close engagement with a particular work, Cather's essay, which appeared just one month after the ten-year anniversary of her subject's death, reflects on the core animating spirit that defined Carlyle as both an individual and an artist. This spirit, as Cather sees it, was "thoroughly un-English"; Carlyle's dark cast of mind and "intensely reverent nature" instead links him to a Romantic tradition that strikes Cather as "most decidedly German."[1] Carlyle cherished solitude, study, and reflection. "He loved his books and loved Nature better," as Cather puts it.[2] But for all his apparent introversion, Carlyle also exhibited a passion that enabled him to share in the emotional life of society at large: "His love and sympathy for humanity were boundless. . . . He could understand how the Marseillaise might set men's hearts on fire; the storming of the Bastille, and the revolt of the women [were] pictures after his own heart, in which the hot blood of the old sea kings still raged."[3] Cather's invocation of the raging blood of sea kings and the incendiary passion of French revolutionaries might at first seem overwrought, but only if we fail to notice her subtler modulations of historical tone. Recalling her emphasis on Carlyle's Germanic comportment, her language riffs on the subjective expressionism of eighteenth-century *Sturm und Drang* aesthetics. Yet Cather's riffing in the key of German Romanticism also comes immediately after she sounds the key notes of Enlightenment philosophy ("humanity") and Victorian aesthetics ("sympathy," "earnestness"). From Cather's perspective near the fin de siècle, then, Carlyle was a figure who integrated the best of the European eighteenth and nineteenth centuries, at once a reflective genius and an accomplished artist. But it is Carlyle's range and intensity of feeling that remains paramount; it coexists with and also rises above any other aesthetic or philosophical ideals.

Already in her first major publication, Cather had isolated the foremost principle that would continue to ground her writing and thinking throughout the next four decades: the principle of *vitality,* which she also referred to as *the life force.* Although references to vitality appear frequently in Cather's early writing as a university student, nowhere did she realize her vision for artistic greatness more clearly than in that first essay on Carlyle. Indeed, much of what Cather would later write on the subjects of living, feeling, and art-making could be seen as extensions of lessons she derived from Carlyle's biography. Perhaps the most potent of such lessons appears in Cather's reflection on the vital relationship Carlyle enjoyed with the Scottish landscape: "He went far out into one of the most desolate spots of Scotland, and made his home there. There among the wild heaths, and black marshes, and grim dark forests, which have remained unchanged since the time of the Picts and the Saxons, he did his best work. He drew his strength from those wild landscapes; he breathed into himself the fury of the winds; the strength of the storm went into his blood."[4] The impressionistic style Cather employs here links the wildness of the Scottish moors and the intensity of Carlyle's artistic vision. She imagines Carlyle being literally inspired by "the fury of the winds" and fed on "the strength of the storm." His is truly a life lived in vital relation to the "elemental things." Cather also presents Carlyle as having existed in continuity with "the time of the Picts and the Saxons," that brutal era of conquest and survival when life balanced along the edge of death, trembled on the threshold of myth. Cather's invocation of a historical imaginary of conquest gestures toward a vital continuity that persists between the great warriors of old and the great artists of the present century.

The importance of Cather's writing on Carlyle derives in part from the links it draws, first, between art and vitality, and second, between the art–vitality matrix and the metaphor of conquest. The above passage also proves valuable for the way it attaches these links to the concepts of *landscape* and *home* that appear in the first sentence: "He went far into one of the most desolate spots in Scotland, and made his home there." Even more than suggesting a conceptual intimacy between landscape and home, Cather figures a notion of landscape *as* home, a site of primal, affective embodiment. For despite the grammar of making a home *in* the Scottish landscape and living *among* its "wild heaths, and black marshes, and grim dark forests," Cather imagines a sustaining spiritual-*cum*-physiological bond between Carlyle and the Scottish moor. The very air and soil of the moor incarnate Carlyle's breath and flesh; their organic vitality constitutes him and invigorates his art. And quietly underwriting all of this is the ambivalent echo of historical conquest, ambivalent because it remains unclear how—or to whom, or to what—this echo relates. Is Carlyle "conquered" by the spirit of place, thus making his art into nature's mouthpiece? Or is the Scottish landscape "conquered" by Carlyle, who galvanizes its spiritual energy for the purpose of feeding his work? Without ever resolving it, Cather would remain preoccupied with this precise ambiguity for the rest of her long career.

The present chapter explores how Cather's fiction brings notions of vitality, landscape, home, and conquest into an uncertain intimacy, one that wields a metaphysical power over the subject who finds him or herself caught up in its matrix. As she imagines it for Carlyle, making one's "home" in landscape enables a subject to channel a profound sense of (re)invigoration. *Being at home in landscape,* then, becomes a conceptual shorthand for a vitalist ontology in which landscape, fashioned imaginatively into a home, enables one to feel more alive. And yet to sustain this invigorating state of being in the American contexts Cather typically privileges, the settler subject must disavow the histories of conquest and dispossession that enable their imaginative claim on the landscape in the first place. This disavowal is structural and submerged rather than conscious. Indeed, Cather's landscapes achieve their home(l)y status through narrative means that, when carefully attended to, reveal a foundational dependence on coloniality. The remainder of the chapter elaborates how Cather's interest in home and landscape come together through a complex preoccupation with architecture, and how the interrelations between these concepts yields a metaphysics of landscape that is, I argue, fundamental to settler colonial homemaking. The centerpiece of this analysis is Cather's 1925 novel *The Professor's House,* which features an intricate narrative architecture whose function is to gather—or, in narratological terms, to focalize—landscape into a metaphysical home.

CATHER'S VITALIST ONTOLOGY OF ARCHITECTURE

Narrative homemaking stands at the center of Willa Cather's novelistic project. From the pioneer mythos that occupies her Prairie Trilogy to Jean-Marie Latour's imaginative acts of homecoming at the end of *Death Comes for the Archbishop* (1927), Cather fashions home(l)y spaces where her protagonists may fully grow into their most vital selves.[5] In this sense, Cather shares with Heidegger a deep interest in recovering something of a primordial home in which human being may properly "dwell." As the previous chapter argued, the ultimate horizon for such dwelling in Heidegger's thought lies within the World (*Welt*), and even more particularly within landscape (*Landschaft*). Cather's visions of home likewise rely heavily on landscape description, and this link between landscape and homemaking has drawn both praise and criticism. Whereas Cather's earliest reviewers celebrated the aptitude of her painterly renderings of the Nebraska prairie, more recent critics have linked her landscape descriptions to the troubling politics of "homeland" thinking.[6] The final section of this chapter will return to the critique of Cather's politics. For now, however, the more pressing issue relates to the role narrative plays in constructing Cather's politics in the first place, which we can only understand by stepping back yet further to examine the philosophical foundations of her novelistic art.

What other critics have largely missed when subjecting Cather to ideological critique is the ontological significance of her home(l)y rhetoric. Judith Fryer's spatio-feminist approach to Cather's fiction represents one important exception.[7] Fryer characterizes Cather's spatial imagination in terms of Gaston Bachelard's concept of "felicitous space," which, in its simplest sense, refers to any space that allows one "to be at home." This four-word phrase encapsulates Bachelard's thesis in *The Poetics of Space,* where the author develops a phenomenology of intimate places organized around the home, a felicitous space that "concentrates being within limits that protect."[8] To *be at home,* then, means to stand more fully in being by inhabiting a space of radical belonging. For Fryer, whose notion of imaginative structures ranges across a spectrum of both physically and conceptually spatial entities (including but not limited to architectural, geographical, and social structures), felicitous space does not primarily refer to an actual home (i.e., a *house*) or to images thereof. Fryer echoes Bachelard's own desire to speak to a more essential understanding of the home: "All really inhabited space bears the essence of the notion of home."[9] For both, then, homemaking happens whenever space is gathered into a site where one feels centered and safe. Whether explicitly architectural or not, "home" fosters a sense of heightened ontological density—a "concentration of being."

Fryer's work has had a notable influence on Cather studies, inspiring numerous scholars to draw on Bachelard's phenomenology.[10] In their emphasis on Bachelard's principles of home and dwelling, however, none of these scholars (Fryer included) has explicitly connected the home(l)y essence of dwelling to Cather's first principle of art: vitality. This connection can only be made explicit if we attend to the way Cather uses the figure of architecture to frame her understanding of the vitality of art.

Cather's early writings during her literary apprenticeship at the University of Nebraska consistently emphasized the primacy of "the life force"—that is, the vitality that endows the artist with a singular vision. Vitality innervates the imagination. It provides a mysterious, quasi-spiritual impetus that enables the artist to manifest work that embodies a living form and hence bears the mark of its maker's own life force. The "vital artist," as we might call them, marries depth of passion with organic intuition. Such an artist also possesses an ability to create through subtlety and suggestion—to show rather than tell. When it comes to fiction, for Cather, vitality manifests most prominently in three aspects: character, scene, and narrative form. The first of these is also the most obvious. Primary characters in fiction succeed to the extent that they exhibit sufficient emotional complexity and undergo plausible—if not always obviously consistent—development over the course of a story. Vital characters thus possess what E. M. Forster termed "roundedness," and what James Wood describes as a vividness capable of surprise.[11] The vitality of character must be matched by the vitality of scene. For Cather, a scene is more than a mere backdrop for action; it provides a three-dimensional living world. The livingness of scene flows directly from the vitality of character: "When a writer has a strong or revelatory experience with his characters, he unconsciously creates a scene; gets a depth of picture, and writes, as it were, in three dimensions instead of two."[12]

Further linked to the vitality of character and scene is a certain organic liveliness of narrative form, or what I shall call "narrative architecture." The vital artist's ability to create atmosphere through subtlety and suggestion enables narrative architecture to evolve from the emotional demands of the story. By contrast, the nonvital artist tends to create, as Cather says, by "mere construction," which entails the imposition of a structural design for the sake of cleverness. Such formal rigidity often exhibits a certain coldness and lack of feeling: "Too much symmetry kills things."[13] Thus, to achieve a sufficient level of organicism, the vital artist must experiment with narrative in ways that reconcile the apparent opposition between art and life. Cather herself sought to generate narrative forms that could remain "pliable enough to preserve the spirit and vitality of the recollections that emerged in the process of composing."[14] Cather's idea of pliability finds expression in *Death Comes for the Archbishop* and *Shadows on the Rock*, both of which exhibit disjointed yet expansive episodic structures that conjure the temporal spaciousness of legend. Despite its highly structured form, a certain expressive pliability is also on display in *The Professor's House*, discussed in detail below. Each of these works embodies a unique form that Cather arrived at organically and left more roughly hewn than cleanly polished. For Cather, such unburnished form in a literary work signals a creative vitality that springs from a deep well of passion. And despite emerging through process of composition, narrative architecture retains a level of primacy for Cather, as it manifests an imaginative space that can infuse both character and scene with their constitutive vitality.

Yet to understand the crucial link between (narrative) architecture and vitality in Cather's thinking and artistic practice, it is also necessary to situate Cather vis-à-vis two thinkers she herself read and admired: John Ruskin and Henri Bergson.[15] Each of these thinkers unites notions of form and vitality, albeit in rather different ways. Ruskin, for instance, was a critic of art and architecture who placed a premium on the liveliness of form, yet he also represented the moral probity of Victorian society. By contrast, the work of Henri Bergson marked the emergence of a revitalized philosophy that dispensed with the dead-end thinking of mechanism. His notion of *élan vital*, or vital impetus, as the creative force behind evolution located a living force within the material, physiological structures of organisms. Although very different from one another, both Ruskin and Bergson contribute to Cather's theory of art a sense of living structures, thereby illuminating her interest in the vitality of narrative architecture and the characters and scenes such architecture brings into focus.

John Ruskin's writing on painting and architecture qualified him, in Cather's estimation, to be "perhaps the last of the great worshippers of beauty."[16] Ruskin had a powerful influence on Cather during her literary apprenticeship, and particularly in the period between 1893 and 1896, when she contributed frequently to the *Nebraska State Journal*. Near the end of this period, in a column dated 17 May 1896, Cather praises Ruskin for his commitment to emotional intelligence over unfeeling erudition: "He knew that to know is little and to feel is all."[17] Cather's language here most closely echoes a passage that appears in Ruskin's magisterial work *Modern Painters*.[18] However, her overriding interest in vitality directs us to another important work, *The Seven Lamps of Architecture*, in which Ruskin more explicitly works out the importance of vitality to art. He asserts that vitality matters to architecture "more than [to] any other [art]; for it, being especially dependent . . . on the warmth of the true life, is also peculiarly sensible of the hemlock cold of the false: and I do not know anything more oppressive, when the mind is once awakened to its characteristics, than the aspect of a dead architecture."[19] Ruskin understands architecture to exist in dynamic relation to a life principle, which depends in turn on the actions of inhabitants. In order for architecture to remain "alive" it requires inhabitants to pass in and out of it—to cut through, (re)orient, and (re)constitute its spaces. Architecture is as much about human interaction as it is about geometry and design. As another architectural theorist puts it, "there is no architecture without program, without action, without event."[20]

For Ruskin, then, the vitality of architecture depends on the interaction of form and event, though he adds a moral inflection (as well as a gender bias) when he defines architectural vitality in terms of the "true" and the "false" life, respectively distinguished by the principles of activity and passivity. Ruskin describes the true life as "the independent force by which [one] moulds and governs external things; it is a force of assimilation which converts everything around him into food, or into instruments; and which . . . never forfeits its own authority as a judging principle."[21] The true life represents a traditionally masculine ideal achieved by establishing a definitive link between intention and mastery, and never allowing anything or anyone to obstruct one's way. By contrast, the false life takes on the conventionally feminized principle of passivity, and it emerges when one lacks the self-possession necessary to realize individual intention. In the false life, then, "we do what we have not purposed, and speak what we do not mean, and assent to what we do not understand. . . . [The false life] becomes to the true life what an arborescence is to a tree, a candied agglomeration of thoughts and habits foreign to it, brittle, obstinate, and icy, which can neither bend nor grow, *but must be crushed and broken to bits, if it stand in our way*."[22] Whereas the true life harbors an imperialist sense of conquest, bending the external world to its will, Ruskin casts the false life in the role of the conquered, powerless and marked by the forces of oppression and assimilation. Extending this logic, Ruskin implies that without the vitality necessary to animate the "event" of architecture, an architectural entity itself falls out of sync with the life principle; it grows false, it "dies." And the relationship between architecture and vitality is reciprocal: just as, in the false life, inhabitants no longer properly inhabit their own lives, dwellings cease properly to house their inhabitants. Despite his objection to "the troublesomeness of the metaphysicians,"[23] a metaphysics not unlike that of Bachelard's felicitous space and Heidegger's dwelling resonates at the heart of Ruskin's understanding of architectural vitality. The true life infuses architecture with a transformative liveliness such that a house (a physical structure) becomes a home (ontological ground). Within this paradigm, the one who lives the true life feels more fully alive, more fully present in the world, and more fully in being.

Henri Bergson's psychological approach to evolution has less of an ontological thrust than Ruskin's architectural thinking, but it retains a similar focus on life force. Throughout his writings of the late nineteenth and early twentieth centuries—including *Time and Free Will* (1889), *Matter and Memory* (1896), and especially *Creative Evolution* (1907)—Bergson maintains a link between vitality and form. But in place of Ruskin's "spatial" emphasis on the moral structure of this link, Bergson spotlights its "temporal" unfolding. Ruskin's yoking of intention and realization remained bound by teleological thinking. By contrast, Bergson dispenses with the folly of "finalism" and its assumption that "things and beings merely realize a programme previously arranged."[24] He instead introduces his notion of *durée*, or duration, which expresses his belief in the radical nature of change. For Bergson, change is continuous, such that "there is no essential difference between passing from one state to another and persisting in the same state."[25] The continuity of change does not enforce presentist thinking; instead, it indicates the ever-present significance of the past, of memory: "For our duration is not merely one instant replacing another; if it were, there would never be anything but the present—no prolonging of the past into the actual, no evolution, no concrete duration. Duration is the continuous progress of the past which gnaws into the future and which swells as it advances."[26] The past thus continuously structures the development of the present through the unfolding of an inertial force.

Bergson gives this inertial force a name: *élan vital*, or "vital impetus" in Arthur Mitchell's translation. The *élan vital* indicates a "psychological" element that plays an active role in organic evolution. Again resisting teleology, Bergson's vital impetus represents less of a striving toward than an emerging from; it designates a "limited force" that is "always seeking to transcend itself" and that "always remains inadequate to the work it would fain produce."[27] Instead of realizing a set intention, then, the *élan vital* truly is an *impetus* in that it creates movement from the push of the past—"the great blast of life."[28] Bergson argues that individual organisms (and especially humans) realize their vital impetus by establishing a conduct that allows them to "ripen" into themselves organically rather than to "counterfeit" alternative modes of being. The ripening process continues "indefinitely . . . without ever reaching its goal. . . . Such is the character of our evolution; and such also, without doubt, that of the evolution of life."[29] For Bergson, then, the relationship between vitality and form is a temporal one: the vital impetus *endures* such that it is the past that structures present and future forms of life. The evolution of the self depends less on setting specific intentions (as Ruskin argues) and more on a creative impulse that initiates movement "from behind," as it were, and thereby retains the openness of the future.

The influences of Ruskin and Bergson converge in Cather's narrative depictions of landscape. On the surface, Cather's interest in landscape most obviously connects her to Ruskin. In the first volume of *Modern Painters*, Ruskin famously asserts the aesthetic and moral superiority of modern landscape painters based on their rejection of Renaissance pictorial conventions in favor of more realistic representation. As Joseph Murphy points out, however, "Ruskin did not view painters as slavish transcribers of nature's iconography; according to his romantic typology, the landscape's inherent meanings are discovered through the artist's imagination and arrangement."[30] Landscape's "meanings" are therefore not immanent in how they appear to the eye but in how painters like J. M. W. Turner and John Constable organized them within the frame of the canvas. Cather's landscapes likewise derive their meaning from their particular position within a larger narrative framework. As I will later show with regard to *The Professor's House*, it is not a naturalist depiction of scene but rather the structural positioning of landscape description within the narrative architecture that proves crucial to the development of vitality and meaning in the novel at large.

And yet Cather's use of landscape is not entirely Ruskinian. Whereas Ruskin subscribes to "a theological conception of history" that "suspends

[its] inhabitants between type and fulfillment,"[31] Cather avoids the moralizing and teleological thrust of such a paradigm. Instead, echoing Bergson, Cather's landscapes invoke history and memory in order to create the dynamic atmosphere she considered necessary for the lively ripening of both character and scene. As she puts it: "When we have a vivid experience in social intercourse, pleasant or unpleasant, it records itself in our memory in the form of a scene; and when it flashes back to us, all sorts of apparently unimportant details are flashed back with it."[32] Thus, while Ruskin provides a model for the vitality of (narrative) architecture and its framing of landscape, it is this latter, Bergsonian swelling of the past into the present that further invests Cather's fictional landscapes with their particular meaningfulness as ontologically significant dwelling places.

LANDSCAPE IN THE HOUSE OF FICTION

Given the symbolic significance of the home in her novels, it comes as little surprise that Cather frames her theory of fiction in terms of a house. Cather's well-known essay "The Novel Démeublé" takes part in a tradition of literary interpretation structured on architectural metaphor, from Ruskin's *Seven Lamps of Architecture* to Walter Pater's "literary architecture"[33] to Henry James's "house of fiction."[34] The commonplace interpretation of Cather's attempt to "unfurnish" (*démeubler*) the novel views her domestic metaphor as a parable of novelistic selection. In this sense, the unfurnished novel at once repudiates the overstuffed realist narratives of the nineteenth century and reminds modern novelists that art finally resides, as Nikolai Gogol says, in the art of choosing. Rather than populating narrative with a catalogue of objects, one must choose only "the eternal material of art" and toss out all the rest.[35] For Cather, this means distilling elemental passions in order to conjure the essence of a vital life force. The house of fiction must therefore remain uncluttered and open for the play of grand, human emotions.

Despite the elegance of this essay and its own "unfurnished" prose style, Cather's housecleaning metaphor does not fully disclose by what means she intends to tidy up the novel. In an oft-cited passage she calls for novelists simply to "throw all the furniture out the window,"[36] but this rhetoric of physical jettisoning proves misleading, even logically opposed to her emphasis on art-making as a matter of selection. As she suggests in her discussion of Leo Tolstoy, rather than defenestrating the literary "furniture," the novelist must instead make strategic use of it to evoke the essence of character: "the clothes, the dishes, the haunting interiors of [Tolstoy's] old Moscow houses, *are always so much a part of the emotions of the people that they are perfectly synthesized;* they seem to exist, not so much in the author's mind, as in the *emotional penumbra* of the characters themselves."[37] In this sense, the proper method for unfurnishing the novel entails something like the absorption of mise-en-scène into subjectivity, fusing character and setting into an integral union. This method gathers a character's environs into a context of significance and emotional resonance—not an "outward" projection as with the pathetic fallacy, but an "inward" synthesis of place into the shadowy realm of a character's "emotional penumbra." The latter model proposes an alternative to Cather's own frequently invoked metaphor of fiction as a stage completely purged of scenery and bereft of all but "one passion, and four walls."[38] Rather than allowing the setting to be a static backdrop against which passion unfolds, scenery itself must manifest the play of emotional drama. What this means is that, for Cather, scene and character coexist in an unusually intimate relation to one another; they are, so to speak, topological inversions of one another, with character "housing" scene and scene "inhabiting" character. And as these paradoxical inversions suggest, the reciprocal relationship between character and scene plays a central architectural role in Cather's narratives of home(making). The relationship is, as John Stilgoe puts it in another context, "the armature around which the work revolves."[39]

How are we to decide between these two contrary readings: *démeubler* as the imaginative clearing of novelistic space, and *démeubler* as radical absorption of environment into character? It would be a mistake to read Cather's essay as a blanket rejection of all realistic description, and particularly of setting. Some interpreters have found this tack tempting and have argued that Cather makes such narrative digressions obsolete by relying on readers to conjure their own images.[40] These interpreters reflect a common tendency to take Cather too much at her word, typically emphasizing her powerful idea that only what is felt but not named can be considered "created" as such. Such an interpretation is not wrong in its point of emphasis. Indeed, Cather repeats such sentiments in her essay "On the Art of Fiction," where she once again stresses that art should simplify: "Any first-rate novel or story must have in it the strength of a dozen fairly good stories that have been sacrificed to it."[41] Furthermore, this reading has the advantage of bringing Cather's theory into conversation with other modernists such as Ernest Hemingway, whose so-called iceberg theory of omission entails a similar philosophy of the unwritten.[42] However, such an interpretation privileges readerly practices rather than asking how Cather meant to actualize such absent presences in her art. And indeed, Cather's novels are full of celebrated passages of landscape description, many of which do important narrative work that requires closer examination. Rather than present a concretely rule-bound understanding of Cather's literary practice, I read "The Novel Démeublé" as a call to unfurnish the novel not simply by avoiding description but, more fundamentally (and abstractly), by gathering the necessary scenic elements into that which, for Cather, represented the living passion of human emotions. The novel *démeublé* interiorizes the external world as the expression of a character's "emotional penumbra"—it brings "outland inland," as Deborah Karush puts it.[43]

It is precisely through this spatial process of bringing the outside in that Cather fashions her literary landscapes. Using explicitly architectural terms, Cather intimates such a process in a letter from 1938, where she explains the experimental narrative structure of her 1925 novel *The Professor's House*. She begins by describing a literary device commonly employed in early French and Spanish novels, where the authors insert a *nouvelle* into the middle of a *roman*. This is precisely what Cather has done in her own novel: the independent novella "Tom Outland's Story," which narrates the title character's youthful adventures in New Mexico, constitutes the centerpiece of a three-part narrative, framed on either side by the story of an aging midwestern professor of history. In spite of this literary precedent, however, Cather admits that "the experiment which interested me was something a little more vague, and was very much akin to the arrangement followed in sonatas."[44] Curiously, Cather refrains from developing the musical example and proceeds directly to a discussion of painting, which in turn elaborates an architectural metaphor:

> Just before I began the book I had seen, in Paris, an exhibition of old and modern Dutch paintings. In many of them the scene presented was a living-room warmly furnished, or a kitchen full of food and coppers. But in most of the interiors, whether drawing-room or kitchen, there was a square window, open, through which one saw the masts of ships, or a stretch of grey sea. The feeling of the sea that one got through those square windows was

remarkable, and gave me a sense of the fleets of Dutch ships that ply quietly on all the waters of the globe—to Java, etc.[45]

Cather emphasizes the way these Dutch paintings dramatize an intimacy between inside and outside. The "warmly furnished" interiors domesticate the exterior world by organizing it into a view—a seascape—bringing the volatile waters beyond the windowpane into harmless proximity, where their aesthetic and symbolic qualities may be contemplated in comfort. Cather even briefly projects herself into a drawing-room like those depicted in the paintings, complete with a window that affords a view of ships on the sea. The window does more than simply usher the seascape into the domestic interior; it also inspires Cather with grand feelings she associates with narratives of oceanic exploration and seafaring adventure. Whereas the Dutch ships Cather imagines seeing through the window are most likely engaged in goods trade fueled by imperialism and its attendant violence, the window itself frames the view and compresses it into something more artful, even poetic: a fleet that will "ply quietly on all the waters of the globe." The outside world enters the interior as a framed and hence domesticated pictorial representation that, in turn, invests the inhabitant with a flush of vital feelings.

Whereas the windows in the Dutch paintings seem to remain closed, the revitalizing relationship Cather conjures between architecture and landscape functions most powerfully when windows remain open to the world: "In my book I tried to make Professor St. Peter's house rather overcrowded and stuffy with new things . . . until one got rather stifled. Then I wanted to open the square window and let in the fresh air that blew off the Blue Mesa, and the fine disregard of trivialities which was in Tom Outland's face and in his behaviour."[46] Cather translates the architectural premise of the Dutch paintings into the narrative architecture of her novel—namely, by allowing the fictional window in the Professor's house to double as a narratological window in *The Professor's House*. Thus, the landscape brought into the fictional home simultaneously becomes a fixture in the house of fiction itself. Furthermore, leaving these fictional and narratological windows open invites an enlivening breeze to permeate both imaginative spaces. For Cather, the figurative "fresh air" ushered in along with the landscape vision (in this case, the landscape of New Mexico's Blue Mesa) helps to charge her fictional spaces and the characters who inhabit them with an enhanced sense of energy and passion. In other words, the narrative architecture of *The Professor's House* acts as a kind of focalizing device that brings the vital landscape of the U.S. Southwest into focus, infusing the novel with that landscape's revitalizing energy by way of what I call a "home(l)y metaphysics."[47]

AT HOME IN LANDSCAPE

Cather's protagonist in *The Professor's House* is Godfrey St. Peter, a fifty-two-year-old history professor who feels permanently exiled from the present. As a historian, it is the past that enlivens him rather than the commercial frontiers of the modern world; he trades in intellectual artifacts rather than the kinds of material commodities preferred by his own family. Even his name places him in the company of historical pioneers, linking, as it does, Godfrey of Boulogne (conqueror of Jerusalem) and Saint Peter (founder of the Roman Church). The Professor has likewise enjoyed a successful career as a scholarly pioneer, spending his academic life chronicling early American imperialism in his multivolume masterwork, *Spanish Adventures in North America*. With the recent publication of the history's final volume and the impending conclusion of his career, however, St. Peter faces the closing of an intellectual frontier; his connection to the past dwindles. The pain of this transition is exacerbated by his family's move into a new house, which, for St. Peter, means abandoning the old house and hence the last-remaining ties to his own personal history. The novel opens in the midst of these changes: "The moving was over and done. Professor St. Peter was alone in the dismantled house where he had lived ever since his marriage, where he had worked out his career and brought up his two daughters."[48] The pondering length of the second sentence gives the lie to the blunt brevity of the first. The move may be complete, but St. Peter is far from renouncing his own past. He holds fast to a secret vitality.

My insistence on the Professor's "secret vitality" may at first seem strange, especially given that critics tend to emphasize his existential desuetude. Many scholars see St. Peter as a stuffy academic whose occasional patriarchal and even imperialist tendencies have left him a lonely exile, alienated from his own family within a deteriorating old house. For such critics, he appears little more than an old-fashioned historian who inexplicably clings to a dying architecture as he faces the closing of his career, the slow dissolution of his family, and the sense of his own impending death. But in contrast to this depiction of St. Peter as a desiccated and vestigial figure, I see the Professor as an artist in the Catherian sense—that is, an individual of singular vision who is animated by grand passion. As such, St. Peter's attic study serves a secondary function as an artist's studio. St. Peter considers his attic separate from the rest of the otherwise "dismantled" building, and he reckons uncertainly with "the feeling that under his workroom there was a dead, empty house" (7). Included among the "dead" and "empty" rooms below is St. Peter's "show study," located more formally on the main floor and complete with "roomy shelves where his library was housed, and a proper desk at which he wrote letters" (8). This study may appear more appropriate for the prestige of intellectual work, but it is little more than an attractive sham, as hollow and dead as the rest of the all-but-abandoned house. By contrast, and despite its architectural informality, his dark "cuddy" at the top of the house serves as the real center of his creative practice. The attic retains the vestiges of life.

More than just the site of his scholarly writing, the attic study serves as St. Peter's main hub for intellectual adventure and creative ferment: "There had been delightful excursions and digressions; the two Sabbatical years when he was in Spain studying records, two summers in the South-west on the trail of his adventurers, another in Old Mexico, dashes to France to see his foster-brothers. *But the notes and records and the ideas always came back to this room*" (16; emphasis added). If, as Bernice Slote has suggested, Cather's thinking about the creation of art finally comes to "bear on [the] ways by which an artist might bring disparate elements into a whole,"[49] then as a scholar who has spent years organizing decades of research into his history of the Spanish explorers in the Southwest, St. Peter undoubtedly qualifies as an artist, and his attic study is the site where the real work of gathering disparate elements into a unified whole has taken place: "It was here [that his notes] were digested and sorted, and woven into their proper place in his history" (16).

In addition to providing the space for St. Peter's creativity, there is another way in which the study embodies the vitality of art creation: namely the window that provides "the sole opening for light and air" (7). Aside from its functional purpose as a release valve that keeps the attic from filling up with noxious gas from the house's furnace, this window also affords a view of Lake Michigan, a picturesque vision of "the inland sea of his childhood" that had dazzled him as a boy: "The great fact in life, the always possible escape from dullness, was the lake. . . . [It] ran through the days like the weather, not a thing thought about, but a part of consciousness itself. . . .

[As a boy] he didn't observe the details or know what it was that made him happy; but now, forty years later, he could recall all its aspects perfectly. They had made pictures in him when he was unwilling and unconscious, when his eyes were merely wide open" (20–21). This description evokes a youthful state of existence in an unreflective present, a life *lived* rather than "a thing thought about." Yet the "unconscious" process elaborated here, in which environmental forces silently impress images on St. Peter's "unwilling" consciousness, also bespeaks an attachment to this landscape that was forged as a boy and traumatically ruptured before he came of age: "When he was eight years old, his parents sold the lakeside farm and dragged him and his brothers and sisters out to the wheat lands of central Kansas. St. Peter nearly died of it" (21). The trauma of this move haunted him throughout his youth, so that even during his idyllic student years in France, "that stretch of blue water was the one thing he was home-sick for" (21).

The deeply imprinted "pictures" of Lake Michigan eventually lead St. Peter to accept a faculty position at Hamilton, a moderately prestigious university near his childhood home. This homecoming restores to him the long-lost vitality he had enjoyed as a boy, particularly through the view of the lake furnished by his attic window: "The sight of [Lake Michigan] from his study window these many years had been of more assistance than all the convenient things he had done without would have been" (22). More than simply offering a sense of proximity to the lake, the window enlivens the room with a generative energy vital to the quasi-artistic creation of his scholarly masterpiece. The landscape of St. Peter's childhood home enters the attic interior and infuses it with the essence of dwelling.

And yet, twenty years after his homecoming to the lake, St. Peter still feels like a refugee. The sense of exile he has carried with him from the time of his premature departure from the shores of Lake Michigan remains etched in his consciousness, leaving him with a perennial "homesickness for other lands" (6). Despite the importance of the attic as what Judith Fryer, using Bachelard's phrase, refers to as a "felicitous space" of retreat, the line separating the comfort of a refuge from the refugee's unrest remains fuzzy for St. Peter. In the face of this enduring estrangement, he requires a more fundamental homecoming.

If the view from his study window cannot completely satisfy St. Peter's desire for homecoming, the novella installed within the novel presents an opportunity for a more profound return. "Tom Outland's Story" opens up a narratological picture window that frames a vista of the U.S. Southwest, ushering this landscape into the larger narrative architecture. In contrast to "the inland sea of his childhood," this "outland" vista is home to St. Peter's creative spirit. Not only is it the region on which his historical research has focused, but it also serves as the imaginative site of his connection with Tom Outland, a former student who had also been a surrogate son and heir before his premature death in World War I. Tom is a metonymically multivalent figure for St. Peter. He stands in for the southwestern landscape and for the Professor's own lost youth, and he also represents the ideal of an artist-like figure who remains untouched by modern commercialism. Tom is a consignment from St. Peter's past who, as Bergson would say, "swells" into and innervates his present. In order to tap into this revitalizing swell, St. Peter turns to the old diary where Tom recounted his youthful adventures. While the rest of his family vacations in Europe, St. Peter settles himself in the attic and begins transforming the diary into a publishable manuscript. The Professor's commitment to this editorial project at once enables his homecoming to a nostalgic landscape and fosters a return to a former self, a more "authentic" way of being.

Tom's account of his southwestern adventures dramatizes his imaginative (re)possession of the Blue Mesa landscape, which he and his friend Roddy first discover as ranchers in Pardee, New Mexico. From the vantage of their backcountry cabin, Tom and Roddy come into a personal relationship with the Mesa, investing it with meaning: "The Blue Mesa was one of the landmarks we always saw from Pardee—*landmarks mean so much* in a flat country" (165; emphasis added). The geological features that jut out from an otherwise flat and apparently empty horizon jumpstart Tom's imagination, inspiring him to find an entrance into the Mesa, where he comes upon ancient, long-abandoned homes carved into the rockface of a cliff—a place he calls "Cliff City." When he eventually lets Roddy in on his secret, he does so hoping that his companion will share his preservationist impulse, which seeks to protect "those silent and beautiful places" from "vulgar curiosity" (183). From an archaeological point of view, Tom's preservationism seems justified; Cliff City stands as an architectural marvel. When Tom first comes upon the "little city of stone, asleep," his first impression is of sculptural balance:

> It all hung together, seemed to have a kind of *composition:* pale little houses of stone nestling close to one another, perched on top of each other, with

flat roofs, narrow windows, straight walls, and in the middle of the group, a round tower. *It was beautifully proportioned,* that tower, swelling out to a larger girth a little above the base, then growing slender again. There was something symmetrical and powerful about the swell of masonry. *The tower was the fine thing that held all the jumble of houses together and made them mean something.* (180; emphasis added)

The compositional principles of Cliff City mark its difference from other structures Tom has seen in Acoma and Hopi villages. In contrast to the plain functionality of contemporary Native American architectures, "[Tom] felt that only a strong and aspiring people would have built [Cliff City], and *a people with a feeling for design*" (182; emphasis added). According to Tom, the original cliff dwellers were *artists* whose great vision enabled them to organize a "jumble of houses" into a meaningful tableau.[50]

After completing their contracts as ranchers, and animated by a preservationist spirit, Tom and Roddy embark on a scheme to render Cliff City more accessible and prepare it for excavation. As excavation begins, the initial impulse to preserve shades into possessive desire. Reverence for "*their* town" quietly shifts to an expression of pride in "*our* city": "One thing we knew about these people; they hadn't built their town in a hurry. Everything proved their patience and deliberation. . . . But the really splendid thing about our city, the thing that made it delightful to work there, and must have made it delightful to live there, was the setting" (190). The shift in possessive pronoun that takes place in this passage quietly announces an encroaching sense of ownership. The shift in pronoun also discloses a shift in point of reference. As long as Tom and Roddy focus on specific architectural features, Cliff City remains "their town" and belongs to the original, historical inhabitants. Following the initial expression of reverence for the quality of construction, the ellipsis in the passage above conceals a substantial list of architectural elements, including "cedar joints," "poles . . . that held up the clay floor," "door lintels," and "clay dressing . . . frescoed in geometrical patterns" (190). However, once the point of reference shifts from material architecture to the more-than-material splendor of "the setting," "their town" miraculously transmutes into "our city," which "hung like a bird's nest in the cliff, looking off into the box canyon below, and beyond into the wide valley we called Cow Canyon, facing an ocean of clear air" (190–91). This shift from architecture to setting signals a more significant change than the pronominal substitution at first indicates. It marks a move from the admiration of design sense and technical ability—the artists' *skill*—to the admiration of constitution and spirit—the artists' *vitality*. Although Tom and Roddy admire the virtues of the "fine people" who "lived day after day looking down upon such grandeur," the men's evident possessive desire subtly undermines what initially comes across as reverence.

The slipperiness between reverence and possessiveness becomes clearer following Tom's trip to Washington, where he entreats the Smithsonian to support his excavation based on "the beauty and vastness of the setting" (204). But Tom fails to convince the museum curators, and, dejected, he returns to the Mesa only to find that Roddy has sold the excavated artifacts to a foreign businessman. Tom's reaction to the sale illuminates the full extent of his possessiveness. Initially, he claims that Roddy's transaction liquidated a national inheritance: "They belonged to this country, to the State, and to all people. . . . You've gone and sold your country's secrets" (219). Roddy's betrayal is also, and more crucially, a personal one. Following the sale, Tom explains to Roddy that the artifacts "belonged to boys like you and me, that have no other ancestors to inherit from" (219), and he marvels at the idea that these objects "had been preserved through the ages by a miracle, and handed on to you and me, two poor cow-punchers, rough and ignorant" (220–21). Despite his "Fourth of July" talk, then, Tom mourns the loss of the only inheritance to which he, as an orphan, has any claim.

Tom experiences his loss ontologically. When he goes to bed on the night of his return to the Mesa, the very source of his being seems to have been evacuated: "I went to sleep that night hoping I would never waken" (224). Tom's nihilistic drive fades the following night, when, left alone on the Mesa after Roddy's covert departure, he has a transformative experience. He realizes that this "was the first night I was ever *really* on the mesa at all—the first night that *all of me* was there":

> This was the first time I ever saw it as a *whole*. It *all came together* in my understanding, as a series of experiments do when you begin to see where they are leading. Something had happened in me that made it possible for me to *co-ordinate and simplify*, and that process, going on in my mind, brought with it great happiness. *It was possession.* The excitement of my first discovery was a very pale feeling compared to this one. For me the mesa was no longer an adventure, but a *religious emotion.* I had read of *filial piety* in the Latin poets, and I knew that was what I felt for this place. (227; emphasis added)

In this moment, the slow process of Tom's imaginative possession of the Mesa landscape finally comes to completion, focalized through a religious idiom. As he stands atop the Mesa overlooking Cliff City, the landscape converges around him, organizing and "co-ordinat[ing]" itself into a miraculous "whole." The magnificence of this unification inspires in him a "religious emotion," a deep-rooted sense of responsibility to and for that particular place. This responsibility resembles a bond of kinship, one that serves as a kind of ontological recompense for the loss of a material inheritance. As the landscape coalesces around Tom, it forms a unified whole that parallels the radical emergence of his essential self—"all of me was there." The Mesa has become the place where Tom may thrive in the fullness of his *being*; it is the place where he most essentially *belongs*.

Tom Outland's story of imaginative (re)possession and ontological (re)vitalization has a profound effect on St. Peter; it relieves his existential desuetude and brings him home to his childhood self, "the original, unmodified Godfrey St. Peter" (239). Part 3 of *The Professor's House* opens with St. Peter reflecting with melancholy that "the most important things in his life . . . had been determined by chance" (233). He has led a circumstantial life, a "false life" that, according to Ruskin's moralism, has failed to master its own momentum. Even so, Tom Outland's appearance in his life represented "a stroke of chance he couldn't possibly have imagined" (233), a fortuitous event bearing transformative potential. Just as his original encounter with Tom proved expansive for St. Peter, in the present moment he once again embraces the youthful spirit of his onetime companion. And so, as he reconnects with Tom's memory, St. Peter resuscitates an older, more vibrant self and reminds him of a more primal mode of being: "[The] boy who had come back to St. Peter this summer was not a scholar. He was a primitive. He was only interested in earth and woods and water. . . . He seemed to be at the root of the matter, Desire under all desires, Truth under all truths. . . . He was earth, and would return to earth" (241). If St. Peter has lapsed into the Ruskinian false life, he finds some renewal in Bergsonian duration, "the continuous progress of the past which gnaws into the future and which swells as it advances."[51] Revivified through Tom Outland's story and the Southwest landscape it brings into focus, St. Peter returns to a primordial self, the kind of *moi fondamentale* that, according to Bergson, "blazes up from time to time, notably in childhood, in dreams, and in aesthetic feelings."[52] In the end, then, the narratological window that opens onto the Blue Mesa brings the Professor home to himself in a profound way that had only ever been approximated by the attic window that opened onto Lake Michigan.[53]

LANDSCAPE'S HOME(L)Y METAPHYSICS

The foregoing analysis of *The Professor's House* demonstrates how the complex structure of Cather's novel uses architecture as both a thematic and structural tool for framing Tom Outland's imaginative settlement of the Blue Mesa landscape. The novel's complex narrative architecture allows its protagonists to bring the outside inside in a way that recalls an image furnished by Edith Wharton in her 1897 manual on interior decoration, written with the architect Ogden Codman. In a chapter explaining how windows "form the basis of architectural harmony" and "serve to increase the dignity and beauty" of the domestic interior, Wharton emphasizes how they "establish the relation between the inside of the house and the landscape, making the latter what, as seen from a room, it logically ought to be: a part of the wall-decoration."[54] Cather installs windows both literal (St. Peter's attic window) and figurative ("Tom Outland's Story" as narratological window) that allow landscape's revitalizing power to infuse the architecture of the Professor's house as well as of *The Professor's House*.

In Cather's novel it is ultimately landscape that enables a sense of homecoming for Tom as well as for St. Peter, with Tom as proxy. With respect to Tom in particular, his imaginative gathering of his surroundings into a sacred dwelling place has something to do with the metaphysical process of consecration that Mircea Eliade describes in *The Sacred and the Profane*. According to Eliade, from the perspective of someone settling in a new land, unknown and apparently "unoccupied" territory takes on the chaos of homogeneous nonreality and must be worlded through rituals of taking possession. Such rituals "always repeat the cosmogony."[55] Thus, any act of settling a territory doubles as an act of consecration, a symbolic re-founding and re-"cosmicizing" of the world. Tom's particular rituals of taking possession (e.g., survey, excavation) facilitate the rebirth of the Mesa as a landscape that organizes and orients the world into specific configurations of personal (and national) significance. Through this process, Tom consolidates a meaningful landscape; profane space becomes a sacred place of radical belonging.

Significantly, Tom's act of consecration is also an act of settler colonialism. His reworlding of the Mesa landscape enacts what Lorenzo Veracini calls "narrative transfer," which reterritorializes apparently unsettled land by actively unsettling—deterritorializing—Indigenous communities.[56] Settlement demands a vanishing indigene. In the case of "Tom Outland's Story," it also demands the disavowal of history. As an orphan, Tom's imaginative claim to the Blue Mesa landscape is also a claim to a personal inheritance. For St. Peter, who sees Tom as a pioneer and whose own imaginative desire for the New Mexican landscape amplifies this sentiment, Tom's inheritance takes on a national(ist) significance that places its own origins under erasure. Hence, having made a home in landscape, Tom comes to emblematize the ~~Native~~ American spirit of individualism that characterizes the quintessential national spirit of American pioneers and artists alike. As the reference to Eliade suggests, settlement is as much a metaphysical act of taking possession as a physical one. The metaphysical thrust of colonization goes back at least as far as Plato, whose *Phaedrus* dialogue "refers to the relationship between body and soul as 'colonisation': *katoikizein* (specifically: the act of settling a colony)."[57] Just as the soul inhabits an inanimate body, settlers locate their ontological seat in (the myth of) empty land.

And indeed, Tom sets the metaphysical groundwork for his imaginative repossession of the Blue Mesa early in the novel by insisting on the landscape's emptiness. When Tom initially happens upon deposits of pottery shards, arrowheads, and other evidence of an apparently vanished civilization along irrigation ditches coming from the Mesa, he hypothesizes: "There must have been a colony of pueblo Indians here in ancient times" (173). Even as he acknowledges the traces of a previous occupation, however, Tom immediately goes on to ponder the *personal* significance of his discovery: "To people off alone, as we were, there is something stirring about finding evidence of human labour and care *in the soil of an empty country*. It comes to you as a sort of message, makes you feel differently about the ground you walk over every day" (173; emphasis added). Although Tom expresses delight at the discovery that the soil was only *apparently empty*—that in fact it harbors stories of its own that are yet to be unearthed—this moment marks a subtle shift in which Tom imaginatively repossesses these yet-to-be-unearthed stories for himself. What excites Tom is not the discovery of Native American relics but rather the way this discovery holds meaning known only to him. Thus, despite his cursory understanding that the land bears a rich history of inhabitation, he neglects to show any meaningful curiosity about or reverence for an Indigenous community that he automatically consigns to the past, implicitly refusing the possibility of what Gerald Vizenor (Minnesota Chippewa) terms Native "survivance."[58]

Instead, he relies on the landscape's lack of present habitation in order to justify his own exploration of the Blue Mesa and, eventually, his personal claim to its historical legacy.

Settler colonial narratives like Tom's showcase what I call landscape's home(l)y metaphysics. I mean two things by this term: first, a general sense in which the real environment becomes abstracted and hence subject to the kind of imaginative transformation that the Blue Mesa undergoes via the double framing of Tom's and St. Peter's narratives; and second, a more specific sense in which this transmuted reality frames out a "home(l)y" space that secures a sense of being and belonging—an ontologically charged space that Bachelard would describe as "felicitous." Landscape is at once ideological and philosophical; its significance is both political and ontological. In these senses, Cather's narratological use of landscape recalls Heidegger's metaphysical elaboration of *Landschaft*. As explored in the previous chapter, Heidegger treats the figure of landscape as a worlding figure that gathers mere space into a viewing subject's own particular beloved place, installing them more fully within the house of being. And yet, as Mircea Eliade reminds us, and as Tom Outland's story demonstrates, every worlding is necessarily a *reworlding*, an act of colonization. What this means is that landscape's metaphysical power to secure a sense of home—that is, of being-in and belonging-to place—has its roots in coloniality. As with Heidegger, Cather's landscapes broadly recapitulate the coloniality of Nature.

Emphasizing landscape's link to the coloniality of Nature in Cather's fiction adds an essential dimension to previous ecocritical scholarship that has already drawn attention to the imperialist politics inflecting her landscapes. Consider, for example, Carol Steinhagen's standout essay "Dangerous Crossings." Steinhagen's essay is worth examining in detail to see why attending to the coloniality of Nature proves crucial for avoiding the threat of essentialism in settler colonial contexts. Steinhagen begins by warning against equating landscape with Nature: landscape is, rather, "'the land's shape as it is seen from a particular and defined perspective,' both physical and psychic. Landscape is a seen scene."[59] Landscape thus signifies more than distanced perspective; it represents a boundary marker that psychically and physically separates the self (Ego) from the real environment (Nature). "Beyond" this boundary there exists what the author calls "unlandscaped environment,"[60] and transgressing this boundary—making a "dangerous crossing"—involves dissolving the ego into what Herman Melville calls the "'all' feeling" of Nature.⁶¹ Throughout her essay, Steinhagen champions such dangerous crossings beyond landscape: it is tragic that in *My Ántonia* Jim Burden loses this self-dispersive ability, just as it is worth celebrating when St. Peter apparently discovers it. (Though as my argument above should already suggest, St. Peter in fact becomes more firmly entrenched *in* landscape.) In the final analysis, landscape impedes more profound experience; it stymies us from casting our egos to the winds and attending to something like the really real.

If Steinhagen sees landscape as a bad thing, it is only in part due to how it curbs a more primordial experience; it is also because she associates landscape—and the worlding force of *landscaping*—solely with European imperialism, a species of psychic possession that appropriates the environment (which otherwise exists in and for itself) to the ego. Landscape therefore represents an imperialistic betrayal of environmental autonomy. Although I sympathize with this claim, and although my own reading of *The Professor's House* may appear to confirm it, Steinhagen's argument proceeds from a problematic assumption that goes unacknowledged in her essay: namely, it hinges on racialist distinctions she mistakes for essential differences. Early in the essay she writes: "For most westerners this experience of unity must remain temporary, if it is achieved at all, because their sense of nature is shaped by the historical forces that have given rise to the concept of landscape."⁶² Ostensibly, Steinhagen only wishes to comment on "Western" culture here, but this passage silently affirms the commonplace image of non-Western peoples as being in harmony ("at one") with Nature.⁶³ This affirmation returns more explicitly near the essay's end, when she speaks of the "respectful view of the Native American oneness with the natural environment, contrasted with the white man's way of 'assert[ing] himself in any landscape.'"⁶⁴ According to this reading, non-Indigenous peoples remain psychically stymied in landscape, whereas Indigenous peoples (here: Native Americans, broadly construed) enjoy a liberated and exclusive oneness with the environmental expanse.

Again, I do not dispute the link Steinhagen makes between landscape and psychic possession, but she is mistaken in her claim that this link inheres solely in "westerners." Take her example from *Death Comes for the Archbishop*, where Jacinto, a Navajo guide, resists revealing the Laguna place-name Snow-Bird Mountain: "Perhaps he knows that Europeans have already Christianized the land with names like Sangre De Christo. They have landscaped his home with language and he and his people must maintain in silence their sense of the surroundings."[65] Christianization threatens local ways of being, and the introduction of new place-names reinforces this process linguistically. But does this religious and linguistic imposition necessarily imply that the Southwest had not already been "landscaped" before the Spanish conquest? Did Native American tribes like the Navajo not have their own religious associations and toponyms? What fundamentally separates "Snow-Bird Mountain" from "Sangre De Christo?" Does the former not mark an Indigenous "psychic possession?"

The point here is that psychic possession is a trait of the human animal rather than just the Western subspecies. This, of course, does not justify the Spanish conquest of the Southwest, or, indeed, any form of imperialism or genocidal dispossession. But in the rush to castigate the politics of European imperialism, Steinhagen threatens to subordinate a vague Indigenous metaphysics to an essentializing notion of a generalized, Romantic oneness with Nature. What is actually at issue here is two competing metaphysics, one of which is bound up with the coloniality of Nature and its complex narrative mechanics that enact erasure to enable homecoming, the other of which is not. Steinhagen effectively denies Indigenous metaphysics when she asserts that Native Americans exist within an environmental reality rather than a landscape of the mind. In doing so, she unintentionally disappears Native Americans into Nature in a way that repeats the logic of elimination at the very heart of settler colonial narratives like Tom Outland's—and, arguably, like *The Professor's House* as a whole. The constitutive reality of modernity/coloniality means that in any context, and particularly in contexts more explicitly marked by (settler) colonialism, it remains crucially important to think through the relation between landscape's home(l)y metaphysics and the coloniality of Nature. Doing so preserves the sovereignty of Indigenous metaphysics without also relenting on the critique of Western metaphysics and its foundations in the modernity/coloniality matrix.

In the following chapter, which considers the British writer D. H. Lawrence's unsettling depictions of the New Mexican landscape, I explore these precise issues in greater detail.

CHAPTER 3

D. H. Lawrence and the Ecological Uncanny

Do you feel strange when you go home?
—D. H. Lawrence, *St. Mawr*

When D. H. Lawrence first arrived in Taos, New Mexico, in 1922, he did not know that his host, Mabel Dodge Luhan, had already tasked him with a strange challenge. Luhan had written a letter to Lawrence earlier that year while he and his wife, Frieda, were living on Sicily. She had just finished reading *Sea and Sardinia*, which she describes as "one of the most actual of travel books."[1] In her memoir of Lawrence's time in New Mexico, Luhan explains how his book had captured her imagination—how, "in that queer way of his, he gives the feel and touch and smell of places so that their reality and their essence are open to one, and one can step right into them."[2] Though the eminent English novelist could not have known it at the time, it was precisely this ability to evoke a sense of place that Luhan desired to put into service when she invited the Lawrences to New Mexico.

Luhan's emphasis on Lawrence's ability to capture that which is "most actual" about the "reality" of place may seem odd for a reader familiar with *Sea and Sardinia*. For one thing, Lawrence's memoir often functions less to conjure a sense of place than to "[*disclose*] *its author*, giving him an unexpected and at times comical precedence over the matter of the place itself."[3] For another, when he does focus on the matter of place, his descriptions are often tinged with the occult. Whereas the first point typifies travel memoir generally, the second gestures to something more characteristically Lawrencian. A penchant for occult language is on full display in the opening pages of *Sea and Sardinia*, where Lawrence envisions Mount Etna as a concealed presence: "Remote under heaven, aloof, so near, yet never with us."[4] He explains that, although artists ceaselessly attempt to reproduce Mount

Etna's "magical" and "witch-like" allure, they only ever capture those visual elements that are "with us" in "our own world," such as the "near ridges, with their olives and white houses."[5] By contrast, Mount Etna—the *real* Mount Etna—retreats "beyond a crystal wall" and remains locked within the "strange chamber of the empyrean."[6] In order to approach the volcano in "her" hidden actuality, Lawrence claims, "You must cross the invisible border. Between the foreground, which is our own, and Etna, pivot of winds in lower heaven, there is a dividing line. You must change your state of mind. A metempsychosis."[7] Lawrence invokes the traditional rhetoric of landscape painting (i.e., foreground, middle ground, background) only to abandon the visual frame of reference altogether. Crossing the "invisible border" that separates observer from volcano, subject from object, is not a matter of focusing the eye differently; the imaginative transmigration described here requires a shift in one's way of thinking and being. According to Lawrence, such a shift enables one to tap into the concealed actuality, but without ever directly encountering the real. Though hidden behind a "crystal wall," Mount Etna's "witch-like" allure radiates through that wall. Paradoxically, the volcano makes its own utter unknowability known through "strange, remote communications": "one can feel a new current of her demon magnetism seize one's living tissue, and change the peaceful lie of one's active cells. She makes a storm in the living plasm, and a new adjustment. And sometimes it is like a madness."[8] Succumbing to the volcano's demon magnetism, allowing for a storm to brew in one's living plasm—these, for Lawrence, figure an embodied notion of "metempsychosis."

Luhan invited Lawrence to New Mexico with just such a metempsychosis in mind. Captivated by his talent to open himself to otherwise hidden actualities, Luhan wanted Lawrence to do for Taos what he had done for Mount Etna: "Here is the only one who can really *see* this Taos country and the Indians, and who can describe it so that it is as much alive between the covers of a book as it is in reality."[9] Luhan obliquely announced her intentions prior to the Lawrences' arrival. Along with her letter of invitation she enclosed a copy of *The Land of Poco Tiempo*, a study of New Mexico's history and peoples written by Charles Lummis in 1893. On the book's title page she penned the following inscription: "Lawrence!—this is the best that has been done yet—And yet if you knew what lies untouched behind these externals, unreached by the illuminating vision of a simple soul yet! Oh, come!"[10] Once again appealing to the writer's ability to get beyond mere "externals," Luhan lures Lawrence with an appeal to the occult—an appeal she echoes with her accompanying gifts for Frieda, which had been imbued with "Indian magic" meant to "draw them to Taos."[11]

The Lawrences' first trip to Taos lasted from September 1922 to March 1923, and they would return again from March 1924 to September 1925. Lawrence's two southwestern sojourns had a great impact on his life, and though his arrival in the midst of political turmoil complicated his initial feelings about the place (more on this political scene later), his travels in and around the region soon inspired a fascination with the place. Lawrence's writings of the period provide ample evidence of this fascination, and to read them against the background of Luhan's celebration of the Lawrencian spirit of place, it would seem that Lawrence satisfactorily answered Luhan's call. For indeed, during his two stays in the region Lawrence produced some of his most memorable later work. And, as many critics have further noted, Lawrence found something in New Mexico that he never had before or after: made materially manifest in Kiowa Ranch on Lobo Mountain, the only property he would ever own, Lawrence found *a sense of home.* Julianna Newmark contends that Lawrence's time in New Mexico worked on him a transformation that "was not only physical, taking him to a great many places to look for a home for his soul, as it were, but . . . was also deeply internal in his sense of what it means to be *at home* in a place."[12] In discovering the hidden reality of New Mexico and its people, this story suggests, Lawrence also discovered his own place of belonging.

Yet this story of Lawrence's home(coming), only infrequently questioned,[13] neglects to notice a constitutive strangeness that always haunts Lawrence's depictions of New Mexico. Despite his growing familiarity with New Mexico, at no point during his time in the Southwest did he ever claim a special affection for the landscape. As the Danish artist Knud Merrild recalls of his sometime New Mexican companion: "Lawrence generally revolted against the scenery. I cannot recall one single instance where he heartily commended the beauty and grandeur of the landscape. On the contrary, he was very much against it as a whole. 'It is so heavy and empty, it sort of hangs over one. It is very depressing,' he would say."[14] Rather than assert an unqualified sense of comfort, Lawrence consistently points to something more essentially unsettling.

Consider his essay "New Mexico" of 1928. Although often cited for its rapturous praise of the southwestern landscape, Lawrence does not concern himself primarily with beauty. Instead, he advances a claim about a hidden wellspring of vitality—the last trace to survive in a world corrupted by Euro-American modernity, making the Southwest the last bastion of a life force that could reverse the spiritual decay of America and, perhaps, of Western civilization at large. Lawrence frames the problem of spiritual decay in the essay's opening paragraph, where he gestures to the apparently diminished mystery of a world exhausted by the tourist industry. He insists that the world's essential mystery is only *apparently* diminished because, like the ocean, everything important about it remains undiscovered. There exist "terrifying under-deeps" in the land as much as in the sea: "Underneath is everything we don't know and are afraid of knowing."[15] Given such a strange opening, it seems odd to emphasize his descriptions of the landscape's "greatness of beauty" and the vast "splendour of it all." As Lawrence's original emphasis suggests, it was not so much the "beauty" of the landscape that inspired him but its startling "greatness." He reinforces this sentiment when he frames his intimation of "splendour" as a kind of "terror": "Leo Stein once wrote to me: [New Mexico's] is the most aesthetically-satisfying landscape I know. To me it was much more than that. *It had a splendid silent terror,* and a vast far-and-wide magnificence which made it way beyond mere aesthetic appreciation."[16] He describes later in the essay that this splendid silent terror reveals New Mexico as a unique site where "man was to get his life into direct contact with the elemental life of the cosmos, mountain-life, cloud-life, thunder-life, air-life, earth-life, sun-life. To come into immediate *felt* contact, and so derive energy, power, and a dark sort of joy."[17] By the end of the essay, this regional wellspring of vitality and the "dark sort of joy" it inspires takes on a world-historical importance: it signals the final collapse of democracy and the rise of an "old religion" that will "scatter" the relics of industrialized modernity and constitute a new beginning for a "genuine America." Far from an appreciation of the landscape's aesthetic recompense, "New Mexico" describes the concealed power of a place that could, despite its concealment, renew a life-centered ethos in an otherwise spiritually rotten world.

Lawrence's argument in "New Mexico" reveals the mistake Luhan made in her initial evaluation of his writing. His gift was not to reveal "the actuality behind the veil."[18] Instead, it was to acknowledge the otherwise unknowable *and allow it to remain unknown.* And if Lawrence's strange sense of place undermines Luhan's desire for him to unveil the singular, hidden reality of Taos, then it also forestalls narratives of Lawrence's apparent homecoming to New Mexico. In contrast to what I have called "home(l)y metaphysics," which defines "home" as a protective space that revitalizes being and secures belonging, for Lawrence, living in Taos may have stimulated his sense of being, but this increased vitality was not attended by an accompanying sense of belonging. In fact, Lawrence more often expressed the sentiment that living in Taos had an *estranging* quality, though this estrangement proved qualitatively different from the kind of alienation generated within industrialized civilization. "Alienation" implies a radical separation, such as between humans and Nature. Alienation designates a kind of existential exile that compromises being such that one does not feel at home anywhere. For Lawrence, however, the New Mexican landscape does not exactly alienate one; rather, *it turns one into an alien*. In contrast to the state of "being alienated," which implies negativity and passivity, "being an alien" is a positive ontological state. Even if by definition an alien cannot claim to "belong" in a particular place, an alien still fully exists: it *is*. This is precisely the nature of Lawrencian estrangement: being without belonging.

Aldous Huxley captures the Lawrencian sense of revitalizing estrangement perfectly when he announces: "The New Mexican is an inhuman landscape. Man is either absent . . . or, if present, seems oddly irrelevant."[19] According to Huxley, whom Lawrence had befriended around the time of his first travels in America, and who himself lived briefly in Taos in the 1930s, the "irrelevance" of humans in an otherwise "inhuman" landscape proves equally liberating and dangerous. In his preface to Knud Merrild's memoir, Huxley describes the site of Lawrence's ranch on Lobo Mountain as a place where "a new kind of inhuman alienness envelops you. Behind and below you lies the desert; but above is a world that at moments seems positively Nordic. . . . Between this sentimental Teutonic inhumanity [of the mountain valleys] and the ferocious American inhumanity of the desert below, there is no middle term. And in this fact consists, precisely, the charm of the New Mexican landscape—its charm and also its horror."[20] Without a "middle term" to organize foreground and background, the whole landscape takes on a radically different reality; its effects are no longer primarily visual. Similar to Lawrence's description of Mount Etna in *Sea and Sardinia*, the visual frame of reference must be abandoned in New Mexico to allow one to "cross the invisible border" and tap into its hidden actuality. Although Huxley's description of Lobo Mountain as a charming yet horrifying hodgepodge of American and European geographies seems strange, it is appropriately strange; that is, it helps explain "Lawrence's ambivalent attitude towards the country—the dislike that mingled with the love, *the dread that accompanied his homesickness.*"[21] For Lawrence, dread and home go together; home should be a place of revitalizing estrangement.

The remainder of this chapter takes a number of apparent digressions in order to explore how this unsettling conception of home as essentially *strange* and *estranging* emerges in several streams of Lawrence's literary and philosophical thinking and practice during the period extending roughly from 1920 to 1925. The next section begins by reframing Lawrence's well-known notion of "the spirit of place" in relation to a wide-ranging rhetoric of spectrality that appears throughout his work of the period. The following section extends this analysis of Lawrence's spectral rhetoric to his thinking about human–nonhuman relations, which gives way to what he calls the "weird anima" of landscape and what I refer to as "the ecological uncanny." The next section turns to Lawrence's 1925 novella *St. Mawr* to analyze how uncanny depictions of the natural world unsettle the normative idea of a home(l)y Nature. I conclude the chapter by addressing Lawrence's writings about the Taos Pueblo and the Hopi Third Mesa to complicate the claim that his representations of Native Americans essentialize Indigenous difference and recapitulate the settler colonial logic of Indigenous elimination. Without celebrating Lawrence as a model of political correctness, the chapter concludes by suggesting that Lawrence's unique depiction of New Mexico's uncanny spirit of place may nonetheless help unsettle fantasies of settler colonial homemaking.

SPECTRAL RHETORIC AND THE "SPIRIT" OF PLACE

Early in *Studies in Classic American Literature* Lawrence writes: "Every continent has its own great spirit of place."[22] Ecocritics have seized upon this sentence as evidence of Lawrence's ecological vision. To date, however, few ecocritics have interpreted this central concept as anything more nuanced than a basic dialectic or "polarity" between humans and their environment.[23] Such readings reflect a more general tendency among ecocritics to reduce the strangeness of Lawrence's thinking, employing paradigms that either situate Lawrence within conventional categorization of the "nature tradition,"[24] forge an essentialist link between "ecology" and "the primitive,"[25] or harness Lawrence's complex and challenging work to support preestablished values mandated by deep ecology and other modalities of environmental ethics.[26]

Yet if anyone is to blame for the unsatisfying treatment of Lawrence and the spirit of place, it is Lawrence himself. Indeed, he glosses the term only briefly: "Every people is polarized in some particular locality, which is home, the homeland. Different places on the face of the earth have different vital effluence, different vibration, different chemical exhalation, different polarity with different stars: call it what you like. But the spirit of place is a great reality."[27] On the surface, Lawrence's characterization of the spirit of place seems to do little more than reiterate the old Roman notion of *genius loci,* which, though originally referencing actual spirits or gods that inhabited particular places, had long since become a generalized expression designating a locality's distinctive atmosphere. Every place is different, Lawrence tells us, and this difference may emerge from any number of factors. We can give this differentiating factor any name we want, but any name we come up with will only gesture to the *fact* of this difference rather than name its *source,* so (apparently) no need to elaborate further. Lawrence's vaguely neoclassical definition of the spirit of place has made it very easy for his readers to misunderstand his point, oversimplifying the *genius loci* as the basis for a mutually constitutive relationship between peoples and places. But if read closely, one finds that Lawrence does not indicate a straightforward dialectic between people and place. He does write that people are polarized *in* a place, but he immediately goes on to suggest that place itself is more readily constituted by something beyond its relationship to human inhabitants. The "great reality" of the spirit of place thus does not emerge along a single axis between humans and their environment.

The ecocritical understanding of Lawrence's spirit of place needs an upgrade, and since Lawrence himself neglects to elaborate here, we must look elsewhere to develop a fuller understanding. The strategy I propose is to shift emphasis from the ecocritical keyword *place* and focus instead on the more Lawrencian word *spirit.* Emphasizing spirit proves advantageous because of its connection to the spectral rhetoric that appears in other parts of *Studies in Classic American Literature* and throughout Lawrence's later writing. Lawrence's spectral rhetoric no doubt stems from his abiding fascination with messianic faith, the cult of the blood, and the dark life force of the unconscious, and he fed these interests with abundant reading on primitivism, theosophy, yoga, and astrology.[28] However, another strain of thinking emerges from within Lawrence's writing, developing over the ten-year period from 1915 to 1925. From this view, Lawrence's spectral rhetoric arises not solely from his research into the occult, but also, surprisingly, from a chemical metaphor he initially entertains to rethink conventional rendering of character.

By the time of his New Mexican writings of the 1920s, Lawrence had already begun to harness the chemical principle of *allotropy* to experiment with a notion of the self as fundamentally multiple. Allotropy refers to a property of some chemical elements and substances to exist in two or more different forms. Lawrence made his first attempts at allotropic character in his novels *The Rainbow* (1915) and *Women in Love* (1920). Lawrence explained the nature of this experiment in a letter to his editor, Edward Garnett, concerning *The Rainbow*. Instead of seeking "the old stable *ego* of the character," Lawrence insists that when reading his new novel one must look for "another *ego*, according to whose action the individual is unrecognisable, and passes through, as it were, allotropic states which it needs a deeper sense than any we've been used to exercise, to discover states of the same radically unchanged element."[29] Instead of the traditional sense of character as something written or engraved and hence indelible, Lawrence's allotropic ego does not exist in a singular or stable state. But even as the allotropic ego manifests in many forms, something remains hidden from view, under the surface, securing identity across various modes of appearance. Lawrence stays with his chemical metaphor to explain: "Like as diamond and coal are the same pure element of carbon. The ordinary novel would trace the history of the diamond—but I say, 'Diamond, what! This is carbon.' And my diamond might be coal or soot, and my theme is carbon."[30] Less interested in the "history" of appearances (carbon's manifestation as coal or diamond), Lawrence fashions a narrative that passes through the ego's plenitude of allotropic states and gropes blindly for the "radically unchanged element" from which they emanate (elemental carbon itself). It is this rejection of the singular, stable ego that contributes to the peculiar way in which Lawrence's novels of 1915 and 1920 cut straight through the "outer shell" of character to reveal the unchanged yet unstable "center." Throughout these novels, Lawrence frequently diverts away from dialogue in order to narrate how individuals' unconscious selves come into direct relation: "They had looked at each other, and seen each other strange, yet near, very near, like a hawk stooping, swooping, dropping into a flame of darkness."[31] Even as Lawrence attempts to bring forth his characters' unconscious selves, his metaphors proliferate once again and refigure the human as plural and fundamentally other than human: at once like a "hawk" and like a "flame." As with Walt Whitman, Lawrence's "I" contains multitudes—yet it also remains an "I."

Though Lawrence leaves the language of allotropy and the allotropic ego behind, its basic principle of multiplicity remains central to the theory of the novel he develops in a series of essays he wrote on the subject in the early 1920s, both during and after his two New Mexican sojourns. Despite differences in focus and theme, what links these essays is a shared sense of the novel as an inherently multiplicitous literary form that allows instability and relativity to persist within an otherwise interconnected whole. In "The Novel" (1925), for instance, Lawrence argues that the "novel is the highest form of human expression so far attained" because "it is so incapable of the absolute" and instead privileges relativity.[32] Lawrence furthers this point in "Morality and the Novel" (1925), where he defines the novel as a literary form that forges new relations rather than attempting to "pin" things down. In this way, the novel fosters a living and hence imbalanced dynamic that he associates with morality: "Morality in the novel is the trembling instability of the balance."[33] *Immorality*, by contrast, emerges the instant "the novelist puts his thumb in the scale, to pull down the balance to his own predilection," which forces a "stable equilibrium" and thereby "kills the novel."[34] What is "moral" in the novel therefore has to do with how narrative constantly mediates a "subtle" and "changing" dynamic within the characters' "circumambient universe." Navigating this vital dynamic entails danger, fear, and pain, and not for the characters alone. Hence Lawrence's proviso: "to read a really new novel will *always* hurt, to some extent."[35] Lawrence echoes this point in "Surgery for the Novel—or a Bomb" (1923), where he claims that the modern novel is a "monster with many faces" that will claw through the ideological walls that imprison and suffocate us.[36] Horror and liberation apparently go hand in hand in modern fiction.

More than anything else, what makes the novel unique for Lawrence is its ability to register the multiplicitous nature of human being. This is the subject of a remarkable essay of the same period, "The Novel and the Feelings" (ca. 1920), in which Lawrence describes the human individual as an "aboriginal jungle," a dense thicket that grows deep in the heart of "the original dark forest within us."[37] According to Lawrence, all of our "feelings" live within this internal wilderness, where they remain largely invisible to our conscious selves. In contrast to the wildness of feelings, Lawrence introduces "emotion" as the name for those "convenient" affects (especially love) that civilized society likes to "domesticate" and turn into "our pet favourite[s]."[38] The problem with privileging emotions over feelings is that it reduces the self, exhibiting only a curated selection of affects that reduces the individual. Try as we might to tame feelings into emotions, Lawrence insists that we cannot escape the wild multiplicity of our selves:

> What am I, when I am at home? I'm supposed to be a sensible human being.... [Yet] I have a whole stormy chaos of "feelings."... Some of them roar like lions, some twist like snakes, some bleat like snow-white lambs, some warble like linnets, some are absolutely dumb, but swift as slippery fishes, some are oysters that open on occasion.... In the night you can hear them bellowing.... [S]ince the forest is inside all of us, and in every forest there's a whole assortment of big game and dangerous creatures, it's one against a thousand.³⁹

Lawrence's strange bestiary of the self emphasizes how coming "home" to ourselves entails the acknowledgment of two unsettling facts: first, that humans are constituted by an unknowable multiplicity; and second, that we are fundamentally indomesticable. Avoidance of these facts has obscured the truth of our own inner "darkness" and encouraged us to exist "in sheer terror of ourselves," a state of affairs Lawrence says we can resist by actively "listening-in to the voices of the honourable beasts that call in the dark paths of the veins of our body, from the God in our heart."⁴⁰ Such a practice reveals the sheer multidimensionality—and profound nonhumanness—at the heart of the human individual, which literary convention otherwise reduces to the apparent singularity of "character."

With Lawrence's thinking about the nature of novels, characters, and the allotropic ego in mind, it is possible to return to *Studies in Classic American Literature* (1923) and clarify the status of the spectral rhetoric that appears in that work. In its broadest strokes, Lawrence's study of American literature seeks to uncover an emerging spirit of place in the United States as it sloughs off the Enlightenment humanism of the Old World and, like a caterpillar in its chrysalis, secretly forges a radically new "spirit." Lawrence tracks the gestation of this emergent New World spirit through several generations of American writers, beginning with the most quintessentially American: Benjamin Franklin. Lawrence expresses open disgust for Franklin's status as the chief intellectual architect of American freedom and democracy. He casts Franklin as an idealist with a dangerous tendency to celebrate both "liberty" and "the perfectability of man." To Lawrence, these represent incompatible desires: individuals cannot be liberated if they remain concerned with their own perfection—particularly when that perfection is contingent on arbitrary virtues and a quasi-capitalist formulation of American citizenship. Lawrence links Franklin's notion of perfectibility to the latter's famous decree, "Henceforth be masterless!," which is really a call for his fellow Americans to master themselves. The danger of this decree lies in its attempt to reduce "man" to mere parts of himself, and specifically the "cultivated" and "known" parts: "The *wholeness* of a man is his soul. Not merely that nice little comfortable bit which Benjamin marks out. . . . Why, the soul of man is a vast forest, and all Benjamin intended was a neat back garden."[41] Once again advocating for the inherent wildness of the human soul, Lawrence revises Franklin's credo. He claims that we can attain freedom only by relinquishing the desire for self-mastery and by allowing ourselves "Henceforth [to] be mastered!" But mastered by what—or whom? Lawrence's answer is as intriguing as it is unsatisfying: "IT chooses for us," though he quickly replaces "IT" with his preferred term, "the Holy Ghost." This Ghost—which is *not* the Holy Ghost of Christian theology—serves as the source of the self's "deepest promptings"; it is a multiplicitous and unknowable "master" that inhabits us and "drives" us without our direct awareness.[42]

The path I have traced through Lawrence's writing on the novel and back to his study of American literature at last allows me to return to the issue at the heart of this section and make the following claim: in contrast to Franklin's "kitchen garden scheme of things,"[43] Lawrence's notion of the spirit of place conceptually externalizes what he describes in the "The Novel and the Feelings" as the "dark continent" of the self. Just as the individual self becomes a "jungle" populated by thousands of species of indomesticable "feelings," the spirit of place names the totality of entities within a specific locale. Whereas the self represents an "internal" plurality, the spirit of place multiplies this exponentially. It therefore comes to name not just the totality of entities in a place, but the innumerable interrelations that occur among them. Lawrence's ideas of selfhood and place both employ a spectral rhetoric. A hidden "Ghost" haunts the human individual and invisibly drives it, just as the "spirit" of place indicates something more-than-physical that emanates from the totality of human and other-than-human relations. In both cases, Lawrence invokes spectral figures to indicate the strange way in which some hidden complexity reveals itself in its essential hiddenness—that is, makes its inaccessibility hauntingly present.

If this rhetoric of haunting seems needlessly obscure, it bears noting that, for Lawrence, being is simultaneously physical and metaphysical. In his essay "Reflections on the Death of a Porcupine" (1925), Lawrence makes an argument for the materiality of being: "The clue to all existence is being. But you can't have being without existence, any more than you can have the dandelion flower without the leaves and the long tap root.... [Being] is a transcendent form of existence, and as much material as existence is. *Only the matter suddenly enters the fourth dimension.* All existence is dual, and surging towards a consummation into being."[44] Being at once manifests itself to conscious experience and "transcends" that experience, entering "the fourth dimension" that Lawrence elsewhere associates with the unconscious. Even the unconscious, however, has a material seat. Lawrence formalizes this claim in *Fantasia of the Unconscious* (1922), where he argues that the unconscious "mind" is not a consolidated thing so much as a complex physiological system of "plexuses and planes." Like the "Holy Ghost," "IT," the "fourth dimension," and a host of other terms, the "unconscious" serves as a shorthand referring to an irreducible complexity at the heart of things that is at once apparent and withdrawn, material and immaterial. This complexity haunts *all living things,* and not just humans: "Any creature that attains to its own fulness of being, its own *living* self, becomes unique, a nonpareil. It has its place in the fourth dimension, the heaven of existence, and there it is perfect, it is beyond comparison."[45] An individual dandelion is just as capable of achieving a full expression of being as a human or a hawk, and just like any other vibrant being it contains multiple competing forces within itself. Thus, even within a single dandelion seed "sits the Holy Ghost[,] . . . which holds the light and the dark, the day and the night, the wet and the sunny, united in one little clue."[46]

The example of the dandelion demonstrates that, for Lawrence, all living things, and not just humans, are defined by an essential multiplicity. This does not, however, make all entities "equal" in a democratic sense. Elsewhere in "Reflections on the Death of a Porcupine" Lawrence insists on a hierarchy of vitality among living entities, where "higher" means "more alive": "As far as existence goes, that life-species is the highest which can devour, or destroy, or subjugate every other life-species against which it is pitted in contest," and this goes as much for "the dandelion [that] can take hold of the land" as for a "race of man [that] can subjugate and rule another race."[47] But this hierarchy does not relativize being in the same way it relativizes vitality. Instead, Lawrence advocates something similar to Ian Bogost's "flat ontology," in which "all things equally exist, yet they do not exist equally."[48] If entities do not "exist equally" in Lawrence's universe, it is because they do not find themselves in relationships of equal power with other entities. We should certainly object to Lawrence's implication that colonial conquest is a mere matter of "vitality" rather than, say, superior technological capacity as well as a superiority complex. Such a view in this context makes for an uncomfortable alliance with the social Darwinism of Herbert Spencer. Lawrence's examples derived from the dynamics of the food chain offer a similarly violent but perhaps less essentializing view of the hierarchy of vitality: "The snake can devour the fiercest insect. The fierce bird can destroy the greatest bird," and so on.[49] The flow of energy "up" the food chain channels vitality toward the "higher" species. Due to the hierarchy of vitality, there can be no equal relationships between individuals. Hence, "There is no such thing as equality."[50] That said, this does not prevent individual entities from achieving a full expression of being through their "pure" participation in the vital flow of the universe at large: "No creature is fully itself till it is, like the dandelion, opened in the bloom of pure relationship to . . . the entire living cosmos."[51] The absoluteness of being (everything exists) paired with the relativity of vitality (things do not exist equally in relation) results in a dynamic, "living cosmos" that is sustained by a vital flow of energy, yet is simultaneously characterized by a vast complexity that that transpires both *within* and *between* individual entities. For lack of a better term, Lawrence encompasses this overwhelming complexity with an underwhelming name: *the spirit of place*.

"WEIRD ANIMA" AND THE ECOLOGICAL UNCANNY

For many of Lawrence's interpreters, all of this talk about a "living cosmos" sounds an awful lot like animism.[52] But just as I cautioned against reducing the spirit of place to a vague relationship between "people" and "the environment," so would I warn against generalizing Lawrence's interest in vitality as a pseudo-primitivist celebration of animism. The problem with this stance is twofold. First, the typical understanding of animism does not align with Lawrence's emphasis on nonanthropocentrism and the materiality of being. Animism refers to the belief that every entity in the universe, whether "animate" or "inanimate," has a soul or a spirit. As David Skrbina summarizes: "These spirits typically have a human-like nature or personality that exhibit all the properties of a rational person, including intelligence, belief, memory, and agency. Furthermore, such spirits usually are not bound to the physical realm; they are immaterial and supernatural beings."[53] Though Lawrence is clearly invested in a wider sense of being, he neither projects human-like qualities onto nonhuman agents, nor does he consider spirit immaterial. Any generalized claim for "souls" or "supernatural" powers that permeate the universe runs the risk of effacing Lawrence's distinction between the absoluteness of being and the relativity of vitality. Which points to the second issue with animism as an interpretive frame: it threatens to reduce the complexity of Lawrence's thinking to a more romantic, universalizing notion of intimacy with the natural world. Animism can idealize Nature, a folly against which Lawrence cautions in no uncertain terms: "What happens when you idealize the soil, the mother-earth, and really go back to it? Then with overwhelming conviction it is borne in upon you . . . that the whole schema of things is against you. The whole massive rolling of natural fate is coming down on you like a slow glacier, to crush you to extinction."[54] Idealism kills, hence Lawrence's caution: "The world ought *not* to be [considered] a harmonious loving place. It ought to be a place of fierce discord and intermittent harmonies, which it is."[55]

It is precisely this complex, chaotic, even menacing universe that Lawrence conjures when he invokes the spirit of place. In contrast to animism's invocation of supernatural, human-like spirits, Lawrence's spectral rhetoric emphasizes the materiality of being, while still insisting on the withdrawn presence of an unrepresentable, more-than-human complexity. Lawrencian spectrality therefore marks a limit of knowledge. Instead of animism, then, Lawrence's work is better characterized by what, in his essay "Introduction to Pictures," he calls *weird anima*. Although written as a preface for the 1929 edition of his own paintings, much of this essay focuses on Paul Cézanne, whom Lawrence admired for his ability to exhibit the tensions between the bourgeois conventions of nineteenth-century painting and his own frustrated attempts to represent the true nature of physical reality. According to Lawrence, Cézanne's apples represent the first rendering of the "real" in modern painting. Lawrence championed the French artist's talent for producing his reality effect without recourse to the clichés of conventional realism or to the hyperrealism of trompe l'oeil. Deemphasizing the spatial aspects of painting (e.g., perspective), Cézanne instead conjured the temporal dimension, using multidirectional brushstrokes to depict a visual field that slips around the work's subject. Lawrence describes this dynamic as it appears in Cézanne's portraits of his wife:

> When he makes Madame Cézanne most *still*, most appley, he starts making the universe slip uneasily around her. It was part of his desire: to make the human form, the *life* form, come to rest. Not static—on the contrary. Mobile but come to rest. And at the same time he set the unmoving material world into motion. Walls twitch and slide, chairs bend or rear up a little, cloths curl like burning paper. . . . In his fight with the cliché he denied that walls are still and chairs are static. In his intuitive self he *felt* for their changes.[56]

Cézanne's portrait depicts the peculiar way in which all things—animate and inanimate alike—persistently undergo some form of flux. This effect required a new mode of painting that captured a deeper reality than conventional realism could manage; that is, an "uneasy" reality that slips, slides, twitches, and curls before our very eyes. But Cézanne did more in his painting than apply his "intuitive consciousness" to individual objects; he also used it to animate whole landscapes.[57] If his still lifes (*natures mortes* in French) make apparently "dead nature" come alive in its thingliness, then, according to Lawrence, his landscape paintings conjure complete panoramas that demonstrate how the whole world around us seethes vibrantly beneath our gaze: "In [Cézanne's] best landscapes we are fascinated by the mysterious *shiftiness* of the scene under our eyes; it shifts about as we watch it. And we realize, with a sort of transport, how intuitively *true* this is of landscape. It is *not* still. *It has its own weird anima,* and to our wide-eyed perception it changes like a living animal under our gaze."[58]

This idea of a *weird anima* proves very suggestive for an ecocritical reconsideration of Lawrence's "spirit" of place. Just as Cézanne relies on a strange representational strategy to visualize the imperceptible yet immanent flux of material reality, Lawrence relies on a "weird" rhetoric of spectrality—where *weird* has the added advantage of referring both to what "[has] the power to control the fate or destiny of human beings" and to that which is "unaccountably or uncomfortably strange [and] uncanny."[59] As for *anima*, this philosophically resonant term references "the animating principle of living things"—that is, the soul, and particularly in its irrational aspect.[60] The weird anima of landscape, then, not only indicates the vast array of individual entities (human and nonhuman) that collectively constitute a particular place; it also indicates that, taken together as a *spirit* of place, this anima has the potential to be profoundly destabilizing and uncanny.

Perhaps no moment in Lawrence's oeuvre quite captures the uncanny effects of weird anima as the scene in *Fantasia of the Unconscious* when the author recounts an experience he had near the Black Forest in Ebersteinburg, Germany. Unable to work with the screams of an unhappy child inside his rented house, Lawrence escapes with notebook and pencil to a chair he has set up in a copse of fir trees. Instead of finding a salubrious retreat, however, he feels distracted by the trees' menacing presence. "I think there are too many trees," he writes: "They seem to crowd round and stare at me, and I feel as if they nudged one another when I'm not looking. . . . I seem to feel them moving and thinking and prowling, and they overwhelm me."[61] Taken aback as to why the trees should inspire a feeling of menace, Lawrence ventures a theory. He suggests that the trees seem so terrifying to him because, despite their close proximity, there is no way for him to relate to them in a familiar way. Trees "have no hands and faces, no eyes," and yet they still possess a "vast and individual life, and an overshadowing will . . . that frightens you."[62] This thought leads Lawrence to propose a strange conundrum: "Suppose you want to look a tree in the face? You can't. It hasn't got a face."[63] With wry economy, Lawrence at once acknowledges and ridicules the instinctive human desire to see Nature as a mirror that reflects us back to ourselves. Yet he also underscores the strange seriousness of his own proposition. Despite recognizing the absurdity of his proposition, a startling gap of difference opens between himself and the tree. Lawrence has no way to commune with the tree, and hence no way to understand its "vast and individual life" or its "overshadowing will." His is a desire without a real object: it is precisely like trying to look someone in the face when they haven't got a face.

But the point here is not simply that such an experience is shocking. Instead, Lawrence's bizarre discourse on the tree with(out) a face serves as an allegory of how conscious perception reveals precisely *nothing* about the tree's true nature, but rather a fundamental and fundamentally unbridgeable gap of difference between tree being and human being. The tree with(out) a face is neither beautiful nor stately, but creepy, and thus poses a problem for any desire to commune with an idealized and hospitable Nature. For Lawrence, the faceless tree inspires another affect, a strange feeling that arises from the ontological gap that renders tree being permanently unknowable to him. It is his dim recognition of this unknowability that makes the trees towering above him appear menacing: "They have no skulls, no minds nor faces, they can't make eyes of love at you. Their vast life dispenses with all this. *But they will live you down.*"[64] Lawrence acknowledges the tree's radically inhuman agency and vitality, but he also recognizes that his own human consciousness will forever fail truly to comprehend the tree. The persistence of this failure registers in his experience as anxiety. Ultimately, then, Lawrence is not telling us about the trees but about himself and the machinations of his own consciousness, which responds with terror to that which it cannot understand.

Lawrence's faceless tree emblematizes *the ecological uncanny*. Like Lawrence's uncanny tree, the ecological uncanny references the strange and estranging way in which Nature persistently defies our desire for it to act as a mirror. The more the natural world seems to resist our humanizing projections, the more unsettling Nature appears. The ecological uncanny therefore thwarts human desire for a comforting organic holism and instead reveals "Nature" for what it is: an idealist metaphysical construct. It defamiliarizes the natural world, and it does so by tapping into our obscure (i.e., repressed) awareness that what we call "Nature" itself recedes absolutely beyond the horizon of human knowledge and human being. By making this horizon visible at all, the ecological uncanny punctures our belief in the limitless self. It shatters our perception of the world as a seamless whole.

But what does the ecological uncanny have to do with Lawrence's spectral rhetoric of the "spirit" of place and the "weird anima" of landscape? As I described in the previous section, Lawrence's writing of the early 1920s develops a lexicon of spectrality that serves as a shorthand for referencing the immensely complex and unknowable dynamics of a given environment. Terms like "Ghost" function for Lawrence as a placeholder to express something otherwise inexpressible—a hidden multiplicity that at once exceeds perceptual abilities and recedes from the phenomenal world, and yet is not just supersensible or metaphysical but also embodied. But Lawrence's spectral rhetoric also connotes haunting and conjures anxiety, even fear. As such, the rhetoric of spectrality that appears throughout Lawrence's writing of this period is closely linked to the ecological uncanny that arises in *Fantasia of the Unconscious*, which first appeared in 1922. The spectral rhetoric of the "Ghost," "weird anima," and the "spirit" of place recapitulates the logic of the ecological uncanny. This rhetoric summons a sense of terror vis-à-vis the inhuman—not to demonstrate something intrinsically nefarious about nonhuman others but rather to register the profound difficulty of relinquishing an anthropocentric point of view. In fact, when it comes down to it, Lawrence has a rather straightforward take on what "the environment" is: "just living creatures, no matter how large you make the capital L. Out of living creatures the material cosmos was made: out of the death of living creatures when their little living bodies fell dead and fell asunder into all sorts of matter and forces and energies, sun, moons, stars and worlds. So you got the universe."[65] What's terrifying is not the universe or the cosmos itself, which is no singular thing but rather a multiplicity of human and other-than-human selves. What's terrifying is coming to terms with one's own repression of this basic fact, which requires the deflation one's unjustifiably engorged sense of self as well as the acceptance of an upsetting, unsettling view of Nature.

UNSETTLING NATURE AND REDEFINING HOME IN *ST. MAWR*

Nowhere does Lawrence explore the ideological importance of the ecological uncanny more powerfully than in his tour de force novella *St. Mawr*. Written between May and October 1924 on the Lawrences' own Kiowa Ranch, the novella is principally concerned with finding a home for modernity's orphans. The narrative centers on Lou Witt, a twenty-five-year-old American woman who finds herself in Europe, where she feels haunted by "the lurking sense of being an outsider everywhere, . . . at home anywhere and nowhere."[66] The novella follows the itinerary of its protagonist from England to New Mexico in order to chart a transition from the living death of Euro-American modernity to the revivifying wildness that yet remains at the fringes of the American frontier. In addition to narrating a geographical and ideological trajectory from the Old World to the New, the novella is exemplary for offering a mash-up of significant Lawrencian themes, from the negotiation of love and sex between men and women to the significance of animals, the spirit of place, travel, home(lessness), and so on. As such, *St. Mawr* demonstrates "the sure-footed rapidity of Lawrence's late style, which had evolved by now into a kind of hieroglyphic alphabet for transcribing the living world at high speed."[67]

The novella's extraordinary syncretism converges in particular on the final act, where Lou Witt purchases a ranch outside of Santa Fe. The narrative makes a lengthy digression to describe the topography of the ranch and offer a detailed history of its human habitation prior to Lou's arrival. From this engagement emerges a powerful vision of the ranch as a living landscape constantly seething with nonhuman—even *inhuman*—life. As James Lasdun notes, descriptions in *St. Mawr* are certainly "beautifully perceptive as to the surface of things," but they "are never static or merely pictorial."[68]

Indeed, Lawrence's lively descriptions invoke an uncanny spirit of place that, for Lou, proves weirdly animating. The ranch becomes her first real home, and yet it is a home characterized not by stability, protection, and comfort, but by an unsettlingly proliferous wildness.

Lou's ranch was first settled by a schoolmaster who had come out west in a failed search for gold. The schoolmaster developed the land and tried to cultivate alfalfa, but this endeavor failed too, forcing him to sell the homestead to a trader. Exhibiting true pioneer spirit, the trader "tackled [the ranch] with a will," erecting several buildings and ensuring they all had running water—"for being a true American, he felt he could not *really* say he had conquered his environment till he had got running water, taps, and wash-hand basins inside his house" (161). But for all his attempts to "conquer" it, the ranch exuded "some mysterious malevolence"; it "prey[ed] upon the will" (163) and left the trader dispirited.

Whereas the ranch emasculated the trader, his "New England wife" maintained a deep attraction to the land that "intensif[ied] her ego, making her full of violence" (164). The narrator articulates this unnamed woman's egoistic drive through the transcendental refrain of the beyond: "Beyond the pine-trees, ah, there beyond, there was beauty for the spirit to soar in. . . . Beyond was the only distance, the desert a thousand feet below, and beyond" (165). The New England woman desired to inhabit this perfect Beyond, to draw its *"absolute* beauty" around her in a fit of "egoistic passion" (165). Despite her ecstatic response to the beauty of the ranch landscape, the wife dimly registered certain transitional moments when that beauty melted into something more unsettling, as when "the sun would go down blazing above the shallow cauldron of simmering darkness, and the round mountain of Colorado would lump into uncanny significance, northwards. That was always rather frightening" (166). Even so, she denied this "frightening," "uncanny significance" any real power and insisted on the transcendental: "Ah: it was beauty, beauty absolute, at any hour of the day. . . . It was always beauty, *always!*" (166).

Against the New England woman's desire to project herself into the landscape's beautiful "beyond," the narrator offers a corrective vision of the ranch as something "vast and living" and incredibly unsettling. One finds oneself completely enmeshed in this living environment, and yet simultaneously estranged from it: "The great circling landscape lived its own life, sumptuous and uncaring. Man did not exist for it" (166). The narrator further insists: "Even a woman cannot live only into the distance, the beyond. Willy-nilly she finds herself juxtaposed to the near things, the thing in itself. And willy-nilly she is caught up into the fight with the immediate object" (166–67). As the living world converged around her, the wife found herself in discomfiting proximity to a seething environment. The New England woman "had fought to make the nearness as perfect as the distance" (167), but as the narrator suggests, she now contended with the ghostly, hidden realities that lurked beyond her apprehension of even those "near things." The "beyond" she had previously idealized retreated into the entities themselves, and, in the process, the absolute beauty of the distant view swooped into terrifying proximity.

Fully enmeshed in a proximal environment that is itself constituted by a dense assemblage of proliferating entities, the New England wife experienced a crisis of faith. The flourishing vitality that surrounded her on all sides challenged her fragile illusions about universal love and cast her as an alien in her own home: "The very chipmunks, in their jerky helter-skelter, the blue jays wrangling in the pine-tree in the dawn, the grey squirrel undulating to the tree-trunk, then pausing to chatter at her and scold her, with a shrewd fearlessness, as if she were the alien, the outsider, the creature that should not be permitted among the trees, all destroyed the illusion she cherished, of love, universal love. There was no love on this ranch. There was life, intense, bristling life, full of energy, but also, with an undertone of savage sordidness" (168). The narrator further describes the "tussle of wild life" as a botanical survival of the fittest, a "seething conflict of plants trying to get hold" (168). This battleground featured an array of vegetal soldiers, from "fang-mouthed" nettles to "ghostly" mariposas and spiny desert cactuses that lay underfoot. After suffering years of this unrelenting battle, during which "the animosity of the spirit of place" utterly exhausted her, the New England woman retired from the ranch and returned to civilization: "She was glad to come to a more human home, her house in the village. . . . [The ranch] had broken something in her" (170). After her departure, a skeleton crew of Mexican workers tended the land, but the venture still required too much effort to be profitable. They eventually abandoned the ranch at the onset of the Great War, leaving the tussle of wild life to reclaim it.

Lou's haste to purchase the ranch after the narrator's elaboration of its vexed history has struck some readers as ill advised. Donald Gutierrez, for instance, asserts that the novel's ambivalent spirit of place embodies "two hylozoisms, one negative, the other positive," and they "appear to be unreconciled."[69] Lou's zeal therefore seems dangerously misplaced.[70] However, the novella's lengthy engagement with the New Mexican landscape signals that the ranch serves a purpose that in fact exceeds such a reductionist reading. Indeed, Lawrence presents his readers with the more challenging task of remaining with the contradiction in order to discover just what the lack of reconciliation might produce.

One way of reading the ambivalence of the final episode without needing to reconcile Gutierrez's competing "hylozoisms" would be to consider Lou's point of view vis-à-vis her mother's. Over the course of the novella Mrs. Witt's growing dissatisfaction with contemporary masculinity gradually comes into line with Lou's more wide-ranging critique of modern civilization. However, the long journey across the Atlantic and on to Texas and New Mexico saps Mrs. Witt's existential strength; by novella's end she lacks the stamina required to keep striving for a truly new way of living. Thus, when Lou returns to her newly purchased ranch with her mother in tow, the dispirited Mrs. Witt misunderstands her daughter's state of mind. Concerned for Lou's well-being as a lone woman on the ranch, Mrs. Witt appeals to the latent sexual attraction between her daughter and a man named Phoenix. She suggests that erotic love might "be coming into the foreground before long," thereby enabling Lou finally to live: "You've never *lived* yet" (173). But Lou resents her mother's assumption that a conception of living should be directly tied to sexuality: "What do you call life? . . . Wriggling half-naked at a public show, and going off in a taxi to sleep with some half-drunken fool who thinks he's a man because—Oh mother, I don't even want to think of it. I know you have a lurking idea that *that is life*" (173). Lou's rejection of the link between life and sex demonstrates how shallow Mrs. Witt's critique of modern life has always been in contrast with her own. Whereas Mrs. Witt remains narrowly focused on relations between men and women, Lou recognizes that a renewed sense of vitality requires an entirely new model for living. For this reason, she insists to her mother that her life on the ranch represents "the beginning of something else, and the end of something that's done with" (174). With this claim to a definitive break, a "silence that was a pure breach" (174) opens up between mother and daughter, and the novella refuses to close it.

In addition to resisting Mrs. Witt's erotic solution to the problem of spiritual deterioration, Lou is also exemplary for her divergence from her predecessors on the ranch. Unlike both the schoolmaster and the trader, she rejects the ranch as a site for her own (self-)production. And in contrast to the trader's wife, she eschews any desire for the "absolute beauty" of the landscape. Alternatively, she envisions herself as a "Vestal Virgin" who pledges herself "to the unseen presences . . . to the unseen gods, the unseen spirits, the hidden fire" (158–59). As she later tells her mother, though withdrawn and hence invisible to her, these unseen presences represent worlds beyond the human that, taken together, fashion a living spirit of place that she now calls home, even if that home may at times prove unsettling. In a final bid to justify herself to her mother, Lou explains: "There's something else for me, mother. There's something else even that loves me and wants me. I can't tell you what it is. It's a spirit. And it's here, on this ranch. It's here, in this landscape. . . . I don't know what it is, definitely. It's something wild, that will hurt me sometimes and will wear me down sometimes. I know it. But it's something big, bigger than men, bigger than people, bigger than religion. It's something to do with wild America. And it's something to do with me" (175). If Lou's zeal for the ranch at first seems too hasty, this speech reveals that her decision to purchase it is not naïve. The novella closes with an otherwise ambivalent Mrs. Witt congratulating Lou on how "cheap" she got the ranch, "considering all there is to it" (175). But this celebration of cheapness casts Lou's relationship to the ranch solely in terms of ownership, which Lou's speech implicitly rejects. Her point is not that she has become the proud owner of a slice of "wild America" but rather that her purchase (which is merely a socioeconomic arrangement, and hence meaningless in any essential sense) places her in the midst of a complex network of vital relations.

What all of this means is that we can only understand Lou's zeal for this ranch in all of its dangerous splendor once we recognize that the novella's final episode redefines what it means to be "at home" in the first place. Although Lou's description of an unknown "something" or "spirit" that simultaneously welcomes and endangers seems blatantly contradictory, her observations emerge from a deeper recognition that home, understood as a protective enclosure, tends to extinguish life rather than revitalize it. Like Lawrence, she holds herself open to the sensation of estrangement that arises from an environment—the ranch—that alienates through its wild proliferation of life. Lou emphasizes her experience of *being in* a particular place without insisting that being depends on a concurrent sense of *belonging to* that place, or having that place *belong to* her. By dispensing with protective ideological enclosures and remaining open to the unstable dynamics of vitality, Lou settles into New Mexico's unsettling "spirit" of place.

NATIVE AMERICANS AND LAWRENCE'S "GHOST"

Yet for all that *St. Mawr* rewrites the script of ecological homecoming by unsettling normative ideas about Nature as home, it is not immediately clear that the novella unsettles the settler colonial paradigm that these ideas implicitly uphold. Certainly Lou Witt does not enact the same kind of imaginative possession of her New Mexican ranch that Tom Outland does of the Blue Mesa in *The Professor's House*. Even so, as a Texan she is the product of a settler culture, and her act of homemaking still functions through an implicit erasure of Indigenous presence. Thus, to avoid too hastily celebrating Lawrence's invocation of the unsettling spirit of place, it is necessary to reckon with his writing about Native Americans in the Southwest.

Scholars have taken divergent perspectives on Lawrence's representation of Native Americans, with some rebuking his obsession with difference,[71] and others underscoring the importance of Lawrence's desire to stay with the challenges presented by alterity.[72] My purpose here is not to uphold Lawrence as a model of intercultural thinking. But by the same token, the quickness with which scholars have castigated him or else cast aside his idiosyncratic language has too often resulted in fundamental misunderstandings of his work.[73] This final section makes no claims about Lawrence being an exemplary reporter on Native American political affairs or an "authentic knower" of Indigenous cultural traditions; as an Englishman and an outsider, he himself adamantly rejected such recognition, as I'll shortly discuss. I do, however, want to question the argument that Lawrence's representation of Native Americans necessarily recapitulates the genocidal logic of settler colonialism. Instead of sublimating Indigenous presence and disappearing that spiritualized essence into Nature, Lawrence affirms what he (rightly or wrongly) believes to be a Native understanding of the spirit of place as a dynamic matrix of competing forces. By doing so, Lawrence contributes to a discourse seeking to unsettle Euro-American settler fantasies of a home(l)y Nature.

Let us return to the scene that opened this chapter: Lawrence's arrival in Taos at the invitation of Mabel Dodge Luhan. Unbeknownst to Lawrence, Luhan's desire for a unique individual who could "see" the hidden reality of Taos and its original inhabitants emerged from political as well as personal motivations. By the time of the Lawrences' first visit in the early 1920s, the people of the nearby Taos Pueblo were in the midst of a struggle to retain cultural autonomy in the face of Euro-American economic and political encroachments. Debates on this issue reached a boiling point with the introduction in 1922 of the Bursum Bill, which addressed the problem of competing entitlement claims on lands surrounding Indian pueblos. During the previous colonial period under Mexican rule, land grants had allocated nearly 700,000 acres to southwestern Indians. Since 1848, however, a widespread pattern of enclosure had developed in which lands were slowly bought up, encroached upon, or otherwise appropriated, often by monopolistic landowners seeking the best-irrigated portions of the original Pueblo land grants. A 1913 ruling challenged many of the claims that non-Indians had made on Indian lands over the previous sixty-five years. When the Lawrences arrived in New Mexico, however, the newly introduced Bursum Bill sought to reverse the 1913 ruling by (re)confirming nearly all non-Indian claims. In the midst of these cultural and political challenges, Luhan and her husband, Tony, a Tiwa tribal leader, organized a massive effort in defense of the New Mexican pueblos.

Although Lawrence had little if any knowledge of the contemporary political context when he traveled to Taos, he quickly realized that Luhan's invitation implicated him in a situation he did not fully comprehend. His arrival in New Mexico was thus characterized by ambivalence. In December 1922, shortly after coming to Taos, he published a piece in the *New York Times* where he obliquely communicates his frustration with the political crisis into which Luhan had thrust him: "But it's Bursum, Bursum, Bursum! the Bill, the Bill, the Bill! . . . O Mr. Secretary Fall, you bad man, you good man, you Fall, you Rise, you Fall! The Joy Survey, Oh, Joy, No Joy, Once Joy, Now Woe! Woe! Whoa! Whoa, Bursum! Whoa, Bill! Whoa-a-a! . . . [I]magine me, lamblike and bewildered, muttering softly to myself, between soft groans, trying to make head or tail of myself in my present situation."[74] Lawrence conveys his resistance to the political storm through an ironic tone that parodies the prose stylings of Gertrude Stein and implicitly satirizes his host and patron, who had been the subject of one of Stein's famous poetic "portraits" in 1912. Lawrence also forcefully draws attention to his own lack of expertise in the situation, and, wishing to deflect responsibility for taking any political action, he casts himself as a marginal "actor," a mere extra in the "Wild West show" unfolding around him.[75] Part of Lawrence's desire to withdraw from the scene related to the fact that New Mexico confronted him with unexpected realities, such as the sight of supposedly "primitive" Indians participating in the "modern" world of machines, and the fact that the Taos Pueblo community no longer lived "in idyllic independence from 'modern' America but rather under its gross political sway."[76] In another essay, titled "Indians and an Englishman" (1923), Lawrence further remarks with distaste on the "farce" of New Mexico, with its "incongruous" mixture of "wildness and wooliness and westernity and motor-cars and art and sage and savage."[77] Lawrence again expresses his ambivalence by rejecting all claims to expertise and painting a disinterested, even insulting, portrait of the region.

Though the tone Lawrence adopts in these essays appears to result from a disgust with politics, this is not the main source of his concern, which instead stems from his basic distinction between Euro-American modernity as spiritually poor and existentially depleted, and Native cultures as vital and (re)vitalizing. As Carey Snyder has shown, this distinction was commonplace among contemporary anthropologists like Edward Sapir, who posited an opposition between "spurious" and "genuine" cultures.[78] Lawrence adopted a similar opposition that imagined Euro-American and Native cultures as being locked in détente of the spirit, and he believed the settler had the upper hand. According to Ben Conisbee Baer, Lawrence entertained an apocalyptic vision of American history in which Euro-American ascendency would play out against the vanishing of Indigenous peoples—whether by cultural assimilation, forced removal, or genocide. In other words, Baer understands Lawrence as affirming what Patrick Wolfe describes as the settler colonial logic of elimination.[79] It is with this in mind that he reads Lawrence's writing on the Bursum Bill not as a journalistic report on politics but rather as "a proleptic epitaph of the pueblos" intended to register that, "as the Indians die, the process of their incorporation will be the seeding of a post-apocalyptic new birth."[80]

Baer finds justification for this reading in a passage from *Studies in Classic American Literature* wherein Lawrence appears to imagine Native Americans as an absent presence. This passage comes from a chapter on James Fenimore Cooper in which Lawrence discusses the author's vision of the "inception of a new race" emerging from a rapprochement between Europeans and Native Americans. Whereas Cooper stages a dream that moves "beyond democracy" and realizes a "new soul" for America, Lawrence retreats from this fantasy, insisting that "there can be no fusion" between

Europeans and Indians as long as the former hold onto contradictory desires to "extirpate . . . [and] glorify" the latter.[81] Lawrence clearly posits that Europeanness itself stands at the center of this contradiction. Those whites who "would like to see this Red brother exterminated" desire as much "not only for the sake of grabbing his land, but because of the silent, invisible, but deadly hostility between the spirit of the two races."[82] By contrast, whites who "laud [the Indian] to the skies" do so in part because they have "a big grouch against [their] own whiteness."[83] Either way, it is the "white spirit" that proves toxic. Lawrence elaborates this toxicity later in the essay, where he writes: "When you are actually *in* America, America hurts, because it has a powerful disintegrative influence upon the white psyche. It is full of grinning, unappeased aboriginal demons, too, ghosts, and it persecutes the white men."[84]

Baer picks up on the language of this passage and claims that despite his evident critique, Lawrence unwittingly replicates the settler logic of elimination. Lawrence attempts "to realign the dream of the beyond to the 'reality' of America's 'grinning, unappeased aboriginal demons' which at once paradoxically become both the acme of literality . . . [and] mediating figures enabling the white presence to make the transition to a radically new future."[85] This realignment depends on the conflicting figuration of the Indians as simultaneously alive and dead, as "reality" and "ghosts." Baer highlights the paradox of this bifurcation by describing how in Lawrence's text the (ghostly) Indians become sublimated into the land itself: "It is the 'landscape' itself that becomes threatening. Its original inhabitants, having first become demons or ghosts, are finally identified with the earth to the point that it is 'the American landscape' that has 'never been at one with the white man,' and the landscape itself that is 'grinning.' . . . Lawrence elaborates a little on his ghost motif, making of the Indians something which haunts the modernized world."[86] Baer's analysis pivots on a reading of Lawrence's ghostly rhetoric within the framework of Derridean spectrality, such that the Indians operate in the text both as a *historical* haunting of modernity and as a *linguistic* haunting of the "trace" more generally. For Baer, this Derridean reading specifically undermines claims that Lawrence makes throughout his New Mexican writings regarding the nonrepresentational nature of Native American ritual. In response to the violence of the "representational average" made manifest in the ideology of mass entertainment,[87] Lawrence saw in Indigenous religion the possibility of nonrepresentation. For instance, throughout *Mornings in Mexico* Lawrence engages in pseudo-ethnographic description of numerous Native American songs and dances. As opposed to American mass entertainment, Indigenous ritual does not gesture to any sort of universal mind or meaning: "[The ritual] means nothing at all. There is no spectacle, no spectator.... There is none of the hardness of representation. They are not representing something, not even playing. It is a soft, subtle *being* something."[88] Lawrence develops this thesis further in his analysis of the Ghost Dance of the Hopi Third Mesa. According to his interpretation, the Ghost Dance neither operates by some system of representation and substitution (which requires a division between seer and seen), nor does it offer a second-order *re*-presentation of cosmic meaning. Instead, the Ghost Dance emphasizes the direct embodiment of ritual rhythms, what Lawrence often terms "the rhythm of the blood." But for Baer, Lawrence's claim regarding the nonrepresentational nature of Indigenous religion undoes itself in passages like the one depicting Native Americans as grinning ghosts in an unsettling landscape. The trace-like haunting of the ghostly Indians (re-)opens the possibility of representation, implicitly revealing a hidden essentialism that nevertheless "powerfully remains in [Lawrence's] text."[89]

Two problems persist in Baer's reading. Firstly, it makes only a cursory nod to the fact that, in context, Lawrence's figuration of Indians has a bitterly ironic tone, since the point is that Indians only appear as "grinning, unappeasing aboriginal demons" to the Europeans who fear them. The historian Philip J. Deloria (Dakota) underscores this point when he argues that Lawrence is not in fact subscribing to the fantasy of the grinning Indian but rather exposing how "the contradictions embedded in noble savagery have themselves been the precondition for the formation of American identities."[90] Lawrence's apocalyptic vision of vanishing Indians does express the contradictory logic by which American identity is consolidated through the elimination of Native peoples, but it critiques rather than recapitulates that logic.

Secondly, and more crucially, the Derridean thrust of Baer's reading fails to account for Lawrence's more basic point that things can *exist* without being *represented*. Lawrence cannot himself elude this mediating force: his primary medium is writing, so obviously he cannot escape language altogether. But this does not prevent him from using language to gesture indirectly at something that *does* escape it. The genre of horror helps clarify the point. Eugene Thacker argues that horror deals in "the enigmatic thought of the unknown,"[91] and its fearful fascination with the unknown highlights the possibility of the *unrepresentable* as such. In this sense, horror operates through positive negation. Graham Harman makes an analogous argument when he claims that H. P. Lovecraft's greatest stylistic accomplishment, the thing that endowed his fiction with its quintessential weirdness, resided in his ability to create gaps between language and the objects that language attempts—yet obviously fails—to describe, thereby drawing attention to the unrepresentable.[92] Thacker frames this logic of the horror genre as a form of negative capability: "The world-for-itself presents itself to us, but without simply becoming the world-for-us; it is, to borrow from Schopenhauer, 'the world-in-itself-for-us.'"[93]

Like horror, Lawrence's Ghost offers an expression of the inexpressible. As already addressed at length above, repeatedly in his work Lawrence mobilizes a spectral rhetoric to emphasize how the complex, vital exchange transpiring within and between all entities always evades capture. As such, the spectrality implicit in the Lawrence's Ghost does not involve the conventional sense of haunting as a lingering of the past into the present. Although Lawrence does share with Derrida a basic understanding of spectrality as a condition that is proper to being, it would be wrong to assert that Lawrence's work exhibits anything like Derrida's "hauntology." Hauntology emphasizes temporal disjunction and thinks "ontological actuality" solely in terms of the deconstructive trace: "a trace of which life and death would themselves be but traces and traces of traces."[94] By contrast, spectrality for Lawrence is, strangely, about how materiality and immateriality haunt each other across the dimensions of "presence" and "absence," appearance and withdrawal. Lawrencian spectrality recognizes that a plurality of possibilities can and always do (co)exist, and that these possibilities can never be reduced to singularity or certainty.

What this means is that Indigenous peoples are ghostly for Lawrence in the same way that any other living entity is ghostly, and not because they appear in modernity as a "proleptic" haunting of a future in which they have already succumbed to the settler logic of elimination. To be sure, Lawrence does see that apocalyptic logic playing out in the contradictory negotiations of settler identity. But the political, cultural, and existential dangers facing Native Americans notwithstanding, what makes them ghostly is the simple fact that, like all other beings, they "have a little Ghost inside . . . which sees both ways, or even many ways."[95] Contra Baer, then, instead of an attempt to *escape* representation, of which the figure of the Ghost at once marks an attempt and a failure, Lawrence's spectral rhetoric signals the unrepresentable as such—that which escapes language, even if language is required to register its existence.

And what escapes language is, precisely, the nature of Indian *difference*. Baer's attempt to read Lawrence's landscape vision as the product of Indian difference bespeaks a failure to regard Lawrence's broader point that the issue of alterity extends at once deep into the human (e.g., the "allotropic" ego) and far beyond the human (e.g., the "weird anima" of landscape, the "spirit" of place). In his writing on the Taos Pueblo, then, Lawrence does not disappear Indians into the landscape in the way that Baer, echoing other critics, presumes. He does something far weirder. Instead of relegating Indians to a background, Lawrence ushers background into foreground, situating Indigenous peoples both *within* and *as a constitutive part of* New Mexico's vitally unsettling "spirit" of place.

The reading of Lawrence's representation of Native Americans offered here poses a challenge to the usual interpretation of Lawrencian apocalypticism, expressed in its characteristic form by Baer when he claims that, for Lawrence, "the seeding of a post-apocalyptic new birth" necessarily required the disappearance (figurative or literal) of Native peoples.[96] Lawrence certainly saw such an apocalyptic ending to Indigenous life as an urgent possibility given the advancement of Euro-American modernity. But it is a mistake to think he understood this as inevitable, and he certainty did not think it desirable. For Lawrence also entertained an alternative vision in which, as he announced in the essay "New Mexico," the vital "old religion" of the land and its Indigenous inhabitants would prevail over the spiritually corrosive modernity encroaching from the west. The flourishing of this old religion could mark a new beginning for a "genuine America" that, like Lou's ranch, would profoundly unsettle the normative ideal of a home(l)y Nature that sits at the heart of the U.S. settler colonial imagination. And, like Lou herself, Lawrence maintained the possibility that "the spirit can change." Though he insisted that the "white man's spirit can never become as the red man's spirit," he also believed that "it can cease to be the opposite and the negative of the red man's spirit. It can open out a new great area of consciousness, in which there is room for the red spirit too."[97] Though hardly a vision of decolonization, Lawrence's repeated attempts to expose what he called "the myth of the essential white America"[98] may nonetheless count as profoundly decolonial.

EXCURSUS I

Ecological Realism

> The spirit of place is a great reality.
> —D. H. Lawrence, *Studies in Classic American Literature*

D. H. Lawrence's unsettling New Mexican landscapes demonstrate an important principle for ecological thinking. When the idealized Nature and its attendant, colonialist ideology of home fall prey to the ecological uncanny, another view of reality emerges: the "great reality" of that extraordinarily complex meshwork of dynamic, materially embodied being Lawrence referred to as the "spirit of place." Lawrence recognized that the ghostly nature of this reality could not be represented directly. Rather, its obscure, spectral manifestation compels us to contemplate what Timothy Morton has called "the subaesthetic level of being"—a submerged ontological register that "we can't call . . . beautiful (self-contained, harmonious) or sublime (awe-inspiring, awesome)." On the contrary: *"This level unsettles and disgusts."*[1] Morton's claim poses a problem for representation, and as Lawrence's own engagement with the crisis of representation indicates, the unsettling reality that comes into view when the beautiful, mirror-like surface of Nature warps signals the need for a rethinking of realism. This excursus, which leads from the first into the second half of the book, plots a course over the unstable terrain of uncanny ecology toward a literary theory of ecological realism.

Of course, a robust theory of ecological realism already exists within the field of environmental psychology. Initially developed by James Gibson, ecological realism repositions the material environment at the center of perception in a way that emphasizes the immediate reality of *information* rather than the mediating mechanisms of *sensation*.[2] Specifically, Gibson's theory focuses on the "affordances" made to an organism by its environment rather than on the internal mechanisms that determine how an organism perceives its world. In this sense, Gibson does not take a conventional stimulus–response view of perception, preferring to "ask not what's inside your head, but what your head's inside of."[3] Gibson's emphasis on the reality of the in situ environment forms the basis for what has since been called ecological realism: the direct (rather than representational) perception of the really existing and fully populated environments in which organisms find themselves.[4] Like other recent philosophical and scientific realisms, such as speculative realism and quantum realism, ecological realism cannot easily be dismissed as naïve. Edward Reed explains:

> As with other forms of realism, ecological realism takes observation to be a mode of discovery, as a way of finding out what is the case. However, unlike most existing forms of scientific realism, ecological realism can and does take into account complexities in what is the case and the very real limitations of statements about what is the case. *The environment is rich beyond understanding* in what it affords and in the information available within it. It is therefore impossible in principle for an observer to perceive any situation completely.... There is always more to know about the environment is a basic tenet of ecological realism.[5]

In Gibson's theory, then, organisms perceive their worlds and the entities that populate them more or less as they actually are. And yet, despite ecological realism's tendency to see perception as direct and veridical, perception inevitably remains incomplete. This incompleteness is certainly due in part to the limitations of particular perceptual apparatus. Human eyes can see only a small range of the electromagnetic spectrum, fly eyes can distinguish only a couple of colors, and mollusks perceive only the relative presence or absence of light, yet all three organisms accurately perceive real information in their own limited ways. In the case of Gibson's ecological realism, however, perceptual incompleteness has more to do with "the selectivity of perception; we attend to the affordances of things we are interested in, and often do not attend to other aspects of situations."[6] Thus, ecological realism claims that organisms patch their inevitably partial worlds together through the attention they pay to environmental particularities.

Ecological realism's point that attention is always selective implies gaps of perceptual darkness from whence a previously unregistered reality can creep up on one. It is precisely through these gaps in perception that the ecological uncanny surfaces, as it did for Lawrence when his speculation about the radical difference of tree being disturbed his erstwhile forest retreat. Aside from the speculation itself, what felt especially unsettling for Lawrence was how the very question of tree being had remained obscured from his attention even as these great entities surrounded him on every side. Likewise, when we suddenly notice something previously concealed from our noticing—when a hidden entity jumps out, so to speak, from the perceptual darkness—it is unsettling because it was already so close. We register the moment of revelation, when the strange familiarity of that which was previously withdrawn appears, as the sign of the return of some repressed reality. The ecological uncanny thus has a close relation to ecological realism, and this relationship turns on relative levels of attentiveness as well as perceptual bias. The more we learn about gaps in our perception of the world, the more we understand how little we know.

But how might all of this apply to a narrative context? As discussed in this book's introduction, Freud turned to literature for his exploration of the uncanny because readers tend to pay more attention to literary worlds than to their own immediate environments. Literature provides a space to practice attention and thus to develop an understanding of how the uncanny functions across various levels of mediation. Following a similar logic, would it be possible to develop a literary theory of ecological realism that attends more fully to the unsettling nature of other-than-human reality in the ways that Lou Witt exemplifies?

Such an ecological realism would necessarily strain the confines of more ordinary literary realisms. For one thing, it would require an expansion of the social totality that Georg Lukács sees as the real object of any "critical" realism worth its salt. Lukács lauds realism for its critical understanding of the world, whereas he dismisses modernism and its "naturalist" tendencies as naïve—"exclud[ing] critical detachment . . . and stuck to first impressions."[7] Realism's critical acumen should ultimately be directed toward what Lukács sees as the most basic element of literature: its depiction of "the interaction of character and environment."[8] Yet Lukács's use of the term *environment* does not admit anything beyond *human* society: "Content determines form. But there is no content of which Man himself is not the focal point."[9] By contrast, like Gibson's environmental psychology, the kind of ecological realism I wish to develop here can account for the much wider society of entities that populates literary environments. Encounters with such entities are very real indeed, and though mediated, literary descriptions of these encounters cannot be dismissed as merely naturalistic in Lukács's sense. The point is not only to affirm the real existence of these entities, or in this case their presence in a literary text or world, but also to emphasize that our "first impressions"—in "fiction" as much as "reality"—are not naïve so much as incomplete. Like Lukácsian realism, then, ecological realism also stresses reflection: it takes critical distance to come to terms with just how limited our knowledge of literary reality really is, and thereby achieve what James Clerk Maxwell once deemed "thoroughly conscious ignorance."[10] And not just critical distance: we also need creativity to move beyond stale postmodernist truisms that reduce literary realism's "reality effects" to the mere play of dead signs.[11]

For this reason, the literary form of ecological realism I am proposing also necessarily exceeds what Njabulo Ndebele in the South African context has termed *the ordinary*. Ndebele equates the ordinary with a "sobering rationality" that emphasizes the mechanisms of causality over grotesque representations of history's unwanted effects, the latter of which he calls *the spectacular*: "Paying attention to the ordinary and its methods will result in a significant growth of consciousness."[12] Ndebele's contrast between the ordinary and the spectacular revamps Lukács's opposition between realism and naturalism.[13] Like realism, the ordinary is an analytical mode that discloses injustice and points a way forward for the social collective; like naturalism, the spectacular uselessly wallows in the moralism and self-pity of the individual. Ndebele's vilification of the spectacular reprises Lukács's resistance to modernist subjectivism, which in Kafka, for example, gives rise to "the expression of a ghostly un-reality, a nightmare world."[14] Modernist subjectivity also warps time and space such that, in Proust, "the artist's world disintegrates into a multiplicity of partial worlds."[15] Both of these situations result in a form of literary bad faith that disrupts the primary dialectic between "Man and Society." However, as its connection to the ecological uncanny should already indicate, ecological realism will necessarily retain the aspects of ghostliness and fragmentation that Lukács and Ndebele banish from their realisms. As suggested in the chapter on Lawrence, ghostliness does appear as a distortion of reality, but this distortion is produced by the limits of perception—not, as Lukács sees it, by the individual artist's deliberate "attenuation of actuality."[16] Nature is not *intrinsically* uncanny, nor is its uncanny appearance simply the result of warped representation. Rather, it is an effect of our inevitably partial attention, which leaves gaps that, when revealed, produce something like an unsettling or uncanny mood (*unheimliche Stimmung*). Ecological realism registers that ecology often manifests specularly and that it has the ability to unsettle the (human) social totality.

The coming chapters will explore the particularities of this literary form of ecological realism further, but I should like to signal here that this new realism may be a formal characteristic of a text, but it is also, and more importantly, *a new way of reading*. Ecological realism is both a way of approaching a text's worlding mechanisms (its particular "ecology") and a method for thinking about ecological signifiers that appear within a narrative. One might even consider it a species of literary phenomenology. This form of reading is situated somewhere between "symptomatic" and "surface" reading practices. Symptomatic reading emerged in the 1970s and 1980s, when the more widespread application of Marxism and psychoanalysis stimulated the production of interdisciplinary theoretical work. Most emblematic of this work is Fredric Jameson's *The Political Unconscious*. Jameson imagined the text as being shaped by an absent cause, a sort of literary unconscious, and he reframed interpretation as an attempt to identify the latent or "repressed" meaning that hides below the text's manifest surface. Ecological realism does not entirely rely on symptomatic reading because, like in environmental psychology, it takes entities encountered in the world (including the literary world) at "face value," so to speak. We know our perception of these entities is limited, but their reality is immediately obvious to us. For this reason, we needn't always seek out the latent meanings of contradictions that disturb the surface of a text, especially when the repressed content of those contradictions is fulfilled, as it is for Jameson, by History itself. Instead, ecological realism follows Mary Crane in her resistance to Jameson's view of contradiction as a key for unlocking the obscure machinations of history. Instead of being the result of repression, contradiction is an effect of cognition. Crane's theory of the unconscious is closer to Lawrence's than Freud's in the sense that it consists of complex "structures of categorization" that occur too rapidly to be perceived.[17] Rather than seeing meaning as the "allegorical" product of the distance between what is latent and manifest, the cognitive critic takes a more realist approach, locating meaning in the perception, however incomplete, of really existing cognitive structures. Although Crane understands these structures to be truly invisible to the individual, they are not located in the depths of the psyche so much as they sit on the "surface" of cognition.

The out-in-the-open quality of Crane's theory points the way toward what Stephen Best and Sharon Marcus call surface reading. As they explain, "we take surface to mean what is evident, perceptible, apprehensible in texts; what is neither hidden nor hiding. . . . A surface is what insists on being looked *at* rather than what we must train ourselves to see through."[18] Paradoxically—and especially so given the reign of symptomatic reading—what is most manifest in a text has the tendency to hide in plain sight. In their work, Best and Marcus propose a number of different modalities of surface reading, each of which ultimately promotes an understanding of the textual surface in a way that recalls Foucault's characterization of archival research: "rather than dig for 'relations that are secret, hidden, more silent or deeper than . . . consciousness,'" the researcher seeks "'to define the relations on the very surface of discourse' and 'to make visible what is invisible only because it's too much on the surface of things.' [*Surface*] *reading sees ghosts as presences, not absences, and lets ghosts be ghosts, instead of saying what they are ghosts of.*"[19] Once again, the surface of a text presents its readers with a kind of reality that is immediate in spite of its virtual (i.e., literary, representational) character. Within this virtual reality, readers encounter entities directly and must on some level accept them at face value. This applies as much to the surface *of* the text as to surfaces *in* the text.[20] In a descriptive reading practice, figural language has a shocking tendency to become literal. This is the sense in which I take Best and Marcus's allusion to ghosts, which fortuitously recalls the "ghosts" of the New Mexican landscape that Lawrence understood as embodied yet unknowable—though in another register, he too lets ghosts be ghosts.

Yet from the perspective of ecological realism, surface reading is not entirely satisfactory either. Just as symptomatic reading disavows the surface, surface reading disavows depth. As Best and Marcus describe it, "in the geometrical sense, [a surface] has length and breadth but no thickness, and therefore covers no depth."[21] Ecological realism cannot subscribe to an idea of pure surface because ecological reality itself has depth and breadth. Timothy Morton's rethinking of "interconnection" demonstrates why. In contemporary environmental ethics, which adopts deep ecology's philosophy of the earth as a living whole, interconnection is inseparable from its associated principles of harmony, balance, and equilibrium—all concepts seeking a utopian seamlessness that reduces Nature's multidimensionality to a pure surface. Unlike the Whole of Nature, however, what Morton calls "the mesh" of ecology has many holes in it: "Interconnection implies separateness and difference.... The mesh isn't a background against which the strange stranger [i.e., the individual entity] appears. It is the entanglement of all strangers."[22] Entanglement therefore requires depth because it entails an overlapping multiplicity of differential relations.

Here I again appeal to Lawrence's *St. Mawr,* where the "weird anima" of Lou Witt's ranch signals a vast array of human and nonhuman entities that collectively constitute that particular place. Each of these entities has its own phenomenal world gathered around it, what Jakob von Uexküll calls an organism's *Umwelt.* The surface of a text may indicate this multiplicity of worlds, but it cannot contain it. For example, when a character in a novel encounters an ecological signifier—say, the sound of a frog croaking in the distance—that frog's presence is registered on the textual surface. However, being marginal vis-à-vis the protagonist and their story, the frog and its frog world necessarily remain unexplored. But just as a real-world frog has its own froggy existence and froggy experiences that may interact with yet not depend on humans, the frog in the world of the novel can also be said to exist within its own froggy cosmos, at once "inside" the diegetic space of the novel and yet "outside" the realm of human narrative concern. The frog's strange location at the limits of the narrative (or perhaps I should say beyond the horizon of narrative focalization) strains any notion that it exists solely on the textual surface. For this reason, surfaces may suggest a multiplicity of *Umwelten,* but they cannot fully encompass this multiplicity. This is because, as in the case of the frog world, nonhuman worlds are largely invisible in human narratives—they exist as holes in the textual surface, one of many possible gaps that might disturb the narrative totality if interpretation did not work so hard to paper over them and produce the illusion of a pure surface.

More often than not, we readers fail to notice textual gaps in the first place. Wolfgang Iser refers to such gaps as "blanks," and they play a crucial role in his phenomenological approach to reading. According to Iser, literature is indeterminate in two senses. First, literature itself never constitutes a complete world—this is the problem of givenness, according to which literature is mostly nongiven, just as real-world reality is not fully given to consciousness and must be gathered into meaning from the units of perception our partial attention affords us. The reader's inability to hold the whole text in her mind at the same time constitutes the second level of indeterminacy, which must be supplemented by creating "formulations of the unwritten" as much as of the written. Literature communicates through this indeterminacy, which is grounded in *the blank*. The blank "designates a vacancy in the overall system of the text, the filling of which brings about an interaction of textual patterns."[23] The point, then, is that blanks don't necessarily call out for completion—that is, they don't signify a need for the reader to supplement the text from outside. Rather, blanks indicate a need for *combination:*

> It is only when the schemata of the text are related to one another that the imaginary object can begin to be formed, and it is the blanks that get this connecting operation under way. They indicate that the different segments of the text *are* to be connected, even though the text itself does not say so. They are the unseen joints of the text, and as they mark off schemata and textual perspectives from one another, they simultaneously trigger acts of ideation on the reader's part. Consequently, when the schemata and perspectives have been linked together, the blanks "disappear."[24]

Just as Heidegger's phenomenology stresses the gathering together of surroundings into the seamless unity of the World, a phenomenological approach to reading shows that the textual surface is in part a product of human cognition, which strives to gather narrative space into a unified, albeit imaginary, world. In both the real world and in narrative worlds, then, "we witness . . . the collapse of the multi-layered [reality] of nature, the overlapping *Umwelten* which interpenetrate but never touch, into a single spatial pancake."[25]

For all of these reasons, ecological realism demands a mode of reading that neither solely plumbs the text's depths nor exclusively skirts over its surface. Attending to uncanny ecology is crucial for this process because its unsettling mood forces us to pay closer attention to the gaps in perception. Like a black hole that is only visible from the radiation it discharges, the uncanny mood that issues from the holes in the whole helps to detect the presence of hidden gaps in the perceptual world, of unseen depths beneath the textual surface.[26] The ecological uncanny may not fully reveal the veiled operations of a text in the way that symptomatic reading desires, but it certainly has a powerful effect on characters and readers alike. *Affect* therefore seems to be key for ecological realism.

Curiously, affect also turns out to be key for Fredric Jameson's recent rethinking of literary realism. Jameson's theory posits realism as a consequence of the dialectical tension between narrative and scene; that is, between the chronological movement of *récit* and the "eternal present" of the *roman*. In essence, this is a temporal dialectic, one that plays out between the time of "destiny" (teleology) and the time of "unfolding" (phenomenology). According to Jameson, literary realism emerges from the unresolved tension between the two temporalities, which produces a strange optical illusion—a "doubling of perception, in which the aesthetic perspective of the painter does not replace that of the explorer-protagonist, but rather imperceptibly slips in beside it, in a kind of stereoscopic view."[27] Jameson terms this perceptual doubling *affect*: "Affect is perhaps here present as a kind of invisible figuration, which doubles the literal invisibly; a convex that shows through, as though reality itself blushed imperceptibly."[28] For Jameson, affect is at once embodied and beyond language. As for Lawrence, this places *affect* in opposition to *feelings*. Whereas feelings are named emotions that have already entered a linguistic register, "the positive content of an affect is to activate the body" and thus is opposed to language.[29] A unique challenge arises, since affect cannot be described without becoming meaningful, and hence allegorical. Yet as Jameson insists, "it is allegory and the body which repel one another and fail to mix."[30] Thus, just as "characters become the most perfunctory pretexts for what is virtually an autonomous unfolding of sense data," Jameson warns that "there can be no ultimate 'zero degree' in perception, that all such seemingly pure data are still haunted by a meaning of some kind, which is to say an ideological connotation."[31] For this reason, in order to resist vulgarizing affect through allegory (e.g., personification, "naming and nomination"), he argues that "we need a different kind of language to identify affect without, by naming it, presuming to define its content."[32]

As Jameson's own discussion of the strange doubling of perception indicates, the uncanny proves useful for developing a conceptual language for affect that resists defining content. The uncanny evokes a specific kind of mood, one that is located in bodily experience and yet escapes immediate cognition. Jameson points out that, for Heidegger, such a mood, or *Stimmung*, "was neither subjective nor objective, neither irrational nor cognitive, but rather a constitutive dimension of our being-in-the-world."[33] The ecological uncanny defines just such an *unheimliche Stimmung*, an unsettling mood that emerges from a broad range of very real (and mostly material) ecological encounters. Hence, it is not a *feeling* in the sense of being a conscious state; rather, it signals a gap between language and experience, perception and reality. Jameson evokes something very close to the ecological uncanny in his response to a long descriptive passage from *Le ventre de Paris*, in which Émile Zola writes about the "chaotic" and "bewildering multiplicity" of a market full of "masses of vegetables," "heaps of edible flesh and blood," and, "in a kind of delirious climax, the world of seafood in which the category of fish differentiates into pages of monsters and weird otherworldly beings."[34] Jameson's analysis is worth quoting at length:

> It would appear at first glance, and in the light of Zola's remarkable organizational procedures, that what is at stake here is a resolution of multiplicity back into unity, of difference back into identity. The enormous lists and catalogues would seem to be subsumed under generic categories and everyday commonsense universals: from life to the edible, from the edible to plants and animals, from the latter to meats and fish, and so on. In fact, I believe that this impression is at the least ambiguous; and that simultaneously with this first centrifugal movement of mastery and subsumption, of the ordering of raw nameless things into their proper genetic classifications, there exists a second movement which undermines this one and secretly discredits it—*a tremendous fermenting and bubbling pullulation* in which the simplicity of words and names is *unsettled to the point of an ecstatic dizziness* by the visual multiplicity of the things themselves and the sensations that they *press on the unforewarned observer*. The unexpected result is that far from enriching representational language with all kinds of new meanings, *the gap between words and things is heightened;* perceptions turn into sensations; words no longer take on a body at prey to its nameless experience. Finally the realm of the visual begins to separate from that of the verbal and conceptual and to float away in a new kind of autonomy.[35]

In this extended excursus on the unsettling unreality of the real, Jameson confirms the point I made above that a new realism would need to exceed both Ndebele's emphasis on the ordinary as well as Lukács's preoccupation with the (human) social totality. Whereas Lukács saw Zola's writing as politically admirable but ideologically compromised due to its alignment with naturalism rather than realism,[36] Jameson finds in Le ventre de Paris an exemplar of how realism's "antinomies"—narrative and description—reveal the constitutive and often uncanny strangeness of reality. As the chapters in part 2 of this book will show, a literary theory of ecological realism depends on just such uncanny strangeness.

PART II

CHAPTER 4

(Un)settling the Southern African Farm/world

> What is a farm but a mute gospel?
> —Emerson, "Nature"

If part 1 of this book focused on landscape description as a settler home-making practice, part 2 makes a conceptual shift to landscape description as a settler "worlding" mechanism. As chapter 1 argued, in Heidegger's lexicon the term *Landschaft*, or landscape, names an everyday manifestation of what he more famously terms *Welt*, or World. World signifies a total context of meaning that gathers around the human subject, involving it in an inextricable relationship of concern (*Fürsorge*). According to Heidegger, we rarely experience natural phenomena as merely there and awaiting us to assign them a meaning. Instead, we notice only phenomena that are already involved with our concerns, in which case they (always) already have meaning for us.

In *Being and Time*, Heidegger adopts the figure of the farmer and his farm to explain how concern regulates the relationship between Dasein and the World. When the farmer examines his crop or looks to the horizon for oncoming weather, he is inspecting his surroundings for meaningful signs. These signs constitute both the World around him and his understanding of that World. For example, the farmer never experiences the south wind merely as the patterned movement of air particles or the bodily sensation of a cool breeze on the skin. Instead, it is always—and always *first*—a meaningful sign of coming rain. The farmer only encounters the south wind *as* the south wind because it already bears directly on his concern for the well-being of his farm and the livelihood of his family.[1] Meaningfulness (*Bedeutsamkeit*) therefore saturates the World, and as such, signs do not require interpretation so much as the direct bearing of concern. Through the farmer's concern, his farm emerges as a context of meaning—a home(l)y landscape that assumes a World-like structure.

In the interest of thinking landscape as a worlding mechanism, part 2 extrapolates from Heidegger's example of what I call the "farmworld": the topos of the farm as a world in and of itself. Although the following chapters focus on the farmworld as a literary topos, it is also necessary to understand something of its historical pedigree. It should be remembered that even though agriculture is a characteristically human endeavor that has driven the growth of the human species from the Neolithic Revolution to the contemporary system of global agriculture, the (nearly) universal significance of agrarian life in human history should not erase the fact that agriculture is also a multivalent ideological practice that manifests differently in the many social, political, historical, and ecological contexts where it thrives.[2] This is perhaps the most important lesson that postcolonial environmental history has taught us: contest over land is always also a contest over competing land-use ideologies.[3] For this reason, ecological history must always also concern itself with social and political histories.

Heidegger's farmer is indisputably European and deeply influenced by the Old World notion of *husbandry*. As John Stilgoe asserts: "Husbandry is not farming. Husbandry is noble in the eyes of others; it is the avocation of enlightened kings; it is the first work of God himself."[4] In the seventeenth-century English imaginary, husbandry united the husbandman with the feminized earth—"Mother Nature"—in a matrimonial bond. The bond with the land produced fruits like the marital bond produced children, and hence the husbandman's labor served as partial answer to God's decree in Genesis to "be fruitful and multiply." In the New World colonies especially, husbandry also responded to God's command to "fill the earth, and subdue it": "It was the husbandman . . . who made the wilderness into a garden, who civilized a chaos and made it bear fruit and grain of every kind."[5]

Given the particular challenges the colonists faced in the American wilderness, however, the *husbandman* had to become a *farmer*. More engineer than spouse, the American farmer sought to master and transform the land. Early colonists like Hector St. John de Crèvecoeur celebrated the transformational project as essential to the emergent American identity. In the third epistle from his *Letters from an American Farmer* (1782), Crèvecoeur famously asks, "What, then, is the American, this new man?"[6] In answer he encourages his European reader "to examine how the world is gradually settled, how the howling swamp is converted into a pleasing meadow, the rough ridge into a fine field; and to hear the cheerful whistling, the rural song, where there was no sound heard before, save the yell of the savage, the screech of the owl or the hissing of the snake[.] Here an European, fatigued with luxury, riches, and pleasures, may find a sweet relaxation in a series of interesting scenes, as affecting as they are new."[7] Crèvecoeur champions the redemptive capacity of labor. Not only does agricultural labor uproot and displace undesirable members of the ecological community, but it also cultivates a revitalizing "scene." This scene, which is nothing other than the farm, the farmer's home, constitutes the source of a new identity: "The instant I enter on my own land, the bright idea of property, of exclusive right, of independence, exalt my mind. . . . *What should we American farmers be* without the distinct possession of that soil?"[8] Crèvecoeur's farm consolidates his American self by representing the New World in microcosm.

In *Letters from an American Farmer*, Crèvecoeur offers a prototypical vision of the farmworld: a self-contained ideological space that nourishes and protects his sense of being and belonging. In this sense, the farmworld topos emblematizes the home(l)y metaphysics of landscape discussed in chapter 2. Yet whereas that chapter emphasized the formal and ideological mechanics of home(l)y metaphysics, the chapters of part 2 focus on its fragility. D. H. Lawrence's uncanny New Mexican landscapes have already offered a glimpse of this fragility. Like Lou Witt, who sought to break with the Old World idealization of Nature, Lawrence chastises Crèvecoeur for his tendency to idealize colonial domestication. Lawrence even seems to relish the biographical irony of *Letters*, which Crèvecoeur revised in Paris around the same time that his farm, Pine Hill, was burned down during an Indian raid. "H. St. J. de C. tried to put Nature-Sweet-and-Pure in his pocket," Lawrence writes: "But nature wasn't having any, she poked her head out and baa-ed."[9] Lawrence's apparent conflation of "Nature-Sweet-and-Pure" and the Indians who destroyed the farm potently ironizes Crèvecoeur's own outlook in the letter quoted above, where "the yell of the savage" as well as "the screech of the owl" and "the hissing of the snake" all had to be eradicated in order to consolidate his farm. According to Lawrence, the amount of physical and ideological[ ]labor Crèvecoeur required to consolidate his farm belies that farm's basic precariousness.

Just as part 2 makes a conceptual turn from landscape as *home* to landscape as *world*, it also makes a geographical shift from the U.S. Southwest to southern Africa. This choice may seem strange considering the preceding discussion of the farm and its centrality in American history and literature, from Crèvecoeur's eighteenth-century account of the cultivation of the colonies to early twentieth-century elaborations of the pioneer mythos such as Cather's *O Pioneers!* (1913) and O. E. Rölvaag's *Giants in the Earth* (1927). The reasons for making this shift are both historical and literary in nature, and they are more about a difference in degree than a difference in kind. To state the case briefly, the historical and literary farm in southern Africa is much more explicitly bound up with the racial politics of settler colonialism than it is in the United States, where this relationship remains systematically obscured. To this day there remains extremely limited public recognition of the United States' settler colonial status and even less hope that meaningful decolonization will ever happen. This is not to say that the colonialist ideology of the American farmer is not obvious. As already suggested with respect to Crèvecoeur, in the early American context the farm clearly represented the New World in microcosm, emblematizing the colonial ethos of freedom and equality. But just as Crèvecoeur celebrates the (false) idea that he and other American farmers "inherited" this ethos without recourse to violence, so too do many modern Americans continue to see the symbolic inheritance of freedom as equally innocent. The colonial history of anti-Indigenous and ecological violence that made farms like Pine Hill—and the United States as a whole—possible remains largely unacknowledged, even if increasingly subject to scholarly and activist critique.[10]

In southern Africa, however, the history of settler colonialism is much more visible and remains at the forefront of social, political, and historical consciousness. Particularly in the case of South Africa, where the Apartheid regime was founded on twin problematics known as the "Native Question" and the "Land Question," the fates of Black Africans and their relations to both the European minority and the land continue to play out under the auspices of the "New South Africa." Post-Apartheid politics have embarked on the ambitious yet delicate and often uneven aim of decolonizing a settler colonial nation. Unlike in the U.S. context, where the farm typically mythologizes the (European immigrant) farmer's claim to the American Dream of property ownership and the freedom of independence, in southern Africa the farm plays a more visibly complex role. From the historical origins of settler colonialism to present-day decolonization, the southern African farm has emblematized the settler colonial project, functioning as a crucible for its racial politics. What is most crucially important about the farmworld in southern Africa—and especially in European fiction from the region—is that it not only interrogates the founding mythology of the farm/nation, but it also disrupts those foundations. In contrast to the American context, where the farm continues to prop up a political mythology that effaces its own origins, the southern African farm at once reiterates these origins and seeks to dismantle them.

The rest of this chapter moves through a range of intellectual terrains. The following section sketches out the historical and literary genealogies of the farm in southern Africa in greater detail, and the next one offers a more concrete formulation of the farmworld concept through a brief reading of an Afrikaans-language *plaasroman,* or farm novel, by C. M. van den Heever. Whereas these first sections emphasize the ideological foundations of the farmworld, the final sections explore the collapse of these foundations. The second half of the chapter (re)turns to the ecological uncanny and the possibility of an ecological realism, beginning with a section that explores the role ecological realism plays in settler colonial texts, and concluding with the question of the southern African farm novel. All of this material is meant to provide the necessary literary, historical, and theoretical foundations for the chapters that follow, which explore how three English-language farm novels from across the long twentieth century progressively unsettle the farmworld ideology by invoking uncanny visions of ecological reality.

THE EUROPEAN FARM IN SOUTHERN AFRICAN HISTORY AND LITERATURE

European histories of southern Africa have mythologized the farm as the primal scene of white belonging, a sacred site that consolidates and sustains white being. As a symbolic entity, the farm's racialized ideological burden developed throughout southern Africa's long history of settlement, which began, so to speak, with the imposition of a foreign metaphysics of land ownership. When the Dutch first arrived in the seventeenth century and encountered the region's first inhabitants, they failed to grasp local cosmologies of inhabitation and land tenure. The colonists viewed the itinerant existence of the Khoisan peoples as a pastoral mode that rejected all forms of possession. This misunderstanding of San pastoralism made it much easier to imagine the landscape as empty and waiting to be claimed by the plow. According to San metaphysics, however, the land represented what the archaeologist Sven Ouzman describes as "a vast network of relations and obligations between people, animals, places, [and] spirits."[11] Far from being empty, the environments inhabited by the San were teeming with life and populated by ancestral spirits that took various forms, such as wind, rain, and stars as well as flora and fauna. Their relations with these entities provided the San with a strange kind of "title-deed" that confounded the European conception of individual ownership: "The land or network of relations was believed to own *them*."[12] It was for this reason that, among the San, land could not be "owned" in the European sense; instead, individuals "retain[ed], through birth, marriage, and residence, rights to special access to particular territories."[13] As a matter of "special access" rather than exclusive ownership, individuals lacked the one right that was crucial for the narrower European definition of possession: the right of alienation.

The Dutch settlers arrived at the Cape with a very different idea about land tenure, one that privileged the long-term settlement of individuals in the service of agricultural production. If the Khoisan system was founded on a notion of special access defined by a relational metaphysics of belonging, the Dutch system was founded on a concept of possession defined by the metaphysics of natural right. According to the latter metaphysics, human labor established the claim of sovereignty over a particular territory. Locke authorizes this doctrine of natural right in the second of his *Two Treatises of Government*, where he insists that, by way of divine covenant, "Man" has property through his labor: "And hence subduing or cultivating the Earth, and having Dominion, we see are joyned together. The one gave Title to the other. So that God, by commanding to subdue, gave Authority so far to *appropriate*. And the Condition of Human Life, which requires Labour and Materials to work on, necessarily introduces *private possessions*."[14] Labor metaphysically endows land with value in the same way it does for the commodity in capitalism. As in Marx's value-form theory, the true value of the Lockean farm does not reside in its materiality, which is essentially "contentless" (*inhaltslos*). Instead, as a kind of commodity, the Lockean farm accrues a more important form of value through a more ineffable form of content (*Gehalt*). Marx describes this kind of content as a crystallized residue that renders a commodity exchangeable, alienable. This residue is invisible, such that "neither microscopes nor chemical reagents are of assistance." Marx insists, instead, that "the power of abstraction must replace both."[15] Likewise, only by abstraction can we understand the true nature of the Lockean farm's value; it lies not in the physical realm, but in the more-than-physical (*übernatürlich*) one.

Conceptually, the European farm in southern Africa marries this Lockean metaphysics with the home(l)y metaphysics of landscape, which gathers mere surroundings into "felicitous space": a *home*. Home(l)y metaphysics has a parallel history in southern Africa that also reaches back to the beginnings of Dutch settlement. Soon after establishing a refreshment station at Table Bay in 1652 for ships passing between Europe and India, the Dutch East India Company released a small number of its servants to become full-time *boere,* or farmers. These farmers were responsible for stocking the station with fresh produce. However, between the fickle climate and cattle raids perpetrated by malcontent groups of Indigenous neighbors whose livelihoods had been disrupted by European settlement, these *boere* only barely eked out a living. Upset with their precarious position on the frontier, these farmers approached Cape Commander Jan van Riebeeck seeking better payment for their annual wheat harvests. But the Company, which continued to see the "free burghers" as servants, refused to offer more than the bare minimum required for survival. Van Riebeeck managed to relieve the brewing tensions, but he realized that more was at stake than the price of wheat. As he wrote in his diary in December 1658, the burghers "would like to be their own masters and overrule the lawful authorities placed over them."[16]

As Hermann Giliomee indicates in his "biography" of the Afrikaners, who descended from the original Dutch settlers, these early tensions foreshadowed South Africa's long and vexed history of settlement. Many of the conflicts that would erupt in the Cape in the ensuing decades, and indeed much of the expansion in the coming centuries, would play out in the midst of the Afrikaners' search for greater autonomy and sustainability in their agricultural endeavors. For instance, following the gruesome Frontier Wars (1770–1812), the insecurity of farms on the Zuurveld gave rise to a number of emigrations from the colonial territory in the nineteenth century. Most famous among these departures was that of the Groot Trek, or Great Trek, by a group of people who would later be known as the Voortrekkers. Emigrating throughout the 1830s and 1840s, the Voortrekkers emphasized political grievances having to do with a lack of land, labor, and security. With farm life under threat from dangers both external (e.g., Khoisan and Xhosa raids) and internal (e.g., poor governance), as well as a growing sense of marginalization from the increasingly dominant British colonial culture, which increasingly framed the Boers as culturally backward brutes, the Voortrekkers left in search of a stronger, more stable connection to the land.[17] In the midst of these settlements, conflicts, emigrations, and resettlements, the farm became for the Afrikaner an elusive historical and cultural fetish object: universally contested and hence in need of ideological protection as much as physical defense. In the Afrikaner imaginary, then, the farm comes to represent a place that secures an otherwise troubled sense of belonging; it becomes a sacred site that disavows the Afrikaners' own violent displacement of Indigenous Africans by imaginatively transforming "empty space" into a home—a self-enclosed world that sustains Afrikaner being.

Elsewhere in southern Africa—and particularly in the British settler colony of Southern Rhodesia—farming had a different, though similarly powerful lure for Europeans. After the turn of the twentieth century, and following the acquisition of the territories south of the Zambezi River and north of the Transvaal by Cecil Rhodes and his British South Africa Company, Southern Rhodesia became something of a Promised Land for British subjects who felt worn down by life in England. The colony was advertised at the British Empire Exhibition as a land of plenty, replete with large and inexpensive tracts of land where even novice farmers could "Get Rich on Maize." Such promises piqued the interest of disaffected men who, in search of adventure, independence, and economic advancement, uprooted their families and moved to southern Africa.

In truth, the situation in Southern Rhodesia proved more complicated than the Empire Exhibition's image of a Land of Plenty suggested. Much of this complication stemmed from the false narrative of the colony's founding, which stated that the Europeans had been forced to act in order to save the peace-loving Shona farmers from the aggression of their warrior neighbors, the Ndebele.[18] Imperial acquisition was thus framed as a protective measure that would, ostensibly, ensure the future prosperity of the new colony's Indigenous agriculturalists. The British alignment with the Shona makes sense insofar as the Shona land tenure ethos and work ethic approximated the Old World ideal of husbandry, which the colonial government had used to promote Southern Rhodesia as the ideal place for noble (and profitable) agricultural pursuits. According to Lawrence Vambe, the Shona had already formed something of a viable peasantry before the arrival of the British South Africa Company. He writes at length of the prosperous time of the Mutapa Empire, which lasted from the mid-fifteenth to the mid-eighteenth century:

> As time went on, the VaShawasha people and their kindred tribesmen came to measure a man's worth and standing in society on the basis of his

husbandry and industry, as proved by his harvests rather than his boasts, his cunning or physical strength. Thus among the pillars of any Shona tribal grouping were people called *hurudza*. This title, both in the singular and the plural, means an agricultural baron or barons, and was given to individuals who proved themselves to be hard-working, productive farmers.... A *hurudza*, in this context, was a man not only of sustained and conscientious industry. He was also a man of wisdom, of value to the people of Zimbabwe. He was well connected socially and accordingly played an influential role in all Shona tribal and national affairs.[19]

Terence Ranger notes that such *hurudza* also enjoyed the favor of the traditional spirit mediums, such that "productive farming went hand in hand with veneration of the spirits."[20] Clearly, the successful farmer held a privileged place in Shona society. Nor did the initial arrival of the British drastically change this state of affairs. Ranger reports that, in the Shona district of Makoni, oral testimony concerning the period between the 1890s and the 1920s "still refers to but does not passionately resent the loss of land to whites. The reason . . . is that despite very extensive land alienation it was possible in this period to create a viable peasant economy."[21] Many Shona peasants remained on European land and could market their produce, and the Reserve lands that had been established by the colonial government still had enough productive soil for Indigenous agriculture.

Eventually, however, the success and relative independence of the Shona peasantry began to cause problems for European settlers, who found no readily available labor for their own farms. In 1930 the colonial administration passed the Land Apportionment Act, which barred Africans from owning non-Reserve lands except in special cases. A series of Maize Control acts also attempted to curb peasant profits. Nevertheless, ingenious peasant farmers continued to find ways to benefit from loopholes in these acts, and European farmers grew jealous. By 1935, as J. A. Edwards has indicated, British settlement in Southern Rhodesia "carried the mark of disappointment," with the majority of white Rhodesians retreating to the less inhospitable Highveld territory: "By the middle of the thirties the settler knew without a shadow of a doubt that his community would always remain small, a minority living among masses of Africans. It was this fearsome truth that gave the ideal of survival such desperate strength."[22]

During this period of desperation the European farm in Southern Rhodesia underwent its own mythologization: unlike in the South African context, where the European farm grounds a claim of natural right, in Southern Rhodesia the European farm came to ground a narrative of ecological salvation. In south-central Africa, the 1930s witnessed a shift in perceptions about conservation, one that increasingly demonized African agriculture as the driver of erosion and other forms of ecological degradation.[23] By the end of the 1930s, the colonial administration came to see African Reserves as sites of a full-blown ecological crisis, a perception that would escalate tensions between the British and the Shona throughout the 1940s. While Africans in the Reserve suffered from low crop prices and the enforced diversion of peasant labor into government conservation work, the colonial administration saw the crisis in terms of the Africans' methods of cultivation, which had supposedly exhausted the soil's productivity. The British position was hypocritical, for in previous decades the colonial agricultural demonstrators had coerced the African peasantry to abandon traditional cultivation and adopt more intensive methods. Terence Ranger summarizes the irony: "Recently the heroes of progress, converts to the 'gospel of the plough,' they now figured as destroyers of the environment."[24] In contrast to the destructiveness of the African farmer, then, the European farmer became an agent of conservation. In a strange parody of Cecil Rhodes's founding myth, in which the Europeans were forced to save the Shona from their warring neighbors, further land alienation came to be founded on the myth that the soil needed to be saved from the agricultural violence of African farmers. In Southern Rhodesia, then, the European farm legitimized the settlers' claim to the land by becoming a space of ecological salvation.

Literature in southern Africa deeply engages with the figure of the European farm, by turns contributing to and challenging its histories and attendant ideologies. Literary interest in the farm has generated two complexly interrelated traditions of "farm novels" that span the last 150 years of southern African letters. This literary legacy begins with the 1883 publication of Olive Schreiner's *The Story of an African Farm*, a profoundly antipastoral novel that concerns itself as much with European moral philosophy and gender politics as it does with the haunting strangeness of its South African setting. Reversing the dark and tortured ethos of Schreiner's Karoo farm, Afrikaans-language *plaasromane* by iconic writers such as Jochem van Bruggen, Daniel François Malherbe, and C. M. van den Heever emerged in the 1920s and 1930s. These novels sought to remake the farm into a literary figure that mythologized the (male) Afrikaner farmer and naturalized his place in the South African landscape.[25] These *plaasromane* were written at the height of Afrikaner nationalism and promoted an ideology consonant not only with the future Apartheid regime but also with the *Blut-und-Boden* politics enshrined in German farm novels (*Bauernromane*) of the same period.[26] In the later twentieth century, new strains of farm novels emerged to critique these literary forebears. Throughout the 1960s, 1970s, and 1980s, Afrikaans-language novelists such as André Brink, Etienne Leroux, and Etienne van den Heerden penned revisionist *plaasromane* in the postmodern mode that attacked and satirized the ideological foundations of the Afrikaner farm imaginary.[27] During this same period, writers such as Nadine Gordimer and J. M. Coetzee published English-language farm novels (*The Conservationist* [1974] and *In the Heart of the Country* [1977], respectively) in the modernist mode that also sought to disrupt the history and ideology of white land ownership in South Africa. Both of these language traditions have survived into the post-Apartheid era, where Afrikaans novels like Marlene van Niekerk's *Agaat* (2004) as well as English novels like Coetzee's *Disgrace* (1999) continue to deploy the farm as a central figure animating debates about land ownership and racial politics in the "New" South Africa.

Far from seeking to rewrite literary histories of the farm in southern African letters, the second part of this book has the much humbler goal of tracing a single curious motif through a number of English-language farm novels.[28] What previous literary criticism has typically identified as a strain of antipastoralism, I read as a more obscure, more radically disruptive mode. The nature of that mode will become clear in the final section of this chapter. However, as this mode turns on the trope of the farmworld, it is necessary to begin by showing just how this trope functions in southern African fiction.

THE FARMWORLD IN THE AFRIKAANS *PLAASROMAN*

The farmworld enshrines the farm as a self-enclosed space, a world in itself that remains mystically separate from its surrounding environs. This definition recalls Mircea Eliade's understanding of the sacred, according to which any act of settlement also functions as an act of consecration, a symbolic re-founding of the world in microcosm.[29] Likewise, the European farm in southern Africa becomes a comforting symbol of *"smallness* in the midst of *vastness."*[30] As a protective enclosure, the farm ensures the well-being of its inhabitants and promises longevity for the ideological structures that sustain them. It becomes a static realm of cultural reproduction that, over time, enables the fabrication of a myth of origins.

Relatively unchallenged versions of this farmworld ideology have occasionally appeared in English-language farm novels. For instance, Pauline Smith's 1925 short story collection *The Little Karoo* and her 1926 novel *The Beadle* are both set on a quintessentially pastoral farm appropriately named "Harmonie." Harmonie represents one farm within the wider Aangenaam (Pleasant) community, nestled in a valley in the Little Karoo. As J. M. Coetzee points out, this Edenic, womb-like enclosure preserves a rural order of patriarchal feudalism against the incursions of modern capitalism.[31] However, more than simply preserving a particular mode of production, Smith's farm rehearses the pastoral as the figure of her own displacement from South Africa. At the age of thirteen she was sent to Scotland for boarding school, and though she made several extended visits to the place where she grew up, she spent the rest of her life in England. As Smith describes in the short autobiographical essay "How and Why I Became an Author," when she began to write following the death of her father, who had remained in South Africa, she did so "to set down for my own comfort the memories of these happier days."[32] Smith's farm therefore gives form to a certain nostalgia for the Karoo from within her own situated Englishness.

Bessie Head's novels *When Rain Clouds Gather* (1968) and *A Question of Power* (1973) offer similarly idyllic visions of agricultural life in southern Africa. Both of these novels take place within farming utopias in rural Botswana that foster communities of refugees. The earlier novel focuses on life in Golema Mmidi, a village whose name means "to grow crops" and which "consisted of individuals who had fled there to escape the tragedies of life."[33] Among these refugees is the protagonist, Makhaya, who has fled the violent racial politics of Apartheid in South Africa, crossed the barbed-wire border to Botswana, and settled in this farming haven where local farming methods and modern scientific methods collaborate to inspire communal uplift. The later novel follows the cross-border escape of another protagonist, Elizabeth, the product of an illegitimate (and illegal) union between a white mother and a Black father. Finding herself homeless and without a solid identity in Botswana, Elizabeth slips between hallucinations and reality as she recovers in Motabeng, or "the place of sand."[34] Here she takes part in a youth development group where work in a community garden eventually helps alleviate her nervous condition. Though each place has its own specific gender, ethnic, and racial politics, Golema Mmidi and

Motabeng both provide semi-utopian havens that offer protective enclosures for southern Africa's outcasts.

Despite these examples, it is not the English-language farm novel that most fully emblematizes the utopian ideology of the farmworld. Rather, it is the Afrikaans *plaasroman* of the 1920s and 1930s that formalizes the farmworld ideology. In these novels, not only does the fantasy of enclosure imaginatively protect the Afrikaner farmer from natural disasters and economic burdens that threaten to destroy the farm; this fantasy also secures the lineage of landownership and hence safeguards the myth of the farmer as a *natuurmens* (natural man) whose being and belonging are sacredly bound up with the land. J. M. Coetzee notes that the development of the *plaasroman* was coterminous with the emergence of a class of farmers rendered landless due to drought and economic depression. The *plaasroman* gave voice to this difficult time by developing a number of recurring themes, such as the problem of inheritance with limited land, the avarice of land speculators, the migration of dispossessed Afrikaners to the cities in search of labor, and the threat of metropolitan values. During an era when men expected to succeed as independent landowners, it is unsurprising that contemporary Afrikaner writers sought "to dignify the disaster by claiming for the old dispensation an antiquity losing itself in the mists of time.... The farms they carved out of the wilds, out of primal, inchoate matter, become the seats to which their lineages are mystically bound."[35] The myth of the farm as a dynastic seat naturalizes white sovereignty through a metaphysics in which patriarchal lineage is paramount. According to the terms of this myth, in which "it is in some sense sacrilegious to sell an ancestral farm," Coetzee asks whether "it is also sacrilegious to *buy* a farm in whose soil the ancestor of another man lie buried."[36] To this question—ironic given the dubious nature of the Afrikaners' original "purchase" of these lands—Coetzee responds that the only way to secure legitimacy through purchase is to invoke a Lockean metaphysics of landownership; that is, "one establishes one's ownership by signing the land with one's imprint as one signs a legal document with one's mark—a process that may take a lifetime."[37]

Perhaps no body of work captures the ethos of the *plaasroman* better than that of Christiaan Maurits van den Heever. A prolific novelist and a poet, Van den Heever is best known for his farm novel *Somer* (1935), which had raced through sixteen print runs by 1951. Like other of his *plaasromane*, *Somer* expresses deep anxieties about landlessness, impoverishment, and dispossession—anxieties that his novels frequently associate with the rapacity of land speculators and the moral corruption of city dwellers. As a foil to these dangers, the farm represents the romantic centerpiece of the Afrikaners' rural lifestyle. However, the farm's symbolic charge in these narratives emerges through negative circumstances. In the prototypical Van den Heever novel, the farmer only realizes his mystical bond to the land when it's too late. An accumulation of threats provides the preconditions for an epiphany in which "for the first time the farm appears to the farmer in the glory of its full meaning, and for the first time the farmer fully knows himself."[38] Unlike his prototypical *plaasroman,* however, *Somer* is more optimistic in that it never fully thwarts the relationship between farm and farmer. Nevertheless, even without the epiphanic moment described by Coetzee, *Somer* employs other narrative strategies that consolidate the farm as a self-contained world. Specifically, the novel introduces a number of natural and anthropogenic forces that threaten the farm from "outside." In moments when dangers come into the foreground the novel retreats into idyllic landscape imagery, rendering the farm an enclosed, monad-like entity. The invocation of landscape offers symbolic recompense that asserts more than just the farmer's natural right to the land, established by generations of labor; it also instigates the home(l)y metaphysics that cathects the land into the farmer's own particular beloved place. In this light, it seems significant that T. J. Haarhoff's English rendering translates the title not as "Summer" but as "Harvest *Home.*"

Somer centers on a tract of family farmland that has been divided between two sons, Frans and Tom. Frans's farm, Driefontein, stands on the brink of financial ruin. Over the years, and as a result of persistent bad luck, Frans has amassed considerable debt from the Land Bank. To make matters worse, Driefontein falls prey to a freak thunderstorm that obliterates the majority of its crop while leaving neighboring lands untouched. Frans faces bankruptcy in the storm's wake, a disaster that puts him face-to-face with the callousness of Nature. Frans (along with the narrator) casts Nature as an obscure, antipathetic force bent on the farm's destruction. Frans frequently bemoans "the fatal power of the earth beneath them," and whenever misfortune befalls him, he repeatedly exclaims, "old Mother Nature was against me."[39] *Somer* consistently pits Nature against the farmer and, by extension, his farm. The farm exists outside of Nature, a separate entity that must be protected from the latter's tyranny.

If the farm must be protected from Nature, it must also be secured against human threats, which the novel represents through two related motifs. First is the motif of the shrinking farm, which speaks to historical anxieties about land shortages. Lacking the finances and space required to expand the family's holdings, farmers had to subdivide their lands to provide plots for their male heirs. As generations passed and farms became smaller, farmers faced diminishing returns and eventual dispossession, which the novel equates with eventual suffocation and death: "Farms get smaller. I tremble to think of my children's future. We must stick to our lands, such as they are; once we lose them, we're driven to Johannesburg or one of those places. And once you're on the mines, you've got phthisis before you know it—your chest whistles as you struggle for breath. Dig ourselves in here—that's all we can do" (11). The only way for the farmer to sustain his livelihood, indeed his life, is to root himself ever more firmly. Yet this need for rooting is challenged by the novel's second motif, in which the ground itself is figured as unstable. As Frans complains to Tom: "What discourages me so, is that it seems impossible to make a stand—the ground slips clean away under your feet" (82; see also 63). This anxiety about the farm slipping out from under him relates to the very real threat posed by the Land Bank, which could seize Driefontein at any moment on account of Frans's failure to pay off his piling debt.

Even as these threats undermine the stability of landownership, they also endow the farm with a sense of unity. Whenever such threats arise in the novel, the narrative retreats into the idyllic language of landscape. Such a strategy gestures further to the optimism of *Somer,* which the English-language translation frames not as a novel but as "an Afrikaans idyll." For a more concrete example, take the scene when a greedy neighbor, Oom (Uncle) Faan, taunts Tom with his plans to purchase Driefontein in the event of bankruptcy. Initially enraged, Tom's thoughts turn to the twin subjects of landscape and genealogy: "Driefontein in *his* hands, Driefontein that since the Voortrek has belonged to the same family! His eye wanders over the surroundings, he sees the dark-green garden, the ridges quivering in the heat, the land where three generations of his family has [*sic*] toiled so hard.... 'Even if I ruin myself,' he mutters, 'I must help Frans to keep his land—the farm must remain in our hands'" (100–101). Faced with the dissolution of farm and family, Tom seeks solace by appealing to the natural features of the land along with what Coetzee calls "lineal consciousness." And yet, the imperative thrust of his repeated use of "must" intimates a sense of desperation. It also admits a certain frailty in the idea of the farm's unity. Once again the novel figures the collapse of Driefontein as tantamount to death. Tom must risk his own ruin to preserve the land and his kin. Hence, while contemplating his brother and the generations of forebears going back to the Great Trek (in reality less than a century had passed), Tom gathers the farm into his visual field and channels the worlding power of landscape.

Landscape idylls appear frequently in the novel, and not always in direct relation to threat. *Somer* also links landscape to "Nature's unified beauty" (18) in pastoral moments that romanticize the farm and the farmer's labor. For example, the third chapter, which tells of the corn harvest on Tom's farm, centers pastoral iconography that genders the relationship between the white male farmer and a feminized land: "Before [the laborers] lies the yellow corn-field, caressed by the soft compulsion of the morning-breeze. A ripple passes over the surface, the ears bow in stately fashion and are still; there is a whispering in the stalks, a whisper of corn-maturing summer, of mother-earth rejoicing in her fruits, bringing a restful joy to the hearts of the harvesters" (17). The harvest ritual places farmer and farm into a gendered relation of mutual attraction: the male presides over the farm, which embodies the inviting warmth of "mother-earth" as opposed to the cold callousness of "old Mother Nature." The land lies subjugated, spread beneath the farmer's gaze. The narrative perspective aligns this patriarchal gaze with the morning breeze, the "soft compulsion" of which caresses the cornstalks into "stately" bows of subservience. The welcoming earth-mother invoked here derives in part from Old World European thinking about husbandry, a subject that brings the southern African and American settler imaginaries into conversation. Annette Kolodny describes "America's oldest and most cherished fantasy" as "a daily reality of harmony between man and nature based on an experience of the land as essentially feminine—that is, not simply the land as mother, but the land as woman, the total female principle of gratification—*enclosing the individual in an environment of receptivity, repose, and painless and integral satisfaction.*"[40] In the South African context the kind of gendered imagery critiqued by Kolodny additionally derives from the idea of the *volksmoeder,* or "mother of the nation." An important figure in Afrikaner nationalism, the *volksmoeder* embodies the spirit of virtue and self-sacrifice, much like Van den Heever's anthropomorphized cornstalks.[41] Allegorically, the *volksmoeder* trope casts the farm as a microcosm of the Afrikaner nation and contributes to Afrikanerdom's rural myth of origins. However, this invocation also endows the farm with a more "housewifely" sensibility, subjected to the power of the farmer who is traditionally said to be "married" to his land, and whose metaphysical matrimony with the farm is consecrated, above all, by the harvest ritual. Whether understood as the "mother of the nation" or the more rustic "farmer's wife" (*boervrou*), the association between farm and *volksmoeder* renders the farmer's land a warm, womb-like enclosure that protects its inhabitants from the harshness of "Mother Nature."

As *Somer*'s third chapter continues, the narrative adopts Oom Tom's perspective as he presides over the corn harvest. Tom tracks the progression of the sun and searches the skies for inclement weather. In these moments, when the narrative takes on the farmer's gaze and, by extension, his concern for the farm, landscape imagery plays a key role in imaginatively managing the progress of the harvest. When looking for the sun, for instance, the narrator takes note of distant rain: "The East has now become dark red, with thin lines of gleaming light along the edges of the hills" (18). When looking to the west where "thunder-clouds appear"—the very thunderclouds that harbor his brother's coming misfortune—the landscape description becomes more detailed, signaling the farmer's increased concentration: "At first the downy points are just visible beyond the quivering mistiness; then, the white curves swell out, column on column, until the snow beacons majestically fill the horizon and serenely dominate the spacious veld" (19).

Tom's close attention to the shape, texture, and movement of the towering cumulonimbus columns approaching from the west reprises Heidegger's farmer, who constitutes the *Weltlichkeit*, or "worldliness," of his farm through his concern about the south wind. Likewise, in *Somer*, as in other *plaasromane* of the same period, the invocation of landscape imagery—whether in response to external threats to the farm (e.g. storms, land speculators, debt) or to romanticize the rural lifestyle the farm sustains—renders the farm a totalizing metaphor, a microcosm to which the farmer is bound in a relationship that is at once biological and existential, mythical and ontological. The farm becomes a *farmworld:* a home(l)y site that guarantees the farmer's sovereign sense of belonging and secures his deepest sense of being-in-the-world.

UNCANNY REALISM AND (SETTLER) COLONIAL ALLEGORY

For all of its ideological reinforcement, the southern African farmworld remains feeble and in need of undoing. The remainder of this chapter—and indeed the remainder of part 2—thus turns from the home(l)y constitution of the farmworld to its dismantling. This latter project has been underway within Afrikaans literature since the 1960s, when revisionist *plaasromane* began to work against the imaginary established by their literary forebears. In contrast to the idyllic pastoralism embodied by earlier novels, these revisionist *plaasromane* deploy antipastoral tactics meant to undermine an inherited ideology. André Letoit, for instance, penned a postmodern send-up of Van den Heever in his 1985 novel *Somer II*, which introduces a potty-mouthed narrator who "soils the original *Somer*'s idylls with his grimy internal tableaux, his maudlin fantasies of self-annihilation, and his characterization of Van den Heever's beloved *taal* [language] as a 'whore.'"[42] In contrast to the totalizing metaphor of the farmworld in *Somer*, Letoit's misbegotten sequel "proposes schizophrenia and fragmentation as an appropriate response to living in South Africa."[43] In addition to Letoit's schizophrenic aesthetics, other Afrikaner writers have sought to dismantle the figure of the farm through similarly postmodern methods. Etienne Leroux's *Seven Days at the Silbersteins* (1964) presents a surrealist satire that reduces the farm to a bizarre backdrop,[44] whereas Etienne van Heerden's much later *Kikuyu* (1999) uses metafictional experimentation to explore wider sociopolitical issues that overshadow the white protagonist's identity struggles on his parents' Karoo farm. Instead of truly undermining the ideology of Afrikaner pastoralism, however, the antipastoralism of these novels at most affects a shift in tone; that is, it satirizes, bastardizes, and makes a mockery, but it does not pose a radical challenge to the farmworld or the home(l)y metaphysics that sustains it. This is because their antipastoralism focuses solely on the level of (human) politics and (human) history. Unlike Lou Witt, the protagonists of these novels do not pay homage to the "unseen presences" that implicitly challenge human sovereignty and complicate the politics of the settler colony.

One late twentieth-century *plaasroman* that takes a slightly different approach to the dismantling of the farmworld is André Brink's 1982 novel *Houd-den-bek*, translated as *A Chain of Voices*. The polyphonic narrative structure of Brink's novel erodes any sense of a stable Afrikaner settlement in ways that at once encompass and transcend the political injustices of displacement and enslavement. Set during a slave uprising in 1825, the novel links together a diverse range of voices, from Boers and their wives to slave laborers on Afrikaner farms to the quasi-independent Khoisan who live and work among the slaves. This structure allows Brink to establish a narrative counterpoint that places white and Black South African voices in tension.

For instance, in contrast to the biblical language the Boer farmers use to resist British laws impinging on their authority and to support their lineal claim to the land, characters like the San matriarch Ma-Rose produce competing mythologies that queer the language of Genesis: "In the beginning there was nothing but stone."[45] Importantly, Ma-Rose's language is as much geological as it is mythic; it points at once to ecological precariousness and to the ideological frailty of the European farms:

> There is a settled look about the string of farms with their houses and outbuilding and kraals, . . . but don't be fooled by that. One single great gust of wind and it's all gone as if it's never been here. The White people, the Honkhoikwa, the Smooth-haired ones, are still strangers to this part. They still bear in them the fear of their fathers who died on the plains or in the forbidding mountains. They do not understand yet. They have not yet become stone and rock embedded in the earth and born from it again and again like the Khoikhoin. One doesn't belong before one's body is shaped from the dust of one's ancestors.[46]

Ma-Rose's prophecy courts the uncanny in its association of death and belonging: to belong, one must be formed from the bone dust of ancestors who perished in the stony wilds. In this way Ma-Rose's geomythography works to re-place Black Africans at the center of legitimacy. Yet even when legitimacy has been established, precariousness still reigns. As Ma-Rose explains, "our mountains are old, stretching like the skeleton of some great long-dead animal from one end of the Bokkeveld to the other, bone upon bone, yet harder than bone; and we all cling to them. *They're our only hold.*"[47] The instability of this uncanny reality descends on the Boer patriarch, Piet van der Merwe, early in the novel. Upon returning to his farm after learning of a slave uprising on an Irish farm (an event that foreshadows the uprising on his own son's farm), Piet sees his home landscape transformed, "as if a strange presence had touched it all and turned it all transparent, revealing veins and secret organs and the skeleton below."[48]

It is here that the ecological uncanny surfaces, threatening to alienate the Afrikaner farmer. It strips away layers of habitual perception that have otherwise idealized the farm as a home(l)y place. Piet's uncanny revelation is much more extreme than the postmodern antipastoralism introduced by Letoit, Leroux, and Van Heerden. Whereas these revisionist *plaasromane* merely negate the ideology of pastoralism, Brink's novel, albeit only briefly, estranges the pastoral from within. Rather than naturalizing belonging, *A*

Chain of Voices introduces a theme that will prove much more significant in the English-language farm novel tradition, which, as subsequent chapters will show, uproots the idea of a home(l)y Nature and unsettles the metaphysics of natural right.

When the farmworld and its attendant ideology fall prey to the ecological uncanny, another view of reality emerges, one that signals the need for a new form of *ecological realism*. As outlined in the excursus that concludes part 1, ecological realism refers as much to a mode of reading as it does to something that marks or defines a particular text. Realism as a reading practice has something to do with reconfiguring perception by shifting what we pay attention to and reorganizing the hierarchy of attention. Hierarchy, a vertical signifier, becomes here a figure for depth. Ecological realism requires a mode of reading situated somewhere between depth and surface—a kind of *shallow* reading that, as Fredric Jameson suggests in *Antinomies of Realism,* cannot completely protect against the "allegorical" reduction to symbolic meaning but can ward off the most egregious forms of such reduction. In this way, *realism* and *allegory,* understood as modes of reading, stand in tension with each other.

This tension between realism and allegory has largely been ignored in postcolonial contexts generally and in the context of southern African farm novels especially; allegorical reading has absolutely predominated in these contexts. Jameson himself has played a crucial role in establishing the centrality of allegory as a postcolonial paradigm, most (in)famously by asserting that all "Third-World" novels double as national allegories.[49] This claim has been both contested[50] and extended[51] within the sphere of postcolonial studies. More significant (and less controversial) for postcolonial theory, however, has been Abdul R. JanMohamed's writing on "Manichean aesthetics."[52] In this work JanMohamed pushes beyond the analysis of stereotyping in colonialist fiction to something he calls "Manichean allegory," a conceptual schema that metonymically extends racial difference into any number of infinitely flexible binary paradigms (white/black, good/evil, rational/emotional, etc.). In each opposition, the European remains hierarchically superior to the non-European. This schema inheres within the colonial mentality, instituting an "economy" that invests racial difference with moral and metaphysical significance.

By offering a powerful tool for understanding the symbolic economy of colonial texts, JanMohamed's theory provides the basic groundwork for what we might rename as *colonial allegory*. Colonial allegory fixates on racial difference as the key for a proliferating series of self/other oppositions that define the colonizer and the colonized against each other. This paradigm assumes a dialectic of identity and difference that works out, however ambivalently, between a single pair of racial others pitted against one another. Such a mutually defining opposition extends to all elements of a colonial text, carving the fictional world into signifiers of Europeanness and non-Europeanness. Nothing escapes this polarizing allegory—not even absence. For example, Van den Heever's *Somer* repeats a common tendency in the South African pastoral to represent only white labor. It has become conventional to read the implicit erasure of Black Africans from the farm landscape as a protective gesture meant to ward off metropolitan critiques that might see the representation of Black labor as a sign of settler sloth or, more damning, of colonial slavery.[53] However, continuing to read Black absence against white presence in this way can only ever return us to the binary relations of colonial allegory.

Colonial allegory is significant for having identified the characteristic ambivalence of the colonial text and the complex ways in which colonial writers are complicit with the reigning imperial ideology. However, now that the influence of postcolonial studies can be felt throughout the humanities, reading practices focused on colonial allegory have become undesirably formulaic. With imperialism having touched nearly every region of the planet, claims to some form or degree of postcoloniality have proliferated. This is at once a sign of postcolonialism's continued importance and of its need to be complicated. By continuing to place a Manichean schema at the center of interpretive practice, postcolonial critics are in danger not only of reinforcing a binary worldview but also of neglecting the many other possible relations that need not be reduced to a single axis of opposition.[54]

One strain of thinking that has emerged to complicate this paradigm is settler colonial theory, which emphasizes a triangular set of population relations rather than a binary colonial "encounter." In place of JanMohamed's Manichean symbolic economy, Lorenzo Veracini introduces a threefold population economy that defines relations between the settler colonizers, the Indigenous colonized, and various nonsettler arrivants (or "exogenous others," as Veracini calls them). Settler colonialism therefore depends on settlers' ability to domesticate and "biopolitically manage" their domains as well as the various populations within them.[55] In contrast to the binary colonial paradigm, then, settler colonial dynamics are determined by at least three polarities: settler–Indigenous relations, settler–exogenous relations, and Indigenous–exogenous relations. For this reason, the kind of hybridity that some postcolonial theorists have celebrated is much less straightforward in the settler colonial context, where not just one but two (or more) dialectical counterpoints are at play. Identity is more strictly managed in the settler colony, where hybrid forms of identification would "disturb the triangular system of relationships ... and ultimately reproduce a dual system where two constitutive categories are mixed without being subsumed."[56] This is why settler colonial regimes manage racially mixed people by subsuming them into one racial category or another, whether by a "one-drop" rule or by some other mechanism. Veracini's notion of a triangular settler colonial population economy suggests that, against the "Manichean" schematic of colonial allegory, we might consider the possibility of a *settler colonial allegory* in which relations other than that between colonizer and colonized are also at play. Not only would this allegorical form resist the sectioning of fictional worlds according to a single opposition; it would also encourage new ways of reading (post)colonial texts that would consider the "exogenous others" that complicate and exceed settler–Indigenous relations.

Specifically, I am interested in how settler colonial allegory would enable reading for ecological realism. Veracini's population economy does not limit the settler colony to three racial groups: different categories of exogenous others might appear within a single context. I therefore see no reason to limit analysis to a strictly "triangular" model with only three types of actors. Nor do I see a reason to limit analysis solely to *human* actors. After all, settlement by definition implies a relation with the land and its flora and fauna, and compelling evidence has been gathered that demonstrates the complex and often negative ways that settler colonial endeavors affect the natural environment. Alfred Crosby describes the "ecological imperialism" perpetuated by colonizing peoples who, either accidentally or deliberately, introduce populations of animals, plants, and microorganisms that precipitate significant transformations in the ecologies of newly colonized regions and often lead to population collapses in the Indigenous human communities.[57] Given the importance of ecological actors in the histories of settlement, it seems necessary to consider them alongside human agents.

I mentioned at the beginning of this chapter that the shift from the American to the southern African context was motivated in large part by the increased visibility of settler colonialism in the latter. My interest in southern African farm novels is also motivated by the proliferation of ecological actors appearing in these texts. Colonial allegory needs to subsume these figures into one of two Manichean categories in order to function. Often, any and all ecological signifiers are taken metonymically as stand-ins for the land itself and, consequently, for the figure of the Black Africans who otherwise remain marginal or absent. In other words, colonial allegorical readings understand ecological actors not as agents in their own right but as mere symptoms of a displaced indigeneity. Settler colonial allegory, by contrast, has more leeway for considering ecological actors at face value because it has less need to see every object as a symptom of something else. Chapter 6 examines this distinction at length. The sound of cicadas that permeates Doris Lessing's novel *The Grass Is Singing* cannot so easily be reduced to a sign of Mary's psychological collapse in the face of her sexual attraction to (and rejection by) Moses, an African laborer on her Rhodesian farm. The overwhelming reality of the cicadas' presence in the novel exceeds any attempt to reduce it to a displaced Africanness. Colonial allegory needs to make this kind of interpretive maneuver in order to reinstate the usual colonial opposition. Settler colonial allegory, however, does not need to make the cicada into something other than what it is. Much like ecological realism, then, settler colonial allegory is strangely situated between surface and symptom. It allows us at once to acknowledge the presence of ecological actors like Lessing's cicadas, and also to recognize that the complexity of the novelistic world both includes and exceeds those actors.

FARM NOVELS AND VISIONS OF THE WORLD WITHOUT US

The remainder of part 2 emphasizes how ecological actors disrupt farmworlds with their uncanny presence, thereby demonstrating that allegory and realism, understood here primarily as *modes* or *strategies* of reading, begin to shear against one another. Specifically, I will trace a strange motif threaded through three English-language farm novels—Olive Schreiner's *The Story of an African Farm* (chapter 5), Doris Lessing's *The Grass Is Singing* (chapter 6), and J. M. Coetzee's *In the Heart of the Country* (excursus 2)—in which uncanny visions of ecological insurgency depict the destruction of the farmworld. With reference to Alan Weisman's popular thought experiment, I call these "world-without-us visions." Like Weisman's speculative project in *The World without Us*, which richly imagines how the natural world would "take back" the earth in the event of humanity's sudden disappearance, these novels envision the ecological reclamation of southern African farms, shattering the illusion of the totalized farmworld.

At first glance, such world-without-us visions appear to replicate what J. M. Coetzee has called the "dream topography" characteristic of southern African landscape poetics, which imagines the country as "a vast, empty, silent space, older than man . . . and destined to be vast, empty, and unchanged long after man has passed from its face."[58] Postcolonial readings have often (and rightly, as the reading of Cather's *The Professor's House* in chapter 2 should indicate) passed judgment on the politics of erasure that underlie the dream topography. In Olive Schreiner's case, Loren Anthony chastises *African Farm* for a moment in which the character Waldo imagines the geological timescale of the veld: "What should be the recounting of a political and historical moment (the displacement of the Bushmen by the Dutch) is transmuted dreamily into a naturalised image of nature as process, timeless and ageless."[59] However, as the next chapter elaborates, the particular circumstances surrounding Waldo's world-without-us vision provide fodder for alternative readings that do interesting political work without resorting to an allegorical mode and its "metaphysical" revelation of hidden meaning. More than simply replacing a violent history of displacement with transcendental lyricism, world-without-us visions in these novels also sponsor unsettling scenes that resist the home(l)y pleasures of landscape metaphysics and destabilize the fantasy of the farm as a self-enclosed entity—"a complete world unto itself," as Stephen Gray writes of Schreiner's farm.[60] Part 2's overarching argument, then, is that these novels signify on the southern African dream topography by reappropriating the image of Timeless Nature as part of a disruptive logic that, like in Lawrence's *St. Mawr*, divorces being from belonging by attending to the universe of things beyond the horizon of human meaning.

Ironically, given his influential critique of the dream topography trope, it is Coetzee who points the way for such a reading in his 1998 memoir *Boyhood*, where he narrates his early sense of the impossibility of owning (though not of inhabiting) the South African veld. In his third-person, quasi-fictional narrative, Coetzee recalls his love for Voëlfontein, his paternal family farm. He describes this love as the freest and most uncomplicated of his life. And yet, "since as far back as he can remember this love has had an edge of pain. He may visit the farm but he will never live there. *The farm is not his home; he will never be more than a guest, an uneasy guest.*"[61] Part of this pain stems from Coetzee's bitter realization that living on the farm requires ties of belonging: "Is there no way of living in the Karoo . . . as he wants to live: without belonging to a family?"⁶² This thought becomes even more unsavory in the next moment when, by virtue of juxtaposition, "belonging to a family" becomes intimately connected to the possession of the farm as a bounded entity: "The farm is huge, so huge that when, on one of their hunts, he and his father come to a fence across the river-bed, and his father announces that they have reached the boundary between Voëlfontein and the next farm, he is taken aback. In his imagination, Voëlfontein is a kingdom in its own right."⁶³ "Belonging" and the sense of possession it entails dismantle young Coetzee's fantasy of the farm as its own endless world. Soon after this scene, the vexed discourse of belonging returns with a different emphasis—one that invites a sense of the uncanny:

> The secret and sacred word that binds him to the farm is *belong*. Out in the veld by himself he can breathe the word aloud: *I belong on the farm.* What he really believes but does not utter, what he keeps to himself for fear that the spell will end, is a different form of the word: *I belong to the farm.* . . . The farm will never belong to him, he will never be more than a visitor: he accepts that. . . . But in his secret heart he knows what the farm in its own way knows too: that Voëlfontein belongs to no one. The farm is greater than any of them. The farm exists from eternity to eternity. When they are all dead, when even the farmhouse has fallen into ruin like the kraals on the hillside, the farm will still be here.⁶⁴

Coetzee taps into the dream topography of South Africa as an eternally "vast, empty, silent space" that spans a geological timeline extending long before and after human presence. But his version of the trope does not simply erase history by naturalizing it as an anomalous bump in a grander "process": Coetzee's revised dream topography recognizes the injustices of history that he learns about in the early chapters of *Boyhood*, albeit in a counterintuitive way. Rather than advocating for the straightforward restitution of lands, Coetzee entertains a more radical solution, one that advocates a politics of absolute nonownership. Such a politics remembers the violence and dispossession perpetrated by the history of Afrikaner settlement, while rejecting the Lockean metaphysics of land-as-property that history enshrined. Crucially, Coetzee's revised dream topography depends on an uncanny awareness of "what the farm in its own way knows": that it belongs "to no one" but itself. The anthropomorphic suggestion of the farm as a knowing entity in itself lends young Coetzee's epiphany an uncanny edge: it renders the home(l)y pleasures of Voëlfontein *unheimlich*, endowed with an unsettling power that lies beyond his control and undermines any desire for a final sense of belonging.

In the chapters that follow, the uncanny reality of ecology begins to dissolve the home(l)y metaphysics of landscape. Throughout part 2 I link the conflict between landscape and ecology to a more "literary-critical" competition playing out in each of these novels between (colonial) allegory and (ecological) realism. Whereas in this study allegory represents a metaphysical reading practice that, in postcolonial criticism at least, tends to see the farm as meaningful only as long as it is, say, a microcosm of the nation, realism offers a speculative mode of reading that attempts to tap into the unsettling ecological realm. The conflict between allegory and realism therefore pits entrenched ideological structures against the de(con)structive force of ecology.

CHAPTER 5

Allegory, Realism, and Uncanny Ecology on Olive Schreiner's African Farm

> Every day the karroo shows us a new wonder sleeping in its teeming bosom.
> —Olive Schreiner, *The Story of an African Farm*

Olive Schreiner's novel *The Story of an African Farm* enjoyed immediate commercial success upon its publication in 1883, initially gaining fame for the way its stark depiction of a world abandoned by God boldly flouts Victorian religious and social mores. Set sometime in the mid- to late nineteenth century against the spare background of the southern African semidesert region known as the Great Karoo, *African Farm* presents a world where the spiritually unmoored have been left to work through their existential turmoil in isolation and cobble together homegrown agnosticisms. The struggles of mind, soul, and body in Schreiner's unshepherded world engender new modes of thinking and being that the novel introduces via its two central characters: a stubborn and rebellious orphan named Lyndall, and an intellectually curious farmer's son named Waldo. Lyndall, who represents the advent of the New Woman, gives voice to a broadly conceived feminist critique of imperialism that implicates all levels of institutionalized patriarchy, from the empire itself to the church and the nuclear family. By contrast, Waldo, who inaugurates the figure of the New Man, sports a raw intellectualism made possible by liberation from the dark obscurity of theology and initiation into the luminous influence of John Stuart Mill, Herbert Spencer, and Ralph Waldo Emerson.

The tension between the novel's stark outlook and its nascent production of new forms of social being is reflected by a certain murkiness in the novel's overall structure and aesthetics. To stay with the matter of character, for instance, even though *African Farm* follows Lyndall and Waldo as they develop from childhood to adulthood, the narrative generally keeps the two characters isolated from one another. Unlike a *Bildungsroman*, which narrates the process of a protagonist's transformation, *African Farm* renders character in a disjunctive fashion that mirrors the book's discontinuous, episodic form. To complicate matters further, Schreiner's novel exhibits significant generic heterogeneity, incorporating elements from romance and the Gothic in addition to sermons, dreams, letters, and allegories, all jostling together uneasily in a "proto-modernist fashion,"[1] yet also ostensibly united within the unique revision of Victorian realism that Schreiner outlines in the novel's preface. Each of these matters—Lyndall's and Waldo's isolation, the novel's fragmentary structure and generic heterogeneity—has left critics divided on the novel's most central preoccupation. Does Lyndall or Waldo represent the true heart of the novel? And relatedly, is *African Farm* most fundamentally concerned with female revolt against patriarchal society, or with intellectual enlightenment in the aftermath of God's abandonment? In each case the novel is, of course, irreducible.

The Story of an African Farm is irreducible in another sense related to Schreiner's self-description as an "English South African" novelist. Literary history has yet to resolve the question of whether Schreiner's work represents an extension of English literary paradigms in South Africa, or if it inaugurates a new South African tradition altogether. Does *African Farm* respond to European tropes of literary pastoralism, or is it better understood vis-à-vis the "land question" that governed the history and politics of the only home Schreiner ever professed to love? Once again, the novel cannot be reduced to one position or the other. *African Farm* is not merely an English novel that has been transported to the Karoo; some form of transculturation has occurred that roots her work firmly in her South African environment.[2] But even as the novel emerges from the South African reality, it also bears the linguistic and ideological marks of a settler culture. The novel is thus truly a hybrid English–South African text.

The in-between status of *African Farm* reflects the in-between status of its author. Throughout her life Schreiner felt straddled between cultures, an experience that mediated her shifting relationship to South African politics and history. From the beginning, Schreiner's life and writing were animated by the anguish of exile, a lifelong sense of alienation that derived from being "a colonial intruder in a foreign land."[3] Although as a white woman she enjoyed a degree of freedom denied to others around her, Schreiner also felt that her whiteness alienated her from Black Africans. Likewise, her early rejection of Christianity exiled her from her family, and her powerful intelligence set her apart from the "niggardly colonial culture" that otherwise stifled her.[4] Schreiner's in-between status is reflected in her shifting relationship to the racial politics of her time. During the Anglo-Boer War, for instance, she passionately defended the Afrikaners against their inhumane treatment at the hands of the British, who in turn detained her for betraying the cause of her own people. Later in life, Schreiner turned against the Afrikaners and the British alike for their equally brutal treatment of Black Africans. Yet throughout these shifting political allegiances, Schreiner always remained the product of colonial society. She therefore occupies an ambivalent position, at once agitating against colonial ideology while occasionally (and likely unknowingly) replicating it.

Linked to the in-between position of both Schreiner and her writing is yet another constitutive ambiguity, in this case related to *African Farm*'s status as the book that launched the farm novel genre in southern African letters. The farm novel genre is an inherently ambivalent one, strung as it is between the twin concerns of the colonial period: the "land question" and the "native question," which, taken together, articulate the racialized problematics of (dis)possession. As outlined in the previous chapter, the farm novel genre uses the white southern African farm as a stage on which to dramatize a variety of land-related concerns, including agricultural best practices, soil fertility, natural disasters, problems of intergenerational inheritance, the fraught nature of land-use economics, the aesthetic compensation of landscape, and so on. More than anything else, though, the southern African farm novel emerges in relation to the sociopolitical question not only of how to divide up the land but also according to what terms to dispossess Indigenous communities, where to relocate them, and how to redistribute capital and labor to ensure the profitability of European settlement. Generally speaking, the farm novel navigates contradictory positions vis-à-vis the land question and the native question by expressing an aesthetic and ideological allegiance either to a pastoral vision of unproblematic belonging, or to an antipastoral vision of existential insecurity. As the previous chapter explored in relation to C. M. van den Heever's *Somer*, Afrikaans-language *plaasromane* of the 1920s and 1930s invoke the aesthetic ideology of pastoralism, attempting against all odds to idealize rural life and ensure a robust sense of Afrikaner belonging. Yet the pastoral vision to which such novels lay claim also clearly comes at the expense of great efforts of disavowal meant to defend the legitimacy of Afrikaner ownership from all threats, whether natural (e.g., storms, droughts), anthropogenic (e.g., land bank, speculators), or metaphysical (e.g., existential uprootedness). The *plaasroman* wears its ambivalence on its proverbial sleeve.

Long before the advent of the Afrikaans *plaasroman,* however, Schreiner had instigated the genre with a farm novel that leaned more toward the aesthetic ideology of antipastoralism. Instead of a prelapsarian world of perfect contentment, Schreiner's *African Farm* portrays a post-theological world of spiritual anguish. The reader senses as much from the very beginning, when the novel opens on a moonlit Karoo landscape that perhaps alienates more than it enchants, exhibiting as it does "a weird and an almost oppressive beauty."[5] Yet Schreiner's inaugural farm novel is also shot through with an ambivalence that, like the later *plaasromane,* holds existential threats in tension with a Romantic idealism. Against the novel's erstwhile antipastoralism, *African Farm* seeks a simultaneous recompense for the loss of God in the metaphysical unity of Nature. Numerous critics have noted that, though Schreiner often uses land to figure her love for South Africa, it also represents a source of melancholy for her. Unable to reconcile these conflicting emotions, she often emphasizes metaphysical pleasure at the expense of the historical injustice. Anne McClintock, for instance, claims that Schreiner sought redemption from her various forms of exile by taking to the desert, where she found a metaphysical solace "projected onto the steadfast immensity of sky and veld."[6] Yet the retreat into her "longing for the infinite" was not innocent: it "also concealed the very real history of colonial plunder that gave her privileged access to this immensity."[7] This ambivalence becomes particularly evident in *African Farm*, which other critics have castigated for its apparent erasure of history through the figure of empty land. Loren Anthony, for example, critiques the novel's metaphysical investment in the land(scape) as a weak attempt to cover up the "overwhelming intransigence of the land issue, which forces a sacralized, pseudo-mystical response to untenable issues in the novel."[8] This is the "spiritual landscape" where, Susan Horton claims, "Schreiner habitually set up residence and felt most at home."[9] As these critics point out, in *African Farm* land represents a site of disavowed historical violence that must remain empty in order for Schreiner (and her characters) to embrace its transcendent potential.

If the ambivalence in Schreiner's representation of land renders her historical and political allegiances murky, then how is the critic to evaluate *African Farm*'s significance as the model that launched the farm novel in southern Africa? Taking note of the novel's irreducibly ambivalent representation of land leads J. M. Coetzee, for one, to question whether *African Farm* qualifies as a farm novel at all. According to him, Schreiner's farm is "too little distinguishable from nature."[10] Yet at the same time as the farm seems indomesticable as raw nature, the bigotry, hypocrisy, and idleness of its human inhabitants lend the farm an urban aura. In this sense, "Schreiner's farm is an unnatural and arbitrary imposition on a doggedly ahistorical landscape."[11] Coetzee thus sees the farm as having a "twofold being," at once as country (natural) and city (unnatural): "These two aspects merely coexist, juxtaposed. They form no synthesis."[12] He ultimately views this unresolved contradiction in the farm's twofold being as a sign of Schreiner's anticolonial vision, one that differs greatly from that of Pauline Smith and C. M. van den Heever, whose farms remain continuous with the venerable Old World farm ideology. Unlike these, Schreiner's farm "is never accepted as home."[13] Influential as Coetzee's reading has been, it does not fully account for the paradox of Schreiner's farm: (anti)colonial ambivalence no longer seems satisfying as a catchall explanation for the novel's challenging ideological murkiness, and postcolonial readings of the novel have not moved beyond this essential position. Instead of being the final word on the subject, Coetzee's reading raises another question: How can Schreiner's farm at once be "too little distinguishable from nature" yet also represent a microcosm of civilization, "a tiny society in the middle of the vastness of nature"?[14] How can the farm be perfectly *natural* and perfectly *unnatural* at the same time?

The sections that follow unfold an alternative reading of *African Farm*'s constitutive ambivalence. This reading links the novel's contradictory treatment of land(scape) and/as Nature to another, perhaps even more constitutive tension between the novel's competing allegiances to allegory and realism. The struggle between allegory and realism plays out most visibly in moments when the novel's metaphysical sympathies come up against the unsettling actuality of the Great Karoo. Yet Schreiner never allows the former fully to yield to the latter. As such, the fragmentary realism she calls into being in the preface to her novel never fully dismantles the allegorical devices meant to unite the text into more singular meaning. Even as other-than-human ecological realities creep in through the cracks of the narrative and give voice to the low hum of an *unheimliche Stimmung,* the metaphysical overtones of a redemptive Nature continue to sing out.

THE METAPHYSICS OF ALLEGORY AND THE STAKES OF REALISM

In his classic work on the subject, Angus Fletcher defines allegory as a symbolic mode that "says one thing and means another."[15] As a procedure of linguistic encoding, allegory fosters a curious doubleness of vision. Such doubleness, we could say, produces a surface that conceals a hidden but symbolically resonant depth. This definition pays heed to the metaphysical origins of allegory as "a human reconstitution of divinely inspired messages, a revealed transcendental language which tries to preserve the remoteness of a properly veiled godhead."[16] Allegory therefore serves both obfuscatory and revelatory purposes. The tension between these conflicting purposes makes allegories "far less often the dull systems that they are reputed to be than they are symbolic power struggles" that complicate the task of interpretation.[17] As Anne McClintock notes, the allegorical mode "both solicits and frustrates the desire for original meaning." This is because, etymologically, allegory establishes an occult relationship between words and things—it purports "to speak in public of other, or secret things."[18]

The understanding of allegory as a metaphysical conceit—one meant (partially) to reveal hidden depths—produces a symbolic structure that is at once disjunctive and unified, irreal and hyperreal. It is worth quoting Fletcher at length on this point:

> [The] visual clarity of allegorical imagery is not normal; it does not coincide with what we experience in daily life. It is much more like the hyperdefinite sight that a drug such as mescaline induces. It is discontinuous, lavish or fragmentary detail. . . . [Furthermore,] its so-called "illustrative" character is more than merely "tacked on" to a moral discussion. Allegorical imagery must be illustrative, because its discontinuous nature does not allow a normal sense world to be created. . . . An allegorical world gives us objects all lined up, as it were, on the frontal plane of a mosaic, each with its own "true," unchanging size and shape. Allegory perhaps has a "reality" of its own, but it is certainly not of the sort that operates in our perceptions of

the physical world. It has an idealizing consistency of thematic content, because, in spite of the visual absurdity of much allegorical imagery, the relations between ideas are under strong logical control.[19]

Allegory produces a distorted version of reality, one that is at once decidedly unrealistic and yet incredibly real. In Fletcher's powerful description, allegory forsakes realistic perspective and instead brings all objects onto the same visual and thematic plane. From a certain point of view, this rejection of naturalistic perspective appears strange and discontinuous. And yet, the "strong logical control" according to which the thematic content of the allegory is organized also functions as a form of unification. A new regime of meaning is consolidated, one that is less "real" but also more "true" than the surface reality.

As this last point suggests, allegory privileges depth over surface in a way that recalls Jameson's symptomatic reading. Yet Fletcher also points out that allegory does not entirely discount the surface. In fact, "The whole point of allegory is that it does not *need* to be read exegetically; it often has a literal level that makes good enough sense all by itself."[20] Even so, an allegory's surface "suggests a peculiar doubleness of intention" that invites interpretation and renders the text "much richer and more interesting."[21] By defining allegory as a surface that "suggests" a hidden depth, Fletcher effectively opens any and all texts to be treated as allegories. Fletcher's treatise appeared in 1964, at the height of structuralist discourse, and the generalization of allegory his book facilitates appears to have become radicalized in the 1970s and 1980s, when poststructuralist critics like Paul de Man located allegory in the critic's own "process of reading in which rhetoric is a disruptive intertwining of trope and persuasion."[22] No longer confined to the likes of *Pilgrim's Progress* or *The Faerie Queene*, it would seem that any text could be understood as an allegory in some looser sense, as long as we are willing always to understand the surface of a text as concealing hidden depths of meaning.

The purpose of this brief excursus on allegory has been, first, to spotlight its metaphysical origins and, second, to suggest that allegory is as much a symbolic mode (that is, a literary figure) as it is a critical decision to "read into" a text. Allegory therefore seems unavoidable; any intimation of meaning is always, as Jameson reminds us, "already an allegory."[23] However, if we recognize that allegorical modes of reading always have a frustrated relationship with a text's surface reality, there is an opening for the critic to shift their perspective and begin not with a gesture toward unified meaning but rather with an investigation of the "broken" text's discontinuous surface.[24]

African Farm enables just such a shift. Schreiner has long been known for her interest in allegory, which she considered the highest mode of symbolic art. She has received both negative and positive attention for this fondness. Many early readers (including Schreiner's own editor, Dan Jacobsen) criticized what they saw as her digressive use of allegory within the context of longer narratives like *African Farm*. As Gerald Monsman notes, however, more recent critics have begun to recognize Schreiner's "adroit use of the symbolic fable, of the religious quest, and of social and political allegory to illuminate the failures of colonialism," particularly in works like *Trooper Peter Halket of Mashonaland,* a short novel that offers a powerful critique of Cecil Rhodes and the violent, profit-driven ideology of his campaign for Rhodesia.[25] These conflicting critical positions repeat the split highlighted above between the discontinuity of allegory as a narrative mode and the totalizing tendency of interpretation.

One useful case for understanding this split is the allegory of the hunter, which first appeared in *African Farm* and was later republished as a self-contained story in Schreiner's 1890 volume *Dreams*. In the novel, a wandering stranger stops at the farm and tells Waldo a story about a man who abandons society to pursue an elusive bird of truth. The hunter spends his life clambering up a precipitous mountain path in search of the bird. In the end, just as he realizes that his own harrowing and apparently fruitless journey has in fact forged a path for other truth seekers, he receives a single feather as a reward. The deployment of such an allegorical digression proves rather complex, at once a complete story in its own right and hence speaking for itself, and at the same time providing a key for allegorical readings of the broader text. Irene Gorak, for instance, suggests that a digression like the hunter's allegory, which is "set like a capstone half-way through the book," may disrupt the narrative, but it also serves to "[crown] the younger character's colonial hopes and terrors in a story of a man who leaves conventional society and its pieties."[26] Formally, then, allegory tethers the narrative together, allowing the reader to fashion a conceptual unity from disparate parts. This function, Gorak concludes, ultimately makes *African Farm* a "test case for the recuperability of allegory in progressive social narrative."[27]

Upon closer inspection, however, allegory has a difficult relationship with "reality" in Gorak's reading. She claims that throughout *African Farm*, "Schreiner empties the world of 'reality' only to saturate it with stories told by alien, potentially treacherous guides."[28] And yet, in order for these self-contained stories and allegories to apply symbolically to the larger narrative, they simultaneously depend on the novel establishing a foundational sense of reality. This claim provides the basis for what Gorak, following the French painter Gustave Coubert, calls a "real allegory," which features "fragments of artifice set in a realistic frame."[29] If "self-conscious art challenges our faith in realistic representation," then real allegory "turns back the challenge," first establishing the order of the "real" to offer "rich opportunities for moral and political digression."[30] Unfortunately, Gorak leaves the nature of real allegory's "reality" obscure. Nor does she clarify how an allegorically inclined critic should understand an allegory's "realistic frame" when otherwise invested in the metaphysical revelation of occult meaning. How is it possible to perceive this realistic frame when the allegorical mode "both solicits and frustrates the desire for original meaning," creating an intrinsic ambiguity that continually "threaten[s] to undermine its [own] intelligibility"?[31] Where does the real reside in a text so richly "intercalated," as Gorak says, with allegorical visions?

To pursue an answer, it is perhaps best to turn to Schreiner's own preface to *African Farm*, which, despite her predilection for the allegorical mode, frames the novel as a realist narrative. Schreiner confesses to her reader that, because her novel deals "with a subject that is far removed from the round of English daily life, it of necessity lacks the charm that hangs about the ideal representation of familiar things" (xxxix). Schreiner goes on to explain that whatever her implied British reader might find unfamiliar in her work could not be made familiar without also being decontextualized and idealized. For Schreiner, such a process of decontextualization and idealization is inherent in the conventional approach to the realist novel. She describes this approach as the "stage method" of representing human life, a method that offers readers an illusion of "completeness" and satisfying predictability: "We know with an immutable certainty that at the right crises each [character] will reappear and act his part, and, when the curtain falls, all will stand before it bowing" (xxxix). According to Schreiner, however, the stage method of conventional realism is profoundly *unrealistic*. It corresponds not with life as it is actually lived, but "with a social evolutionary narrative that attempts to construct the illusion of seamless causality, in which arbitrary, discontinuous, and incidental 'facts' find no place."[32]

In opposition to the stage method of literary realism, Schreiner introduces "the method of the life we all lead." This method frustrates the desire for predictability: "Here, nothing can be prophesied. There is a strange coming and going of feet. Men appear, act and re-act upon each other, and pass away. When the crisis comes the man who would fit it does not return. When the curtain falls no one is ready" (xxxix). Schreiner's realism bears some resemblance to that of Charles Dickens in *Dombey and Son*, which she read while drafting *African Farm* and which, according to Raymond Williams, developed a narrative style that was able to register increasingly complex and arbitrary social relations.[33] Not only does this unconventional realism offer a more realistic representation of the "strange coming and going of feet" characteristic of "the life we all lead," but it does so without submitting real life to the idealizing and unifying strategies normally required to transform it into art and endow it with structured meaning. Schreiner's own method allows for disjointed aesthetics, unresolved conflicts, and a fragmented and episodic narrative structure; she develops a realism that thwarts any attempt to see *African Farm* as a reassuring and satisfying whole.

As Schreiner elaborates further in her preface, *African Farm*'s realism required the author's immersion in and close attention to her immediate surroundings. For this reason, *African Farm* could never have been written far from South Africa's particular reality. Distance gives way to imaginative flights of fancy, which would have reduced Schreiner's novel to little more than a commonplace colonial text, a "history of wild adventure" (xxxix). Such a novel would "best [be] written in Piccadilly or in the Strand," where "the gifts of the creative imagination, untrammelled by contact with any fact, may spread their wings" (xxxix–xl). By contrast, a work such as *African Farm* requires the writer to reject those "brilliant phases and shapes which the imagination sees in far-off lands" and instead "to paint the scenes among which he has grown" (xl). It is only by attending to "what lies before him" that "he will find that *the facts creep in on him*" (xl; emphasis added). Schreiner's realism is thus a homegrown varietal specially cultivated in South African soil; it does not require imaginative flourishes to conjure "brilliant phases and shapes," because the facts of life in South Africa already provide flourish enough. Schreiner's phrase "creep in" is evocative here, at once innocuous and distressing. She seems to suggest that the "facts" of South

African reality settle in gradually over time—they seep into one's consciousness. Yet the notion of *creeping* also spawns a more sinister vibe. Like a horror villain stalking his victim, South African reality creeps up on you, revealing itself just when it may be too late to get away.

For a novel so often studied through the discourse of allegory, what is at stake in reading *African Farm* as inaugurating a creeping, creepy revision of Victorian realism? Deborah Spillman, for one, argues that Schreiner's realism links *African Farm* to Indigenous African aesthetics, and particularly to the aesthetics of the "Bushman" paintings that figure prominently in the novel. While this link casts Schreiner as a "model home-grown artist," Spillman claims that it is also a "sign of [*African Farm*'s] colonial fetishism."[34] By contrast, just as I want to resist reading the novel's allegory solely in connection to its colonial narrative, I also want to resist reading its realism as a form of colonial fetishism. Neither position is entirely satisfying. Indeed, what makes *African Farm* unique is not its exclusive use of allegory *or* realism but rather its manifestation of a deep conflict between them—a conflict that likely animates Coetzee's contradictory (and unresolved) observation that Schreiner's farm is at once perfectly natural and perfectly unnatural. Whereas Coetzee sees this contradiction as a challenge to *African Farm*'s status as a farm novel, I see it as the key to what makes the novel an exemplar of the genre. Through its penchant for allegory and its metaphysical inheritance, the novel emblematizes what the previous chapter called the "farmworld" ideology, which presents the farm as a mystically unified home. The metaphysics of the farmworld emerges in *African Farm* through the figure of Nature, which Schreiner introduces as a transcendental home for those exiled by the disappearance of God. At the same time, however, the allegorical figure of Nature competes with the novel's realism. This realism operates on multiple fronts, by turns fragmenting the novel into discontinuous episodes, infiltrating human narratives with nonhuman points of view, instigating unsettling encounters with strange natural phenomena, and enabling disquieting moments of perceived interobjectivity. In each of these ways, Schreiner's subtly uncanny ecology works to unsettle Nature.

HOME(L)Y METAPHYSICS VERSUS THE CREEPING OF UNCANNY ECOLOGY

Postcolonial scholars often critique Schreiner's treatment of land for its tendency to seek solace in the metaphysical at the expense of history, but their critiques generally neglect to acknowledge a curious slippage between Schreiner's personal investment in land, her literary discourse about landscape, and her philosophy of Nature. This slippage persists throughout all of Schreiner's work, and it exhibits an unstable relationship between her investment in the social, political, and ecological actuality of "real life" and her simultaneous retreat to metaphysical ideals.

For my purposes here, it is Schreiner's collection of political essays, *Thoughts on South Africa*, that most clearly showcases a slippage between land, landscape, and Nature. In the opening chapter, Schreiner attests to the importance of understanding the land, where *land* refers to geographical and ecological features as well as to South Africa's sociopolitical history and the country's unjust land distribution policies. In order to understand "the land," then, Schreiner offers a lengthy survey of South African topography, passing through various ecologies and noting geological, anthropogenic, and historical features. At various points in her survey she stops to conjure the grandeur of different natural scenes, giving the reader something like a typology of South African landscape aesthetics. For Schreiner, however, the true comprehension of South Africa's land requires a further level of abstraction, one that traces a "certain unity" running through the variety of landscapes peculiar to the country. She describes the country's natural unity in terms of "a certain colossal plenitude, a certain large freedom in all its natural proportions."[35] The key to understanding South Africa, then, is not a scientific or historical mapping of its *land*, nor an awareness and appreciation of its various *landscapes*. The issue is Nature writ large: "If Nature here wishes to make a mountain, she runs a range for five hundred miles; if a plain, she levels eighty; if a rock, she tilts five thousand feet of strata on end; our skies are higher and more intensely blue; our waves larger than others; our rivers fiercer. There is nothing measured, small nor petty in South Africa. . . . It is this 'so much' for which the South African yearns when he leaves his native land."[36] Schreiner eventually translates this abstract unity-in-vastness from Nature into the social sphere. She argues that even more important to social cohesion than economic or political unity is a yet more abstract "vital unity" that would allow for "the production of anything great and beautiful by our people as a whole."³⁷ Over the course of the first chapter of *Thoughts on South Africa*, then, the land problem gets transformed, first, into a typology of distinct landscapes, then into an issue of Nature's unity, and, finally, into a matter of social unity. A skeptical reading might venture that Schreiner's dream of metaphysical holism belies an anxiety about the impossibility of any such cohesion. Late in the chapter Schreiner repeatedly deploys a rhetorical strategy whereby she emphasizes various challenges facing South Africa, then roundly refutes any seeming impossibility.³⁸ The oscillation between apparent hopelessness and the rhetorical recuperation of hope suggests an optimism curbed by anxiety—an ambivalent stance that emerges clearly in Schreiner's assumption that, should South Africa ever achieve this higher unity, "no man now living will see the final solution."³⁹ By chapter's end, Nature's metaphysical solace no longer seems a sure bet.

Similar to how the land problem in *Thoughts* transmutes into a question of South Africa's "natural" unity, existential crisis in *African Farm* resolves into the transcendental unity of Nature. Nature performs much of its metaphysical work at the very heart of the novel. In the first chapter of part 2, "Times and Seasons," *African Farm* shifts dramatically in tone as the narrator adopts the inclusive third-person plural "we" and charts how the disenchantment following from Waldo's crisis of faith eventually, though painfully, finds reprieve upon awakening to the deep enchantment of Nature. It is at this point that Schreiner's Nature becomes home(l)y, offering Waldo a way back to himself following his theological exile. The depiction of Nature in this section of *African Farm* adopts elements from two writers of particular importance to Schreiner: the American philosopher-poet Ralph Waldo Emerson, beloved to Schreiner through his *Essays*, and the British polymath Herbert Spencer, whose social Darwinist-inflected *First Principles* she devoured rapturously. Nature in *African Farm* represents a modified Emersonian transcendentalism stripped of its divinity and infused with a secular, more Spencerian sense of reality's absolute unity: "Not a chance jumble," as Schreiner writes, but "a living thing, a *One*" (118).

But just as Schreiner's insistence on spiritual unity in *Thoughts* intimated an unspoken anxiety, so, too, does the resolution of Nature into a miraculous Whole suggest something that haunts *African Farm* at its core. Why, we might ask, does Schreiner need to trade one transcendentalism (God) for another (Nature)? What compensation do the figures of "Universal Unity" and "Universal Life" (260) offer for characters like Waldo and

Lyndall? In spite of the metaphysical recompense furnished by Nature in God's absence, the trade still comes at a cost. To see why, I turn to specific passages from "Times and Seasons," whose lyricism belies shadowy aspects that are otherwise outshone by Nature's enchantment.

Like a miniature *Bildungsroman* of modern humanity, "Times and Seasons" narrates "our" passage from crisis to redemption. Waldo stands in here as the modern everyman. Facing the alienating reality of a universe no longer shepherded by God's presence, Waldo experiences a profound existential crisis. The novel figures this crisis as a disturbing moment of self-recognition. Here, "selfhood" happens to Waldo (and to all of "us") like a scene from some Hegelian horror film in which the self becomes divorced from its surroundings as part of a violent initiation into self-consciousness: "One day we sit there and look up at the blue sky, and down at our fat little knees; and suddenly it strikes us, Who are we? This *I*, what is it? We try to look in upon ourself, and ourself beats back upon ourself. Then we get up in great fear and run home as hard as we can. We can't tell anyone what frightened us. We never quite lose that feeling of *self* again" (103). Self-consciousness comes upon us suddenly and obscurely, and inspires an enduring anxiety. Schreiner captures this anxiety rhetorically in her recursive gesture, "ourself beats back upon ourself," which turns on the uncanny doubling of an alienated self. We could read this strange moment as the novel's existential take on Hegel's *Philosophy of Mind*, in which the individual "reflect[s] itself into itself out of its immersion in the external world."[40] But Schreiner's passage also suggests the possibility for a Hegelianism stripped of the potential for self-knowledge. In the face of self-inquiry ("This *I*, what is it?"), the self may resist—that is, "beat back" upon—itself. Self-consciousness therefore does not resolve the mystery of selfhood or of consciousness; Schreiner refuses to chart a dialectical path toward (self-)transcendence.

Within the narrative logic of "Times and Seasons," Nature appears as a metaphysical balm to alleviate the trauma of existential awakening. What was lost with existential awareness and the crisis of faith is regained through a clarity of vision that no longer sees Nature through a glass darkly:

> From our earliest hour we have been taught that the thought of the heart, the shaping of the rain-cloud, the amount of wool that grows on a sheep's back, the length of a drought, and the growing of the corn, depend on nothing that moves immutable, at the heart of all things; but on the changeable

will of a changeable being, whom our prayers can alter.... Was it possible for us in an instant to see Nature as she is—the flowing vestment of an unchanging reality? When a soul breaks free from the arms of a superstition, bits of the claws and talons break themselves off in him. It is not the work of a day to squeeze them out. (114)

By imagining Christian theology as some fell creature that sinks its talons into the believer and keeps her locked in a false conception of reality, this passage also implies the pain involved in extricating those sunken talons. With the lifting of the theological veil, the revelation of Nature comes all at once, but the shock of its sudden apparition does not fade so swiftly. And the *suddenness* is indeed at issue here: according to Schreiner's imagery, one does not wriggle one's way out of superstition's grip slowly; one must wrest oneself away abruptly and with enough force that the claws break off in the flesh.

According to "Times and Seasons," when one breaks from the talons of theology, one rushes straight into the embrace of Nature and opens one's eyes to her wonders. The turn from God to Nature brings a strange and unanticipated world into focus. The apparent disorder resolves into a miraculous internal order: "And now we turn to Nature. All these years we have lived beside her, and we have never seen her; now we open our eyes and look at her. The rocks have been to us a blur of brown; we bend over them, and the disorganized masses dissolve into a many-coloured, many-shaped, carefully-arranged form of existence.... We have been so blinded by thinking and feeling that we have never seen the world" (116). With its celebratory lyricism, this passage narrates a swift movement away from the pain of theological disillusionment and toward the wonder of Nature's reenchantment and eventually (despite the critique of "thinking") to the miracle of Enlightenment rationality. Indeed, this passage goes on to reveal a number of other "wonder[s] sleeping in [the Karoo's] teeming bosom," including "that wonderful people, the ants," "that smaller people ... who live in the flowers," as well as ground spiders, horned beetles, green flies, and spotted grubs (117). Motivated by a newfound curiosity, we will even adopt a disposition of scientific objectivity that allows us to "look into dead ducks and lambs" and dissect drowned ganders (117–18). Nature offers miraculous wonders, and its mysteries of life and death animate scientific curiosity.

What is significant about these passages is that they move a bit too quickly toward rational illumination, papering a bit too neatly over both the pain of disillusionment and the unsettling possibilities of a Nature that so suddenly reveals itself as having always already been right under "our" noses. If removing the broken-off talons of superstition "is not the work of a day," then neither is coming to terms with a completely new and unfamiliar order of reality. The celebratory embrace of Nature therefore carries with it a hint of repression. As Susan Horton has noted, Schreiner herself often experienced the monochromatic Karoo where she set *African Farm* as a kind of "repressed melancholy." Schreiner's daughter and biographer, Lyndall Gregg, retells her mother's sardonic joke that God had fashioned the Karoo at night, and "when he saw it in the morning, . . . was so angry that He threw stones at it. Anyone who knows the Karoo and those ironstone kopjes," Gregg goes on to explain, "can well understand the deep depression they will induce."[41] The South African–born historian Noël Mostert has also emphasized the repressed melancholy that quite literally hides within the Great Karoo landscape: "an 'arid emptiness' where the fossils of creatures that walked two hundred million years ago 'are embedded now in pale outline in the reddish, bluish and green shales of its mesas.'"[42] In a true Freudian sense, such repression signals an unconscious recognition of Nature's alien reality. Mostert's gesture to the eerie temporality of fossils is surprisingly on point in relation to *African Farm*. Indeed, what the narrator of "Times and Seasons" wishes to see as the Karoo's arid emptiness turns out to be crowded with fossilized creatures: "There on the flat stone, on which we so often have sat to weep and pray, we look down, and see it covered with the fossil footprints of great birds, and the beautiful skeleton of a fish. We have often tried to picture in our mind what the fossilized remains of creatures must be like, and all the while we sat on them" (116). The final sentence subtly belies the tranquility of Nature's beautiful revelation. On the one hand, it plays on the trope woven throughout the novel in which theological blindness gives way to the clear vision of Nature. On the other hand, and despite the narrator's tone of wonder, this sentence offers a literally unsettling image in which the seated, praying believer stands up and turns her eyes from the sky to the ground to see that a more immediate reality has lurked underneath her all along. Nature suddenly appears closer than she ever thought possible, forcing her to trade the dream of history for its physical embodiment in the fossils under her feet.

Although the novel attempts to cast these fossils as specimens of natural beauty, this gesture strikes the reader as a lyrical act of repression that disavows both their strangeness and the suddenness of their unsettling appearance. Allegorical readings of farm novels often interpret such strange phenomena as evocations of the ghost of history—that is, of a kind of recuperative gesture in which the violence of dispossession resurfaces in order symbolically to dispossess the dispossessors. Readings like this tend to revert to politics to resolve narrative weirdness.[43] But something stranger is at work in this context. Here, it is not the ghost of human history that resurfaces. Instead, Nature's revelation represents the ghost of geological *prehistory* erupting into the present. Consider again Mostert's allusion to the repression of the Great Karoo's rich fossil record. Pointing to the Great Karoo's rich fossil record here serves as a reminder of how densely packed with life this apparently empty landscape actually is. Doing so also conjures the inhuman timescales of geology. Jeffrey Jerome Cohen writes, "A fossil is a time traveler and a spark, an interpenetration of epochs," and when we encounter a fossilized specimen, whose organic matter has been slowly colonized by stone over millennia, we feel "conveyed by its historical summons to a perspective in which human relations are not the only measure of reality, to worldly and temporal expansiveness."[44] If the vastness of geological time and the vertiginous "interpenetration of epochs" weren't unsettling enough, close attention to the fossil record reveals other uncanny realities. Take, for example, the discovery of the American mastodon in the nineteenth century. The French paleontologist Georges Cuvier had the fossil remains shipped from New York to Paris, where he could compare them with similarly confounding fossils from elsewhere in the world. His method of comparative paleontology eventually led him to the conclusion that the remains belonged to a creature that had once walked the earth but had since completely died off. Thus was the theory of extinction born. The study of Nature's past therefore offered a strange and upsetting view into a future when any species—humans included—might be wiped off the face of the planet.[45]

Clearly, something obscure and potentially disquieting hides beneath the wondrous surface of Nature, and the narrative of "Times and Seasons" itself seems to repress it. As the episode with the fossils suggests, what hides beneath Nature's surface is the proximal reality of the Great Karoo itself, the unsettling complexity of which threatens to disrupt the reassuring unity of Nature yet never fully succeeds in doing so. Indeed, the narrative transition from God to Nature is too forceful; it disavows any pain involved in such a transfer of sympathies, preferring to bask in the miraculous new transcendentalism of Nature's "Universal Unity."

SCHREINER'S ECOLOGICAL REALISM AND WALDO'S WORLD-WITHOUT-US VISION

The apparent stalemate in *African Farm* between transcendent Nature and unsettling ecological reality reflects the tension described earlier in this chapter: that is, the tension between an allegorical mode that seeks to draw connections across an otherwise disjointed text and organize it into a conceptual unity, and, on the other hand, an uncanny realism that works against this holistic desire by insisting on the strangeness of ecological actuality. At first (and even second) reading, neither allegory nor realism appears to win out. In the end, however, realism has a slight edge on allegory. To see why, I must once again start by turning to some peculiar examples of landscape visions from Schreiner's later work, *Thoughts on South Africa*, for comparison.

As already described above, the opening chapter of *Thoughts* presents a lengthy geographical survey of South Africa. Within this survey, Schreiner offers several "closer" looks at particular environments that feature unsettling encounters with natural phenomena. She describes, for instance, the dramatic compression of distance made possible by the Karoo's vast openness: "In the still, clear air you can see the rocks on a hill ten miles off as if they were beside you; *the stillness is so intense that you can hear the heaving of your own breast.*"[46] The landscape is literally breathtaking, both beautiful and startling: stunning, "intense." In another example, Schreiner includes a brief biological survey, explaining how imitative adaptation has allowed plants and insects to survive in the Karoo's exceptional, arid conditions. She tells of one "curious little plant" with "sharp-pointed green leaves" whose "resemblance to the lichen growing on the rocks, besides which it is always found, is so great, that not till you tread on it, and your foot sinks in, do you discover the deception."[47] Similarly, she describes a "large square insect with hardly any power of flight [that] protects itself by lying motionless on red stones, which it so exactly resembles in color, having even the rough cleavage marks upon it, that it is impossible to detect it, though you know it to be there."[48] In both cases, Schreiner not only expresses wonder at the principles of imitative adaptation at work in the Karoo, but she also intimates the startling effects they have on human travelers passing through: both the prickly plants and the red insects make their otherwise invisible presence known suddenly—and painfully.

In her survey, Schreiner also indulges in more traditional landscape visions. These visions read like traveler's guides meant to help explorers frame an optimal landscape experience. Schreiner's guide-like passages begin by emptying out the landscape, only to reveal it as a "teeming bosom" of life that was never "empty" in the first place. In one instance, detailing how to view the Karoo, Schreiner recommends that the reader set out from "some solitary farmhouse" and travel twenty miles or so, until she can "ride without seeing a living thing, nor passing even a herd of sheep or goats, or a korhaan or mierkat [sic]." With no other creature in sight, Schreiner tells her reader to settle down "in a narrow plain between two low hills" where "the horizon is bounded by a purple mountain thirty miles off." Once the reader has reached an optimal site completely empty of human habitation, she can remove her horse's saddle, take a seat, and enjoy the vastness of the Karoo that stretches to the far horizon. But then, just as the reader has settled in, Schreiner describes a sudden contraction of perspective that retreats from the vast emptiness to reveal the startling proximity of life: "In the red sand at your feet the ants are running to and fro, carrying away the crumbs that may have fallen from your saddle-bag; and in the stillness you can hear your horse break the twigs from the bushes as he feeds; he moves further off, and you cannot hear even that. Then you notice on the red sand a little to the right, at the foot of a Karoo bush, a scaly lizard, with his head raised, and his belly palpitating on the red sand, watching you."[49] Schreiner's overall vision involves a meticulous emptying out of the vast landscape, only to reveal that this apparent emptiness teems with life—and, most importantly, that *this life is right next to you, watching your every move.* As with the view of the far-off stones that drastically compresses distance, this vision and others like it undermine the "imperial eyes" of the traditional landscape viewer by collapsing vast distance (and solitude) into cloying closeness (and intimacy).[50]

In all of these cases, however, despite the sudden revelation of proximity-in-vastness, the unsettling qualities of these phenomena pass, and the traveler swiftly returns to self-reflection. The "oppressive, weird, [and] fantastic"[51] closeness of distant stones; the plants and insects hidden in plain sight; the secretly watchful lizard—all bring the traveler's thoughts back to herself. Schreiner still revels in her solitude despite the disquieting proximity of these ecological strangers. Indeed, she insists that such encounters do not induce anxiety so much as self-awareness: "It is not fear one feels, with that clear, blue sky above one; that which creeps over one is not dread. It was amid such scenes as these, amid such motionless, immeasurable silences, that the Oriental mind first framed its noblest conception of the unknown, the 'I am that I am' of the Hebrew."[52] *Thoughts* retreats to higher metaphysical ground, fleeing uncanny encounter by entering the undiscovered country of the "I." The landscape's vastness empties to become a dwelling for the solitary self.

Like *African Farm*, *Thoughts* seeks a higher home in abstraction. Unlike the novel, however, *Thoughts* offers a more peaceful "return to the self" that eschews any sense of alienation; retreating to the "I" is not only noble but also grounding. In *African Farm*, by contrast, the Hegelian nightmare that decouples the self from the world proves terrifying and painful and must be soothed by the (re)turn to Nature. Crucially, however, in *African Farm* the subject's return to itself does not ultimately solve anything: Lyndall's homegrown feminism leads her to a fatal impasse—pregnant, in ill health, and refusing the warmth of others—whereas Waldo's search for himself in the world at large leads to more profound disenchantment and the conclusion that "the ideal [is] always more beautiful than the real" (226). Far from the transcendentalism of Nature featured in "Times and Seasons," and despite the novel's brief eleventh-hour appeal to "the Divine compensation of Nature" (267), *African Farm* ends with existential tragedy. As Waldo proclaims in his unfinished letter to Lyndall, who dies before he can send it: "Ah, all dreams and lies! No ground anywhere" (259).

Furthermore, although the novel ends with human tragedy, it is the ecological uncanny that has the last word. *African Farm*'s final scene depicts Waldo in contemplation while sitting in the sunshine on the farmhouse *stoep*. As he sits, he reaches out to a group of chickens clucking around him, seeking a moment of connection. The chickens refuse to come to him, and, in this moment of refusal, Waldo dies. These events unfold unbeknownst to Lyndall's cousin, Em, who brings Waldo a glass of milk and finds him in apparent slumber. She places the glass next to him and reenters the house, consoling herself that he will wake soon and will be happy for the milk. The novel ends, however, with a grim subversion of Em's perspective: "But the chickens were wiser" (270). In the novel's final moments, Waldo's death happens like a secret kept not only from Em but also, it seems, from the narrative itself, which insufficiently (or perhaps just ironically) titles this final chapter "Waldo Goes out to Sit in the Sunshine" and which otherwise emphasizes an idea of Nature that enfolds one in the beneficence of a "balmy, restful peacefulness" (266). Waldo does not merge with the landscape or with Nature in this moment, as the novel's final chapter might encourage us to believe. Instead, he dies in disconnection, and in a disturbing rejection of the human desire for intimacy, the chickens wait until Waldo has passed before claiming his body and roosting on his still-warm corpse: "One stood upon his shoulder, and rubbed its little head softly against his black curls; another tried to balance itself on the very edge of the old felt hat. One tiny fellow stood upon his hand, and tried to crow; another had nestled itself down comfortably on the old coat-sleeve, and gone to sleep there" (270).

The chickens' appropriation of Waldo's body could be read as an uncanny realization of a prophetic vision he has near the beginning of the novel, when he speculates with Lyndall and Em about an Indigenous rock painting:

> "Now the Boers have shot them all [i.e., the Bushmen], so that we never see a little yellow face peeping out among the stones." He paused, a dreamy look coming over his face. "And the wild bucks have gone, and those days, and we are here. But we will be gone soon, and only the stones will lie on here, looking at everything like they look now. I know that it is I who am thinking," the fellow added slowly, "but it seems as though it were they who are talking." (16)

Following Coetzee, Loren Anthony has castigated Schreiner for this passage, arguing that it trades the violent reality of displacement for the dream of "the great transcendental sweep of time"—history for Nature. In doing so, Schreiner allows history to speak, but only on the order of the repressed: hence it is the Bushmen speaking through the stones, and thus "the sign of history cannot be fully erased."[53] Though Anthony's postcolonial reading of repressed history's irruption back into the text is powerful, particularly given Waldo's upsetting characterization of the Bushmen as "so small and so ugly" (16), I also think that it sweeps Schreiner's strangeness under the rug. The figure of the talking stones, for instance, is not a momentary reference so easily and singularly connected to the repressed history of displacement. Indeed, Waldo speaks insistently and at length about the stones and the message they bear:

> "Lyndall, has it never seemed to you that the stones *were* talking to you? Sometimes," he added, in a yet lower tone, "I lie under there with my sheep, and it seems that the stones are really speaking—speaking of the old things,

of the time when the strange fishes and animals lived that are turned into stone now, and the lakes were here; and then of the time when the little Bushmen lived here . . . and used to sleep in the wild dog holes, and in the 'sloots,' and eat snakes, and shot the bucks with their poisoned arrows." (15–16; emphasis added)

Rather than speaking of a repressed history extending back only two centuries or so, the stones speak from the inhuman perspective of geological time, endowing Waldo with a vision that stretches long before any human presence and (in the passage quoted above) projects forward to another time after human habitation in the Karoo comes to an end. More than just an image of Nature as a timeless process, this moment indulges in a compelling, even disturbing, vision of what Alan Weisman, in his popular thought experiment, has called "the world without us." Waldo's vision implies more than the erasure of history: it represents a more encompassing recognition of the disparity between the brief timeline of human history and the vast span of geological time. What makes Waldo's world-without-us vision more disturbing is its arrival in a moment of perceived interobjectivity, when he seriously entertains the possibility that he has confused his thoughts with the speech of the stones. Even after admitting the absurdity of the thought, he again asks, "Has it never seemed so to you, Lyndall?" (16).

Waldo's world-without-us vision might be easier to dismiss as an allegory of repressed history if the narrative of *African Farm* did not so frequently adopt nonhuman perspectives. At several points in the novel, the narrative focalization shifts suddenly in and out of animal consciousnesses. Curiously, much like Waldo's passing among the chickens, these moments all occur in circumstances surrounding death. A gray mouse boldly skirts around the room in the quiet that ensues upon Otto's death (62). Waldo's dog, Doss, nervously waits as Lyndall tells her lover she cannot marry him and that she "cannot bear this life" (209), a comment that portends her own death. Although Waldo's interobjective experience with the stones may seem absurd, it is not exceptional in the novel. Indeed, his unsettling vision of Nature's reclamation obscurely prophesies the chickens' appropriation of Waldo's body at novel's end.

And yet, despite this reading's attempt to suture disparate parts of the novel into a unified meaning, the insertion of nonhuman points of view ultimately reasserts the novel's disjunctive structure and aesthetics. From a certain perspective, such flirtation with nonhuman focalization may seem more connective than disruptive, zooming as they do in and out of different points of view in such a way that, as Timothy Morton says of a similar passage in Virginia Woolf's *Mrs. Dalloway*, brings the "environment as such . . . to the fore."[54] But Woolf's careful control of indirect discourse allows her prose to shade smoothly between perspectives within a single sentence. Schreiner's passages, by contrast, emphasize disconnection: they inhabit the minds of these animals at much greater length, and they do so, as I have mentioned, at moments in the narrative when death ruptures connection, or at least threatens to do so. It is in this sense that instances of nonhuman focalization invoke the uncanny effects of Schreiner's ecological realism and shatter Nature's metaphysical unity.

Returning to Waldo's world-without-us vision, what is more important than its potentially disjunctive function in the narrative is its implicit testament to the impossibility of any final human claim to ownership of the Karoo. This is precisely the kind of vision that causes Coetzee's semifictionalized avatar in *Boyhood* to abandon the dream of inheriting the family farm: "Voëlfontein belongs to no one. . . . The farm exists from eternity to eternity."[55] In *Boyhood*, the boy John initially sees Voëlfontein as a self-enclosed "kingdom in its own right," only to realize the unsettling truth that the "farm is not his home; he will never be more than a guest, an uneasy guest."[56] Similarly, any possibility for Schreiner's farm to provide a home is consistently destabilized by the persistent ecological undercurrent that runs throughout the novel, allowing the uncanny force of life to creep in through the narrative cracks—much like Schreiner describes the miraculous teeming of life in the desolate Karoo, where even in "the crevices of the rocks little flowering plants are growing."[57] Although in the end neither the unsettling force of Schreiner's ecological realism nor Waldo's world-without-us vision fully undermines *African Farm*'s metaphysical sympathies or its allegorical predisposition, the novel nevertheless conjures a disruptive ecological realist vision—one that will continue to occupy the English-language farm novel genre into the twentieth century. As the remainder of this book will show, Waldo's vision of the world without us and its prophecy of ecological reclamation returns more forcefully in Doris Lessing's *The Grass Is Singing* (chapter 6) and J. M. Coetzee's *In the Heart of the Country* (excursus 2). Whereas in Schreiner the uncanny work of ecological realism only begins to creep into the narrative, in Lessing and Coetzee it infiltrates with an increasingly apocalyptic vengeance.

CHAPTER 6

Doris Lessing's Ecological Realism

Meanwhile the world churns, bubbles, and ferments.
—Doris Lessing, *A Small Personal Voice*

Doris Lessing spent her childhood on a farm in Banket, a small town in Southern Rhodesia some fifty-five miles from Salisbury in the heart of Mashonaland. Her father, Captain Alfred Tayler, moved his family to southern Africa in 1925 after attending London's Empire Exhibition, which had sold him on the dream of getting rich quick on maize in the far-flung colony. The Taylers arrived in Rhodesia hopeful, and while the rest of his family remained in Salisbury, Alfred set out to find the perfect plot of land. His scouting excursion led him to settle his family in the maize-growing district of Lomagundi in the northeastern part of the colony.

Lessing reflects that soon after her family had landed in southern Africa, it became obvious that the promise of wealth had been a false lure: "[My parents] must have realized by [then] that the enticements of the Empire Exhibition had little to do with reality. Fortunes had been made out of maize during the war, but were not being made now. But maize was what [my father] wanted to grow. And that area was still being 'opened for settlement.'"[1] Yet the impracticality of the situation was not limited to the diminishing returns of colonial maize farming. The particular site her father had chosen was also problematic, perched as it was atop a hill, requiring a team of oxen to transport materials up and down the steep slopes. Adding to the difficulty was the fact that many of the family's "virgin" acres had yet to be cleared of trees and stones. Before cultivation could begin, then, a great deal of labor was required. It took years to make the Tayler farm even partially operational, and although Lessing's parents continued to work the land for the next two decades, it never became profitable.

What the farm lacked in economic and agricultural viability it made up for in picturesqueness. As Lessing recounts, "It was the beauty of the place, that was why my father chose it, and then my mother approved it. . . . a wide sweep of land [that] ended with the Umvukves, or the Great Dyke, where crystalline blues, pinks, purples, mauves, changed with the light all day."[2] No doubt part of the property's visual appeal was its remoteness and the surrounding vacancy. At the time of the family's arrival the Lomagundi region was "very wild and [had] very few people in it."[3] The apparent emptiness of this vast space gave a young Lessing the impression that "'our' land went on indefinitely."[4] Yet despite this imaginative expansiveness, Nature was always very close: "The real bush, the living, working, animal-and-bird-full bush, remained for twenty years, not much affected by us in our house[;] . . . you might startle a duiker or a wild cat or a porcupine only a few yards down from the cleared space."[5] This play of scales—the vastness of space and the proximity of the living bush—contributes to Lessing's sense of her childhood home as an "Eden" imbued with a mythic quality. "Every writer has a myth-country," she writes in her memoir *African Laughter,* and she insists that her myth-country is "the bush I was brought up in, the old house built of earth and grass, the lands around the hill, the animals, the birds."[6]

Considering her youth in Rhodesia and her deep attachment to her southern African surroundings, it is unsurprising that Lessing would have been drawn to the writing of Olive Schreiner. Lessing first read Schreiner at the age of fourteen and immediately felt a kinship with her literary "elder sister."[7] As she records in her afterword to the 1968 edition of *African Farm,* Lessing initially responded to the novel's "magnificent" sense of the African setting. Schreiner's South Africa was the same Africa that Lessing knew intimately from her own childhood and felt was equally "[hers] and everyone's who knows Africa."[8] The immediacy of her identification with the novel and its peculiar mise-en-scène allowed it to "become part of me, as the few rare books do."[9]

Lessing's praise of the magnificence of Schreiner's South Africa—and, moreover, of the home(l)y comfort it provides—may seem surprising given the uncanny ethos of the landscapes in *African Farm,* as explored in the previous chapter. And yet, despite the strange (and estranging) form that Schreiner's realism takes in that novel, Lessing understands well its fidelity to the strange (and estranging) reality of life in the southern African bush. Indeed, when she writes that her family's Rhodesian farm is her "myth-country," she does so with more than romanticized nostalgia. Lessing understands the difficulty and danger of life in the bush. Furthermore, as her many uncompromising critiques of colonial and neocolonial barbarity indicate, she is fully aware of the historical and ideological strata that supported her childhood privilege. Thus, when writing about these experiences, or when recasting them in fictional encounters with the "living bush," Lessing emphasizes that her gesture to the mythic "does not mean something untrue, but a concentration of truth."[10]

Something of this "truth" about the southern African bush appears midway through Lessing's 1962 novel *The Golden Notebook*, where the protagonist, Anna Wulf, recalls a scene from her Rhodesian youth. Walking along a rural road, she and her companions encounter "a festival of insects," replete with butterflies and grasshoppers that "seemed to riot and crawl" after the early-morning rain. While "a million white butterflies with greenish-white wings hovered and lurched" over the low grasses, "on the grass itself and all over the road were a certain species of brightly-coloured grasshopper, in couples. There were millions of them too."[11] In her recollection of this episode, Anna places these species in opposition. The butterflies are "extraordinarily beautiful. As far as we could see, the blue air was graced with white wings. And looking down into a distant vlei, the butterflies were a white glittering haze over green grass."[12] At once distant and fluttering softly, the butterflies have a calming presence. By contrast, Anna and her companions find the riot of mating grasshoppers obscene:

> It was grotesque. . . . In every direction, all around us, were the insects, coupling. One insect, its legs firmly planted on the sand, stood still; while another, apparently identical, was clamped firmly on top of it, so that the one underneath could not move. Or an insect would be trying to climb on top of another, while the one underneath remained still, apparently trying to aid the climber whose earnest or frantic heaves threatened to jerk both over sideways. Or a couple, badly-matched, would right itself and stand waiting while the other fought to resume its position, or another insect, apparently identical, ousted it. But the happy or well-mated insects stood all around us, one above the other, with their bright round idiotic black eyes staring.[13]

In contrast to the butterflies fluttering idyllically in the distance, the grasshoppers appear unsettlingly close; Anna's narration even zooms in on individual couples and their "grotesque" acts of sex and violence. Despite the group's youthful radicalism, the millions of "bright round idiotic black eyes" gazing back at them upset their moral and aesthetic sensibilities: "'Much better watch the butterflies,' said Maryrose, doing so."[14] But in this scene, the orgy of innumerable insects (note the gawping repetition of "millions") affords no escape into romanticism. Paul insists on the truth of the matter: "But my dear Maryrose, . . . you are doubtless imagining in that pretty way of yours that these butterflies are celebrating the joy of life, or simply amusing themselves, but such is not the case. They are merely pursuing vile sex, just like those ever-so-vulgar grasshoppers."[15] Through Paul, Lessing refuses to privilege an aesthetics of prettiness over a grotesque reality.

This episode from *The Golden Notebook* demonstrates how close Lessing's view of southern African reality is to Schreiner's realism. For Lessing, as for Schreiner, the potentially uncanny reality of Nature is linked to the process of existential awakening. And much like her literary forebear, Lessing claims to have "lost religion in a breath" when still a young girl in the bush.[16] Yet unlike Schreiner, who remained hesitant to abandon the transcendental recompense of Nature, Lessing is more willing to accept the unsettling implications of her environmental reality. In *Under My Skin*, for example, Lessing tells the story of a disturbing encounter she had as a young girl with a chameleon in the bush. While pulling apart leaves and closely examining their pores and veins, the sudden appearance of a lizard shocks young Doris: "On the bush is a chameleon. I watch it creep with its slow rocking motion up a branch. And then suddenly . . . I rush screaming up the hill to my mother, sitting in her chair, beside my father, looking out over the bush."[17] Doris leads her mother back to the chameleon, which has moved "quietly a little further up the branch, its eyes swivelling about":

> I am in shock, it is like a dream. I saw the chameleon's insides come out and . . . it happens again, and I scream. "Shhh . . ." says my mother, holding me tight. "It's all right. It is catching flies, can't you see?" I am shuddering with disgust and fear—but with curiosity too. I stand safe inside her firm grasp. "Wait," she says. The club-like tongue of the chameleon darts out, a thick fleshy root, and disappears back inside the chameleon. "Do you see?" says my mother. "It's just its way of feeding itself." I collapse into sobs, and she carries me back up the hill. But I have acquired adult vision; when I see a chameleon, part of my knowledge of it will be that it darts out its enormous thick tongue, but I won't really see it, not really, ever again, not as I saw it the first time.[18]

Two things interest me about this encounter. First is that, even in her retrospective account, Lessing is careful to leave out the detail that initially frightened her and sent her running to her mother. Even when they return to the chameleon, the cause of her terror still remains somewhat obscure: young Doris must first learn that what she initially sees as the chameleon's "insides" is actually just its tongue. The postponement of this detail not only emphasizes its traumatic impact but also speaks to how the child's uncanny experience of the completely alien (at first entirely inexpressible; then, vaguely, "insides") gets revised into banal, everyday knowledge ("tongue"). This leads to the second point: adult vision in all its banality cannot see the "real" chameleon as it actually is. Lessing writes that after this moment she will never *really* see it again. Lessing figures this newfound impossibility as an irreversible side effect of the transition between childhood and adulthood, between the immediacy of experience and the social mediation of knowledge. Socialization, she suggests, offers ready-made categories (here made manifest in the linguistic reduction of "insides" to "tongue") that render the otherwise unfamiliar familiar. Such categories irreversibly program a perceptual gap between ourselves and the alien reality of what Timothy Morton calls "strange strangers." And of course, once we put an entity in a box, we find ourselves just "looking at the box" and "not at the strange strangers" at all.[19]

Unlike Schreiner, Lessing retains the strangeness of the uncanny encounter. Even in the parts of *African Farm* where unsettling phenomena avail themselves, Schreiner's narrative reverts to lyricism, leaving the book awkwardly situated between, on the one hand, her anti-Victorian, proto-modernist aesthetics and, on the other hand, her fidelity to Romantic tropes. Romanticism always wins out for Schreiner. In a letter to Havelock Ellis, for instance, she ventriloquizes Percy Bysshe Shelley to celebrate childlike genius: "Genius does not invent, it perceives!," she writes, before going on to say that this adage "agrees with the true fact . . . that men of genius are always childlike. A child sees everything, looks straight at it, examines it, without any preconceived ideas which hang like a veil between them and the outer world."[20] Like Schreiner's ideal child, Lessing looked straight at that chameleon and examined it without any preconceived ideas. But the encounter fundamentally changed her. Whereas Schreiner insists on the possibility, via genius, of continuing to see Nature as it really is, Lessing's tale of the chameleon admits its impossibility. And yet, though Lessing is fully cognizant both of how the immediately perceivable world comes to be mediated by social and ideological conventions, her story gestures to something that escapes the reduction of the social filter, a reality that remains hidden by the "veil" of human knowledge despite being tantalizingly—if also disturbingly—close. Lessing's continued insistence on a reality that exceeds human perception marks her investment in a form of ecological realism.

LESSING'S REALISMS

The assertion about ecological realism may seem surprising given that Lessing's literary legacy remains most closely associated with her commitment to social realism. The ideological thrust of Lessing's realism can be traced throughout her literary career, from her intellectual awakening in the Communist Party in the 1940s and 1950s to her "psychological" fiction of the 1960s, to her experiments with science fiction in the 1970s and 1980s, and onward. Throughout each of these major periods in her artistic development, Lessing's commitment to social realism reflects her desire to liberate the individual from self-delusion and come to an understanding of how he or she relates to the social totality as a whole. In this sense, penetrating knowledge of the self has a transformational potential for society.

Lessing's commitment to social critique began when she was a young woman in colonial southern Africa. She dropped out of an all-girl Catholic school at the age of thirteen and began her self-education in earnest at fifteen, when, as a nursemaid, she learned about politics and sociology from reading materials furnished by her employer. In 1937, at the age of eighteen, she returned to Salisbury, the capital of Southern Rhodesia (now Harare, Zimbabwe), to work as a telephone operator. After a marriage, two children, and a divorce, Lessing's continued interest in social and political issues drew her to a community organized around a British socialist publishing group, the Left Book Club. By 1942 she had become involved with the Communist Party, and in 1943 she married Gottfried Lessing, one of the founding leaders of the Southern Rhodesian Communist Party.

As she relates in *Under My Skin,* Lessing's relation to Communism was always tenuous. Although drawn by Communism's social vision, the "crudities" of the party's language limited the potency of this vision, reducing it to political theater: "We were playing a role. The play had been written by 'History'—the French Revolution, where the language was first used, and the Russian Revolution—and we were the puppets who mouthed the lines."[21] Lessing was too interested in the complexities of the individual psyche to mouth party pieties. In Gottfried's estimation, his wife's fascination with tales, legends, myths, and fairy stories marked her as more Freudian than Marxist, which in turn made her "unsuitable material as a Communist cadre."[22] They divorced in 1949, and Lessing left Southern Rhodesia the same year with the manuscript of her first major work of fiction—and the one major artifact of her "Communist" period—in hand. *The Grass Is Singing* was published in 1950, one year after Lessing arrived in England. Though the novel reflected a concern with the barbarity of colonial ideology and its entrenched racism and sexism, Lessing's commitment to Communism already was fading: "I joined the Communist Party in, I think, 1951, in London, for reasons which I still don't fully understand, but did not go to meetings and was already a 'dissident,' though the word had not been invented."[23] She left the party five years later and began her literary career in earnest.

The year 1956 marks the beginning of Lessing's "psychological" period, when she began to experiment with literary forms that could support her social vision. In her 1957 essay "The Small Personal Voice," Lessing attests her allegiance to literature's "highest point" in nineteenth-century realism. She privileges the realist novel for its commitment to a common "climate of ethical judgment," a climate that modernism, with its "confusion of standards" and "uncertainty of values," had all but compromised.[24] Realism, by contrast, defines an "art which springs so vigorously and naturally from a strongly-held, though not necessarily intellectually-defined, view of life that it absorbs symbolism."[25] Lessing believed that a strong investment in a particular ethics and worldview would relieve literature from modernism's erstwhile need for a symbolic economy, and thereby support a total social vision that would resist modernist aesthetics of fragmentation.

These themes of fragmentation and unity structure the narrative experiment of Lessing's first major work from this period, 1962's *The Golden Notebook*. This novel oscillates between accounts of Anna Wulf's life as recorded in her variously colored notebooks: black for her experience in southern Africa, red for her political life as a disillusioned Communist, yellow for her semi-autobiographical novel, and blue for her personal diary.

In her 1971 introduction to the novel, Lessing describes the novel's twin themes in terms of "breakdown" and "unity." The very structure of the novel breaks Anna down by compartmentalizing her life and experience, and as various personal, social, and political pressures converge on

Anna and lead to a mental collapse, she attempts to consolidate her sense of self by unifying her separate notebooks into the eponymous golden notebook. Anna's act of self-reunification enables her to move past her political disillusionment and gain a renewed sense of her place in the social totality. This theme of self-consolidation within a total social vision also drives the other major achievements of Lessing's "psychological" period—her epic Children of Violence series (1952–69), as well as her dystopian semi-autobiographical novels *Briefing for a Descent into Hell* (1971) and *Memoirs of a Survivor* (1974).[26]

The psychological emphasis of Lessing's work became more complicated in the late 1960s, when she came under the tutelage of the preeminent authority on Sufism, Idries Shah. Lessing discovered in Sufi thought a spiritual holism that she had found so painfully lacking in Communism. According to Shah, Sufism must not be understood as a rigidly codified system of beliefs; instead, it is a constantly evolving set of spiritual tools meant to assist in gaining an objective perspective on psychological limitations and to develop techniques for transcending them. Paraphrasing Shah in a review of his book *The Sufis*, Lessing explains that Sufism registers "a central transforming force [that] is always at work in the world—the force of evolution itself."[27] Recognizing this evolutionary force has important implications for both the individual and society, a lesson Lessing emphasizes when she uses a quote from Shah's *Caravan of Dreams* as an epigraph to her autobiography *Under My Skin:* "The individual, and groupings of people, have to learn that they cannot reform society in reality . . . unless the individual has learned to locate and allow for the various patterns of coercive institutions . . . which rule him."[28] Shah's words succinctly reflect the thematic concerns of breakdown (of the individual) and unity (of society) that had occupied Lessing in *The Golden Notebook*. And yet, the mystical insight of Sufi wisdom also takes these themes to a place beyond psychological realism. As Lessing herself puts it, and in terms that uncannily echo D. H. Lawrence, Sufism reveals that "Man is not alone; is not a glorious individual—or not in the way he thinks. His 'personality,' what he ordinarily knows of himself, is an assembly of shadows, of conditioned reflexes; his real individuality is hidden and will emerge slowly during the process of learning."[29] The self-knowledge that develops during mystical practices is therefore not merely a matter of understanding a psychologically unified individual but rather a manifestation of a greater, constantly evolving whole: "Sufism, then, emphasises . . . that to want something for oneself before one has reached the stage of being able to see oneself as part of something greater, and which must grow as a whole, is self-defeating."[30]

With her epic series of novels known collectively as *Canopus in Argos: Archives* (1979–83), Lessing turned from "inner space" to "outer space." Lessing began to experiment with science fiction in the late 1970s, attracted to its ability to expand the human mind. Not unlike Sufism's evolutionary perspective, science fiction requires a simultaneous backward- and forward-looking gaze that aims to provide a clear view of the social present. She writes of science fiction writers: "These dazzlers have mapped our world, or worlds, for us, have told us what is going on and in ways no one else has done, have described our nasty present long ago, when it was still the future and the official scientific spokesmen were saying that all manner of things now happening were impossible."[31] Perhaps predictably, Lessing's shift to science fiction dismayed mainstream critics like John Leonard of the *New York Times*, who complained that "[Lessing] now propagandizes on behalf of our insignificance in the cosmic razzmatazz."[32] But just as she'd previously dismissed claims that *The Golden Notebook* propagandized on behalf of women's liberation, Lessing also rejects critics' stereotyping of science fiction: "What they didn't realize was that in science fiction is some of the best social fiction in our time."[33] Indeed, throughout *Canopus in Argos*, Lessing sponsors an intergalactic vision that privileges the social and cultural aspects of future worlds over functional developments in science and technology. Like her inner space fiction, then, Lessing's outer space fiction retains her powerful commitment to social realism.

No matter how apparently speculative her work becomes, Lessing always labors to narrate self-transformation in the interests of social evolution. Such efforts exhibit Lessing's ongoing commitment to a social realism that clearly aligns with that of Georg Lukács, who championed realist fiction as a critical mode for articulating the relationship between individual and society. But if Lessing firmly situates her work within *social realism*, then what can be made of my earlier proposal regarding Lessing's *ecological realism*—a realism that would, of necessity, gesture beyond the *human* social totality?

In seeking to answer this question, it should be recognized that Lessing is a child of the twentieth century, whose unprecedented violence marked her indelibly: "I used to joke that it was the war that had given birth to me, as a defence when weary with the talk about the war that went on—and on—and on. But it was no joke."[34] Born in 1919, Lessing's early childhood played out in the aftermath of World War I, and twenty years later her political coming-of-age unfolded against the background of World War II. Her ongoing struggle to imagine total social transformation thus moves in counterpoint with society's ever-more-total destructive capacity.

This counterpoint emerges most clearly in Lessing's early essay "The Small Personal Voice." As described above, this essay announces Lessing's artistic commitment to literary realism, which, in her words, offers a renewed "climate of ethical judgment" to banish the "uncertainty of values" that haunted modernism. The kind of social commitment Lessing advocates in this essay appears at first to represent an unshakable faith in humanity and its future. And yet, on closer inspection, this commitment follows a negative logic. Rather than insist on any intrinsic, cohesive force that links all of humanity, Lessing stresses that the twentieth century has generated a universal sense of fear and anxiety that unites the species in shared vulnerability. As "the traditional interpreters of dreams and nightmares," artists have a special role to play in bringing this second-order unity to light; indeed, it is their social responsibility to do so, a duty they would be avoiding if they "refused to share in the deep anxieties, terrors, and hopes of human beings everywhere."[35]

For Lessing, who penned this essay in the shadow of World War II, nothing produces a sense of universal vulnerability so much as the atomic bomb. Reflecting on the devastation of Hiroshima and Nagasaki, she writes: "Yesterday, we split the atom. We assaulted that colossal citadel of power, the tiny unit of the substance of the universe."[36] It is here that we can detect the first rift in Lessing's social vision, which she defines negatively against the possibility of universal destruction:

> Everyone in the world now, has moments when he throws down a newspaper, turns off the radio, shuts his ears to the man on the platform, and holds out his hand and looks at it, shaken with terror. . . . We think: the tiny units of the matter of my hand, my flesh, are shared with walls, tables, pavements, trees, flowers, soil . . . and suddenly, and at any moment, a madman may throw a switch, and flesh and soil and leaves may begin to dance together in a flame of destruction. We are all of us made kin with each other and with everything in the world because of the kinship of possible destruction.[37]

This passage operates according to a strange logic. It begins by proposing a unity that goes beyond the social and includes material reality as well, linking animal, vegetable, and mineral. But this unity emerges only negatively as a secondary response to the threat of universal devastation. Rather than being an intrinsic order, "the kinship of possible destruction" is *made*. The social fabrication of this kinship begins with a recognition that the source and the scale of atomic devastation lie absolutely beyond human comprehension: the atomic bomb represents an unsettling encounter between unimaginably minuscule operations of material reality and unimaginably wide-ranging effects of terror and sorrow within the social cosmos. The discontinuity between such a tiny reality and the colossal amount of energy it contains opens up a horrifying gap. Hence, at the same time she proposes a unity between material and social strata, the atomic event that fabricates this kinship also explodes it. Given the catastrophic nature of her example, Lessing's vision of the social totality seems like a weak form of recompense: unity only comes in response to the fear of being blown to bits. As with Schreiner's turn to a unified Nature following God's disappearance, Lessing's turn to social commitment in the face of atomic devastation simply papers over the abyss of terror that continues to yawn despite the soothing gesture toward universal kinship.

This fragile logic of committedness invites us to read Lessing beyond the social realism both she and her critics believe her work to exemplify. As Katherine Fishburn rightly insists, "Doris Lessing has never truly been the realist (we) critics thought her," and the assumption of her commitment to realism "has had the unforeseen consequence of deflecting critical attention away from those very qualities of her fiction that serve to undermine and de(con)struct realistic texts."[38] Fishburn focuses in particular on how Lessing's fiction works to undo the realist marriage plot as well as other "language systems, be they political, social, or mythic," and in light of this dismantling project, she claims that Lessing might be better understood as a "meta-fictionist" and a "meta-physician."[39] While I agree with Fishburn's assessment that Lessing's realism contains the deconstructive power of metafiction, I wish to move in the opposite direction and suggest that Lessing's fiction may in fact sponsor a profoundly antisocial vision—one that does not deconstruct the novelist's realism so much as it undermines her insistence on the metaphysical unity of the *social*.

Appropriately enough, Lessing's antisocial vision has an apocalyptic edge, and while destruction in Lessing is often accompanied by the potential for regeneration, what is intriguing is that such potential rarely promises *human* resurrection. Take, for example, Mother Sugar's comments to Anna near the end of *The Golden Notebook:* "Consider the creative implications of the power locked in the atom! Allow your mind to rest on those first blades of tentative green grass that will poke into the light out of the lava in a million years' time! . . . [It] is possible after all that in order to keep ourselves sane we will have to learn to rely on those blades of grass springing in a million years."[40] In proposing that Anna should imagine far-future regeneration as a meditative practice to soothe her present anxieties, Mother Sugar advocates for a shift of consciousness, but one that moves decisively away from the social cosmos. The continuity of life in her fantasy of destruction resides outside the human sphere, meaning one must "learn to rely on those blades of grass" more than one's own comrades. It is here that the possibility of an ecological realism rises up from below. From the initial crack in Lessing's social realism emerges a single blade of postapocalyptic grass, suggesting the need to think about the future in terms that go beyond the human social totality, which, in itself, is extremely fragile. What this requires is nothing less than a revised ontological perspective. While the language of ontology could not seem further away from Lessing's social(ist) ethos, it is closely linked to the ecological realist vision that emerges in her fantasies of catastrophe.

Nowhere does this link between ecology and ontology feature more powerfully than in Lessing's fiction set in Africa, the primal scene of ontological revelation within Lessing's corpus. Take, for instance, her short story "A Sunrise on the Veld," a tiny parable of ecological realism's uncanny power to reorient one's ontological perspective. The protagonist of this story is a boy who exults in his youth and its intimation of immortality. He rises early one morning and follows a path that leads him beyond the cultivated part of his parents' farm. Alone in the bush, he feels overcome with the "joy of living" and wildly inflates his sense of self: "I contain the world. I can make of it what I want."[41] The boy gives voice to this feeling, singing at the top of his lungs and listening for his song to echo down the river gorge. As he listens, the voice that returns to him undergoes a sinister transformation:

> And for minutes he stood there, shouting and singing and waiting for the lovely eddying sound of the echo; so that his own new strong thoughts came back and washed round his head, as if someone were answering him and encouraging him; till the gorge was full of soft voices clashing back and forth from rock to rock over the river. And then it seemed as if there was a

new voice. He listened, puzzled, for it was not his own. Soon he was leaning forward, all his nerves alert, quite still: somewhere close to him there was a noise that was no joyful bird, nor twinkle of falling water, nor ponderous movement of cattle.... In the deep morning hush that held his future and his past, was a sound of pain, and repeated over and over: it was a kind of shortened scream, as if someone, something, had no breath to scream.[42]

Sobered, the boy follows this other voice until he comes upon what at first appears to be a fantastical creature, a "strange beast that was horned and drunken-legged," whose ragged flesh has raw patches that "were disappearing under moving black and came again elsewhere."[43] As with Lessing's chameleon encounter, it takes a moment for the boy to process what he is actually witnessing: "it *was* a buck." More attentive now, the boy notes that the grass around him is "whispering and alive" and that the "ground was black with ants, great energetic ants that took no notice of him, but hurried and scurried towards the fighting shape, like glistening black water flowing through the grass."[44]

By the time the boy understands what is happening—that the ant swarm is eating the buck alive—it is too late. Just as he picks up his gun to end the buck's suffering, a realization seizes him and swiftly reverses his previous feeling of omnipotence: "All over the bush things like this happen; they happen all the time; this is how life goes on, by living things dying in anguish.... I can't stop it. I can't stop it. There is nothing I can do."[45] For the first time in his life, the boy understands the meaning of fatality—not just *mortality*, that everything dies, but *fatality*, that everything is ultimately helpless in the face of death. This knowledge effects something of an ontological reconstitution of the boy: "It had entered his flesh and his bones and grown in to the furthest corners of his brain and would never leave him."[46]

For my purposes, this story's importance lies not in the boy's facile recognition that this horrific scene represents just "what living is," but rather in the remarkable confluence of affects that links the boy, the ants, and the buck in a web of anxiety, complicity, and biological necessity:

> Suffering, sick, and angry, but also grimly satisfied with his new stoicism, he stood there leaning on his rifle, and watched the seething black mound grow smaller. At his feet, now, were ants trickling back with pink fragments in their mouths, and there was a fresh acid smell in his nostrils. He sternly controlled the uselessly convulsing muscles of his empty stomach, and reminded himself: ants must eat too! At the same time he found that the tears

were streaming down his face, and his clothes were soaked with the sweat of that other creature's pain.[47]

As he witnesses the ants strip the buck to a clean-picked skeleton, the boy at once feels horrified by the brutality of the scene, moved by the buck's suffering, disturbed by his powerlessness to ease his pain, understanding of the ants' need for sustenance, "grimly satisfied with his new stoicism," and newly aware of his own vulnerability. It is this nexus of affects that prevents the boy from reducing his experience to a trite aphorism and—most importantly—gets him thinking. Even at the end of the story, when he tries to brush off the whole experience, the boy admits, "the death of that small animal was a thing that concerned him, and he was by no means finished with it. It lay at the back of his mind uncomfortably. Soon, the very next morning, he would get clear of everybody and go to the bush and think about it."[48]

In "A Sunrise on the Veld," the unsettling voice of ecology speaks through the buck's uncanny scream (*uncanny* because it returns to the boy as the distorted echo of his own voice) and through the eerie susurration of the carnivorous ants in the grass. For the boy in the story, the resonance of this multiform voice is revelatory, transforming his thinking and his sense of being. Such a transformation indicates the power of ecological realism in Lessing's African fiction. Importantly, ecological realism is not the only one at play in this corpus. In other African stories, Lessing explores subject matter more characteristic of social realism, such as colonial ideology and the complex web of race relations.[49] Most interesting for my purposes, however, are those fictions that bring both of Lessing's realisms into play. And no doubt the best example of a text that intermingles both realisms is her first major work of fiction, *The Grass Is Singing*.

GENRE AND NATURE IN *THE GRASS IS SINGING*

Given Lessing's debt to Schreiner, it seems appropriate that her first novel is rather difficult to categorize in terms of genre. Though framed here as a farm novel, like its predecessor, that identification does not account for the several generic transformations that occur throughout the narrative. *The Grass Is Singing* begins, in fact, as a whodunit, with a newspaper clipping announcing a "MURDER MYSTERY." The headline refers to the case of local farmer Dick Turner's wife, Mary, who was murdered at the hands of their Black houseboy, Moses. Yet strangely, at the same time that the novel purports to be a murder mystery, this newspaper report also short-circuits that genre's conventional teleology. Instead of building up to the final reveal of the murderer and his motive, the Special Correspondent already knows "whodunit" and dismisses the need to determine a motive. The clipping reports that "No motive has been discovered," but in truth the white establishment has already judged the event a motiveless crime perpetrated by an unthinking and unfeeling Black African who just happened to be caught stealing from his mistress and reacted badly: "It was thought he was in search of valuables."[50] With the perp in custody and no crime to solve, the mystery hits a dead end on the novel's first page.

Information is clearly missing from the newspaper account. For instance, the clipping fails to mention that Mary and Moses had become entangled in a dialectic of sexual desire and disgust. But salvaging the missing information and revealing the bigger picture requires generic conventions other than those of the murder mystery. At this point, then, *The Grass Is Singing* shifts from a traditional whodunit to a social-realist whodunit. The narrative focus makes a parallel shift from the local newspaper's Special Correspondent to Tony Marston.

Tony is a recent transplant to Southern Rhodesia, who came looking to make it big as a farmer. But farming quickly "lost its glitter to him" (25) as he came to terms with the deep-seated prejudices that govern social life in the colony and realized that he and his fellow newcomers "could not stand out against the society they were joining" (12). Despite the pressure to adopt his peers' bigoted views, however, Tony preserves a sense of justice that refuses to condemn Moses on a whim. With his supposedly naïve perspective intact, Tony becomes a pseudo-detective figure in search of the truth of the event, which he locates in social and historical context: "The important thing, the thing that really mattered, so it seemed to him, was to understand the background" (17). Revealing the "background" takes much more than simply analyzing colonial ideology; it requires understanding how this ideology infects the whole web of social relations, shaping interactions between men and women and between Europeans and Africans. Therefore, and much like Lessing's retrospective account of the chameleon, Tony's investigation must proceed circuitously—and "circuitously it would have to be explained" (17).

The circuitous path of Tony's social-realist whodunit requires the introduction of additional genre conventions—this time from the *Bildungsroman*.

In thinking about Mary's case, Tony "clung obstinately to the belief... that the causes of the murder must be looked for a long way back, and that it was they which were important" (24). The need to examine the details of Mary's personal and social development requires the novel to shift gears yet again, this time leaving Tony behind and adopting a third-person narration that recounts Mary's life story. This story begins with her childhood on a Rhodesian farm, growing up as the only child of an alcoholic father and a mother who succumbed to the stresses of rural life. Despite her depressing origins, Mary is fortunate enough to reach young adulthood during a now-vanished "Golden Age": she receives an education from a good boarding school, works as a personal secretary, lives comfortably in a girls' club, and enjoys an active social life. Yet her extroverted nature belies a rather surface existence in which she spends her spare time either reading sentimental schlock or flirting innocently with young men and refusing their advances. After overhearing gossipmongers speculate about her sexlessness at a dinner party, Mary throws herself into the hunt for a husband. She eventually marries Dick Turner, a struggling farmer still in debt to the Land Bank after five years of profitless cultivation. Dick seems to come right out of the Afrikaans tradition of the *plaasroman*: he exhibits all the characteristics of the farmer as a hardworking *natuurmens*, struggling under debt but still animated by a profound—even ontological—identification with his land: "He worked as only a man possessed by a vision can work, ... *his whole being concentrated on the farm*" (46; emphasis added).

As Mary transitions into country life, the novel makes yet another generic shift, this time to the farm novel. Mary begins her new life on the farm by resolving to adopt her husband's pastoral idealism: "She said to herself, with determination to face it, that she would 'get close to nature.' It was a phrase that took away the edge of her distaste for the veld. '*Getting close to nature*,' which was sanctioned, after all, by the peasant sentimentality of the sort of books she read, was a reassuring abstraction" (50; emphasis added). Faced with returning to the landscape of her childhood, Mary conjures the treacly romanticism of her favorite popular novels as an imaginative homemaking strategy. Yet Mary's longing to "get close to nature" undergoes a series of challenges in the first twelve hours of her life on the farm. After arriving at dusk, Dick vanishes into the house and leaves her alone in the dark. At first comforted by the "the homely sound" of fowls stirring and cackling, her provisional comfort dissipates upon entering the house, a decidedly unhomely space furnished improvisationally with petrol boxes for cupboards and sacking cloth for doors. Dick's house transports Mary back into the farmhouse of her childhood; she senses the presence of her emotionally tortured mother, and she feels "possessed with the thought that her father, from his grave, had sent out his will and forced her back into the kind of life he had made her mother lead" (55–56). Though initially optimistic about her arrival at the farm, Mary now feels overcome with a foreboding sense of being condemned to repeat her mother's desperate and desolate farmwife existence. Morning brings only partial relief; Mary's first daytime view of the farm is at once expansive and confining. The farmland radiates outward in a series of concentric circles of "pale sand," "stretched bush, undulating vleis and ridges, [and] bounded at the horizon by kopjes" (59). Mary feels "closed in" by these circles, but her sense of confinement is outpaced by the landscape's potentially redemptive beauty: "she shaded her eyes and gazed across the vleis, finding it strange and lovely with the dull green foliage, the endless expanses of tawny grass shining gold in the sun, and the vivid arching blue sky" (59). The farm may yet offer some home(l)y comfort.

Mary's new life on the farm initiates a new dialectic in the novel, one that oscillates between her desire for pastoral bliss and the farm's unsettling of that desire. One of the things getting in the way of Mary's attempt to feel at home on Dick's farm is that her point of view remains limited and compartmentalized. However, when Dick gets malaria and she takes up some of his responsibilities, she gains a new, more holistic view of the farm: "She spent several evenings over Dick's books when he was asleep. In the past she had taken no interest in this: it was Dick's affair. But now she was analyzing figures[,] . . . *seeing the farm whole in her mind*. . . . The illness, Dick's enforced seclusion and her enforced activity, *had brought the farm near to her and made it real*" (130; emphasis added). Having a sense of the farm as an agricultural and economic whole—a self-contained entity over which she holds dominion in her husband's stead—gives Mary confidence. And as she becomes more involved with farm affairs, both in the fields and the office, she grows increasingly naturalized to the cyclical temporality of agriculture: "The farm was having the same effect on her that it had had on Dick; she was thinking in terms of the next season" (150).

But as soon as she begins to settle into the rhythms of country life, her sense of the farm's uncanniness returns: "When Mary heard that terrible 'next year' of the struggling farmer, she felt sick. . . . Time, through which she had been living half-consciously, her mind on the future, suddenly lengthened out in front of her. 'Next year' might mean anything. . . . Nothing could change: nothing ever did" (149). The promise of futurity (to the farmer: technological progress, profit-driven longevity) reverts back into the stasis of cyclical time (to Mary: the inescapability of her past, the prolongation of an unchanging present). This revelation wracks Mary with desolation and leads to her first anxiety-induced vision of an eco-apocalypse: "Often in the night she woke and thought of the small brick house, like a frail shell that might crush inwards under the presence of the hostile bush. Often she thought how, if they left this place, one wet fermenting season would swallow the small cleared space, and send the young trees thrusting up from the floor, pushing aside brick and cement, so that in a few months there would be nothing left but heaps of rubble about the trunks of trees" (183). Mary's eco-apocalyptic vision introduces a new kind of problem, one that will return in full force later in the narrative. Given the novel's shifting genre configuration, however, it is not immediately clear how to read such a passage. A conventional reading might suggest that this prophetic experience represents a hysterical response to the stress of rural life—the very stress that sent her mother to her grave. Such a reading makes sense given Mary's evident anxiety about reliving the desperation of her childhood. It also helps explain why it is the cyclical temporality of farm life that precipitates this vision. The sensation of being stuck in a repeating cycle instigates a closed temporal loop—a perverse and self-destructive cyclicality that could be said to drive Mary's vision of the farm turning in on itself and "swallowing" itself into oblivion. This reading neatly situates Mary's vision within a psychological and historical context in a way that helps to explain the events that are yet to unfold. In other words, such a reading fulfills the social-realist mandate laid out by Tony Marston.

In reducing Mary's nightmare vision to an instance of hysteria, however, the conventional reading largely ignores the *content* of this vision, preferring to focus on its psychological structure, or *form*. What, then, can be said about the elements that lie on the "surface" of Mary's experience? For instance, if her prophecy really reflects a deep-seated anxiety about the cyclicality of time and history, then why is it framed in terms of a conflict between the "frail shell" of the brick farmhouse and the "hostile bush" that surrounds it? Why is the vision an eco-apocalyptic one in which the "cleared space" of domesticated farmland is reclaimed by native flora? And what kind of temporality is involved in the image of trees "thrusting up from the floor, pushing aside bricks and cement, so that in a few months there would be nothing left but heaps of rubble"? Is the temporality of the environment's reclamation the same as the cyclical temporality of life on an unproductive farm? The conventional, symptomatic reading neglects to consider any of these questions. Presumably, this is because such questions seem to lie outside the interpretive limits set up by the novel itself, which, from the beginning, organizes a story about the social cosmos of colonial Rhodesia—that is, about *human* relations. To consider the bush *as* the bush or the trees *as* trees therefore seems beside the point, even structurally irrelevant. What it comes down to, then, is an issue of genre and how genre conventions guide our reading. Although the novel has thus far presented itself as a generic chameleon, changing its conventions to suit the needs of the story, in every case—social-realist whodunit, *Bildungsroman,* farm novel—the genre conventions still frame human relations. And is this not just what all literary genres do anyway, "nature writing" included?

What seems more difficult to entertain is the notion that, in a novel where narrative conventions have been so consistently slippery, genre might reveal itself as just another easily dissolvable human construct. Just as metaphysical constructs tend to break down in the face of the mathematics of chaos theory and quantum physics, genre in *The Grass Is Singing* breaks down in the face of the novel's emergent ecological realism. As this chapter's final section shows, Nature is an overdetermined figure in Lessing's novel, always being asked to serve either as a cipher for some repressed aspect of human relations, and race relations in particular, or as a generic ideal, as in Mary's longed-for pastoralism. Either way, Nature remains little more than an allegorical construct, a mask for social phenomena. But when Mary actually "gets close to nature," what she discovers there violates her pastoral idealism. The novel figures the source of this violation most clearly through the incessant and maddening buzz of the cicada. Mary frequently links her sense of oppression on the farm to the cicadas' sonic ubiquity. But the cicada, a constant companion and very *real* presence in southern Africa, refuses to be reduced to a cipher. Whereas Mary (and indeed many readers) attributes symbolic meaning to the cicada's buzzing, it turns out that she is, in fact, tuned in to the uncanny voice of ecology.

ECOLOGY'S UNCANNY VOICE

By the time of her first apocalyptic vision, Mary has strived to establish a sense of home on Dick's farm. She has worked against her sense of claustrophobia in the encircling veld through recuperative landscape visions that feature expansive grasses and lofted skies. She has also worked against her sense of disconnection by developing a holistic view of the farm. But the novel has consistently denied her these home(l)y pleasures through a series of unsettling moments and encounters, and her view of the farm quickly disintegrates from a reassuring sense of wholeness to a terrifying sense of fragmentation.

As the story continues, Mary becomes increasingly entangled in a Manichean push-and-pull of sexual attraction and disgust with her houseboy, Moses, and this colonial dialectic pushes her toward mental breakdown. In a final effort to stave off madness, Mary makes one more attempt to secure a recuperative landscape view. From her bedroom window at dawn, she gazes out upon the farm and briefly revels in a moment of transcendent beauty before the ever-present shrill of the cicada infiltrates it: "The world was a miracle of color, and all for her, all for her! She could have wept with release and lighthearted joy. And then she heard it, that sound she could never bear, the first cicada beginning to shrill somewhere in the trees" (221). Like turpentine sprayed on a canvas, the all-pervading sound of the cicada rapidly dissolves Mary's landscape vision. Colors run into an obscuring haze, and the initial feeling of holistic expansiveness closes in, shrinking the world to the size of her farmhouse prison: "The sky shut down over her, with thick yellowish walls of smoke growing up to meet it. The world was small, shut in a room of heat and haze and light" (222).

The sky and the environs press in, and Mary tries one last time to salvage her sense of peace by imagining herself on the kind of elevated prospect that would offer a quintessential landscape view: "The idea of herself, standing above the house, somewhere on an invisible mountain peak, looking down like a judge on his court, returned" (223–24). But instead of bringing the hoped-for sense of self-consolidation, the view from her imaginary prospect frames a bifurcated vision: "Time taking on the attributes of space, she stood balanced in mid-air, and while she saw Mary Turner rocking in the corner of the sofa, moaning, her fists in her eyes, she saw, too, Mary Turner as she had been, that foolish girl traveling unknowingly to this end" (224). The deflated landscape vision instigates an uncanny double vision that finally shatters Mary's reality, initiating a yet more elaborate vision of her own demise and the destruction of the farm. The very same Nature she had initially hoped to "get close to" spasms in revolt:

> The conflict between her judgment on herself, and her feeling of innocence, of having been propelled by something she did not understand, cracked the wholeness of her vision. She lifted her head, with a startled jerk, thinking only that the trees were pressing in round the house, watching, waiting for the night. When she was gone, she thought, this house would be destroyed. *It would be killed by the bush, which had always hated it, had always stood around it silently, waiting for the moment when it could advance and cover it, for ever, so that nothing remained.* She could see the house, empty, its furnishings rotting. (224–25; emphasis added)

This passage continues at length, describing how rats and beetles would initiate their reclamation, followed by rains that would pound endlessly on the roof and grass that would shoot up through the floors. Branches would break through windowpanes and "the shoulders of trees would press against the brick" (225), pushing until the whole house falls into ruin. After its collapse, "the bush would cover the subsiding mass, and there would be nothing left" (225). Neighbors might look for the hut they knew had stood there, but they would find nothing but partial, decontextualized signs: "a door handle wedged into the crotch of a stem, or a fragment of china in a silt of pebbles. And a little further on, there would be a mound of reddish mud, swathed with rotting thatch like the hair of a dead person, which was all that remained of the Englishman's hut" (226). Architectural and human bodies become conflated, at once sexualized in a bizarre image that juxtaposes the domestic and the wild (the door handle in the crotch of the tree) and weirdly anthropomorphized as a raped and murdered corpse covered in blood-red mud. The sense of sexual violence is palpable: "The house, the store, the chicken-runs, the hut—all gone, nothing left, the bush grown over all! Her mind was filled with green, wet grass, and thrusting bushes. It snapped shut: the vision was gone" (226).

The sexual threat that permeates Mary's apocalyptic vision clearly allegorizes the violence she anticipates from Moses, whom she believes to be lurking in the woods outside, waiting for his chance to rape and perhaps kill her. This, at least, has been the typical critical response to the above passages. As I mentioned earlier, reading the figure of Nature allegorically seems inevitable given that the narrative is a social-realist whodunit turned farm novel that purports to offer a microcosmic view of Southern Rhodesian (and more broadly, of southern African) society. On this view, the apocalyptic vision at novel's end relates, as Jean Marquand argues, to similar visions in T. S. Eliot's *The Waste Land,* from which Lessing gets her title. Like Eliot's vision, Lessing's prophesies the breakdown of the social system: "Lessing sees the farm and by symbolic extension the Southern African system itself as a tottering structure propped up by a slave economy."[51] Accordingly, Mary's emotional breakdown and the vision it sponsors of the farm's devolution is also understood to emerge from the violent play of sexual desire and disgust that undermines the civility of the colonial farmhouse. But such a reading flattens the novel's complexity. Not only does it see Southern Rhodesia as a mere backdrop for the disintegration of the European mind; it also frames the novel in what has become a standard response in postcolonial scholarship, which attends primarily to questions of complicity and sees (unconscious) Manichean allegory as simply reifying white colonial power and dominance.

My own reading of Nature's revolt in Lessing's novel shies away from these flattening tendencies and recognizes that, more than just a backdrop, the Southern Rhodesian environment "is given a *causal* role in inducing the states of the protagonists."[52] As such, my reading echoes Fishburn's observation that Lessing's so-called "realism" does not allow for easy allegorical readings. Yet while Fishburn resists reading the novel as an *unconscious* Manichean allegory, she nevertheless insists on viewing it as a conscious one; that is, she claims that Lessing herself scripts the allegory to suggest how a certain colonial ideology becomes entrenched in the African as much as in the European mind. Hence, Fishburn argues, "rather than murdering Mary because *Lessing* has scripted him as an allegorical dark figure, he murders her . . . because his *society,* which includes Mary, has scripted him as murderous other."[53] In short, the allegory belongs to the metafictional register. In the Lacanian terms privileged by Abdul JanMohamed, on whose work Fishburn draws heavily, *The Grass Is Singing* becomes an "imaginary" rather than a "symbolic" text. The characters are not so much symbols of "good" and "evil," say; rather, their lives play out within an "'imaginary' colonialist discourse" established within and sustained by the novel. The African characters do become symbolically linked to the threatening bush: "for Mary (as other colonials), Africa = native = bush = evil."[54] However, as Fishburn sees it, this imaginary discourse belongs to the world of the novel, from which the author distances herself. That is to say, the threatening wilderness could be read as mere "background" as it is in Joseph Conrad's Congo,[55] but it could also be read as a critique of the symbolic economy that would cause it to be read as such in the first place. In this "metafictional" reading, Fishburn concludes that the novel's real tragedy turns out to be Mary Turner's double bind. On the one hand, Mary honors the social convention of marriage, but her union with Dick Turner compromises her sense of agency. On the other hand, her sexual desire for Moses breaks the racist codes governing white colonial behavior, for which her punishment is madness and death. Tragically, she is "punished both for honoring and breaking illusions."[56]

The problem with Fishburn's reading of *The Grass Is Singing* as a meta-Manichean allegory is that it continues to understand the figure of Nature *only* through its symbolic link to Moses and other dispossessed and maltreated African laborers who are relegated to the novel's background. Postcolonial discourse has scripted such a reading just as inevitably as the (meta-)Manichean allegory has scripted Mary's murder. I therefore propose an alternative reading of Nature "from below," as it were, one that undermines the "metaphysics" of Lessing's metafiction. The issue, simply stated, is that, as an allegorical figure, Nature is entirely overdetermined in this novel. Whether as a symbol for darkness, wildness, evil, or "Africanness" (as in the conventional Manichean allegory), or as part of a colonialist imaginary that makes these symbolic meanings possible (as in the meta-Manichean allegory), or even as an allegorical background symbolizing the psychosexual violence that drives Mary to madness and hysteria, Nature simply has too many roles to play. As a single, homogenized figure, Nature cannot carry the weight of these many allegorical burdens—burdens that are set up as much by the novel as by its critics. What has yet to be entertained is the possibility that Nature appears not as a stand-in at all, *but as itself.*

What does this mean? If, as the previous chapter claimed, allegory and realism were deeply at odds in Olive Schreiner's *The Story of an African Farm*, in Lessing's novel, allegory becomes a kind of social veil that must be lifted in order to disclose reality. This is, of course, an impossible project. As Lessing demonstrates in her account of the chameleon, any encounter with the "real" will always be mediated by social codes, and the only way to suggest even a partial recuperation is to strip back the accumulated layers that tame the strangeness of reality into the banality of common knowledge. This is precisely the effect of Lessing's ecological realism, which attempts to strip back the layers that constitute the allegorical figure of Nature. Although Lessing's ecological realism never fully reveals what lies beneath (again, impossible), it does manage to expose Nature as a mask through which, to echo Lessing's own words, something more mythic may speak with a greater concentration of truth. Specifically, it is the uncanny voice of ecology that we hear speaking indirectly through the social construct of Nature. This claim may seem absurd given that it is through Mary's breakdown that the novel focalizes her eco-apocalyptic vision. However, my argument is not that the natural world surrounding the farm *actually* revolts. Rather, it is Mary's apparent hysteria that endows her with profoundly realist prophetic vision. This vision discloses that what the novel figures as "Nature" does not encompass the totality of all "natural" objects on the farm so much as it represents the unsettling force that has "churned, bubbled, and fermented" beneath the surface of the narrative all along.[57] In other words, Lessing reframes hysteria as a speculative mode that trades Nature's allegory for ecological realism.

This bizarre bait and switch becomes clear in the moments just following Mary's eco-apocalyptic vision. She comes back to herself, standing in her claustrophobic bedroom with all the windows shut. Feeling that "she could not stay in the house," she runs outside and toward the trees that "hated her" (226). This is where she believes Moses is waiting to rape and kill her. What she finds there is not Moses, however, but rather a swarm of cicadas, whose shrill sounds have terrorized her throughout the novel, causing her anxiety and even physical pain. In a strong sense, the cicadas, and not Moses, have been Mary's real antagonists during her time on the farm. Once more Mary finds herself "caught up in a shriek of sound. She opened her eyes again. Straight in front of her was a sapling, its grayish trunk knotted as if it were an old gnarled tree. But they were not knots. Three of those ugly little beetles squatted there, singing away, oblivious of her, of everything, blind to everything but the life-giving sun" (226). This moment presents a very "real" instance of uncanny recognition, as Mary's initial view of the tree resolves into a composite image of clustered cicadas. The encounter sponsors a moment of clear reflection: "She realized, suddenly, standing there, that all those years she had lived in that house . . . she had listened wearily, through the hot dry months, with her nerves prickling, to that terrible shrilling, and had never seen the beetles who made it" (227). Like the moment in Schreiner's *African Farm* where true clarity of vision brings into focus a disturbingly vital world that is—and always *was*—closer than ever imagined,

Mary finally sees what was in front of her all along. The point here is that Mary's flight into the trees contradicts her imaginary fear of Moses and instead brings her near the very real entities that had animated her fear from the beginning. She has gotten her wish: she has finally come "close to nature." But, it bears repeating, the uncanniness of the ecological "real" here is more defined by its power to unsettle than by any qualities intrinsic to it. The cicadas existentially rattle Mary in this moment because, even after she sees them up close and personal, they remain blind and oblivious to her. "Nature" therefore does not rise in revolt in the way Mary prophesies; rather, its uncanny power lies in the cicadas' capacity so violently to shatter her pastoral illusions. As it does for Mary, this state of affairs also provokes an unsettling reversal for the novel's reader. Trained to search for the deep reasons for strange activity on the surface of a text, we find that what lies hidden beneath is, impossibly, right there, out in the open.

But why, if ecological realism prevails, does Moses still end up murdering Mary? Does this act not refute the above reading and tempt the critic to return to an allegorical understanding? Not necessarily. Read in concert with Waldo's world-without-vision from Schreiner's *African Farm*, Mary's eco-apocalyptic prophecy could be read as tapping into a much wider perspective of geological time that registers human habitation as just a blip in the annals of natural history. As a continuation of this trope, which J. M. Coetzee has also termed the South African "dream topography," Mary's vision thus invokes both a past and a future without *any* of us. Many allegorically minded critics of *The Grass Is Singing* have read the farm as a microcosm of southern Africa and Nature (and its revolt) as an allegory for history—that is, as a prophecy of the "ghost of history" being laid to rest followed by the restitution of land to (Black) African hands. My reading, however, suggests that ecological reality's eruption through Nature represents a revolt against all inhabitants who would struggle for ownership. This means that even the oppressed laborers are subject to ecology's uncanny dictates. Perhaps this is why, after killing Mary, Moses retreats to the trees and sits on an ant heap, awaiting his arrest. Rather than being symbolically equated to the "evil bush," Moses, like Mary and all the rest, merely does the *bush's* "evil" bidding.

EXCURSUS II

Exo-Phenomenology

> I would have been far happier under a bush born in a parcel of eggs, bursting my shell in unison with a thousand sisters and invading the world in an army of chopping mandibles.
> —J. M. Coetzee, *In the Heart of the Country*

This book began by investigating the home(l)y discourse that infiltrates eco-phenomenology, both obviously, as in the *oikos* of the *eco*, and less obviously, as in the trope of "return" that eco-phenomenologists adopt from their twentieth-century predecessors. In an attempt to interrogate and ultimately shift away from the discourse of home, part 1 progressively moved toward an uncanny ecology, beginning with an analysis of how the phenomenology of landscape as dwelling is bound up with the coloniality of Nature, and concluding with an unsettling alternative to conventional landscape thinking that decouples being from belonging. The chapters of part 2 applied the lessons learned in part 1 to the southern African context, examining how the trope of the European farm as the mythologized locus of white being-in and belonging-to the settler nation falls prey to increasingly unsettling ecological realisms that demand new forms of reading. The overall arc of the book has thus worked toward a mode of ecological thinking and perceiving that registers the residual biases of the settler colonial imagination.

Yet the analysis of Doris Lessing's realism(s) in the previous chapter has led this argument to a surprising, even politically troubling impasse that persists between the social and the ecological. Recall that Lessing opens *The Grass Is Singing* with Tony Marston, a young white man who wants to determine why a Black servant murdered a white farmwife. Tony invokes social realism as the mode best suited to understanding what led to the tragic event. His invocation occasions a narrative shift that leaves Tony behind and focuses on the life and experiences of the murdered woman. But once it shifts focus to Mary Turner, the narrative never returns to Tony to close the frame. Instead, the novel opens onto a series of increasingly disturbing eco-apocalyptic visions that complicate the novel's erstwhile social-realist orientation. *The Grass Is Singing* thereby marks out a troubling incompatibility between the social and the ecological. Not unlike Olive Schreiner's *The Story of an African Farm*, Lessing's novel slips uneasily from the narrow world of human concern (the "social") to a broader realm that eclipses human affairs (the "ecological") and threatens to brush social, political, and historical affairs aside. But whereas in Schreiner the retreat to Nature enables an escape from existential terror, in Lessing Nature itself invokes terror. Yet to suggest, as I have, that the "evil bush" made Moses kill Mary will no doubt ruffle the feathers of any critic committed to social justice. How can we justify attending to the ecological at the expense of the social, particularly in settler colonial and other contexts marked by ongoing histories of violence and inequity? But by the same token, how can we think the social without recognizing the extent to which all social life unfolds in relation to material contexts that both encompass and exceed the social? The impasse to which Lessing leads us is one that remains an urgent question for contemporary environmental thinking.

To reframe the matter slightly, Lessing's novel highlights the difficulty that arises from emphasizing complicity over contingency. Postcolonial thinking has widely adopted the term *complicity* to reference direct and, especially, indirect forms of involvement in the social, political, and historical violence of coloniality. Settler colonial positionality is always marked by complicity, even when resisting the ideology of the settler state. White writers such as those investigated here "have always been complicit in colonialism's territorial appropriation of land, and voice, and agency, and this has been their inescapable condition even at those moments when they have promulgated their most strident and most spectacular figures of post-colonial resistance."[1] A reading of *The Grass Is Singing* that emphasizes complicity might note that Tony Marston—a newcomer to colonial Rhodesia who, much like Lessing's own family, came to seek his fortune—is complicit in Mary's murder even as he resists normative colonial(ist) views and seeks a deeper truth.

Whereas the discourse of complicity remains rooted in social concerns, *contingency* traverses both the social and the ecological. In the most general sense, contingency is a marker of uncertainty as well as possibility. As a noun, the word refers to a future event or condition that may come to pass, but that cannot be predicted with certainty. Its adjectival form means being subject to or dependent upon chance. Broader in its formulation than complicity, contingency allows for the recognition of a multiplicity of ideological forces that converge on an individual. Yet contingency is also a profoundly ecological principle asserting that evolution follows paths constrained by random historical events. Such events might be part of the human historical record (e.g., meteorites, climate change, urban development), though more often they are hidden from human view, squirreled away in genetic material and splayed out along timescales imperceptible to human experience (e.g., genetic drift, potentiating mutations, population bottlenecks, adaptation, exaption). Alternatively, we might think contingency in terms of the various affect-producing influences that Jane Bennett, following Walt Whitman, has discussed under the term *influx:* that which can "face, flow into, and alter" the human "dividual," in turn leading to a range of thoughts, actions, or incipient stirrings that she terms (again following Whitman) *efflux.* Influx and efflux do not exist in a linear, cause-and-effect relationship; the input conditions a range of possible outputs, which subsequently influence the conditions of inputs. Bennett writes that influences "inhabit and deform the grammatical place of the doer" at the same time as they "release them[selves] from the confinement of being merely the 'context' or 'material conditions' that undergird exclusively human powers of action."[2] The biologist Merlin Sheldrake offers a potent example of such influence when he writes of our "total dependence on fungi" and speculates that "we might dance to their tune more often than we realize."[3] Taken to its extreme, the two-way logic of influx and efflux requires a contingency-based rethinking of all forms of agency. Vicky Kirby indicates as much in her quantum-inflected reading of ecology. She suggests that if by *ecology* we are referring to the dynamism that emerges from the entanglement of space, time, and matter, then every instance of agency (human and other-than-human alike) is in fact an expression of that ever-emergent dynamism.[4] Like others who work on quantum ecologies, such as Karen Barad, Kirby supports a profoundly contingent understanding of "spacetimemattering" as the quantum maker of entangled worlds.

At the very least, the modalities of contingency sketched above present an essential set of tools for thinking beyond the Manichean binaries that dominate (post)colonial discourse. Indeed, they provide powerful ways to think both intrahuman and interspecies relations—that is, the "social" and the "ecological"—together. By no means do I wish to indicate that contingency mediates between the social and the ecological in a way that might ultimately reconcile the two. Thinking contingency may bring social and ecological discourses together, but it does not work out a consensus between them. Thus, contingency enables a more plural mode of thinking, but it does not make it any easier to sort through the complexities and contradictions that inevitably arise. The shift from complicity to contingency leads into very strange, unstable territory whose difficulty to navigate exposes precarious ideological foundations and unveils a need continuously to (re)think politics and ethics beyond the human.

What, then, of eco-phenomenology? After the introduction's initial investigation, eco-phenomenology receded into the background, at once subsumed into and complicated by the increasingly unconventional readings of the foregoing chapters. On the surface of things eco-phenomenology appears to offer an ideal companion to contingency thinking. After all, in its broadest understanding eco-phenomenology attends to relations with those human and nonhuman others that (who) appear within and constitute our individual perceptual worlds. Eco-phenomenologists encourage us to develop our sensuous, somatic relations with the "more-than-human world."[5] Such a practice enables ever-more-nuanced perceptual engagements that reveal just how fully entwined our bodies are with the flesh of the world. We learn to feel how touching always entails being touched; how when we breathe the air, the air "breathes" us back. As the ever-increasing demands of modern life colonize our somatic experiences and perceptions of self and world, phenomenological practices still have a role to play in salvaging a critical sensorium. Yet as the introduction indicated, eco-phenomenology does not yet rigorously think the "ecological" body in relation to the "social" body. Without retroactively learning the lessons on offer from later, more critical environmental philosophies like eco-hermeneutics and eco-deconstruction, any newfangled eco-phenomenology will fall short of the contingency thinking to which I appeal here. Furthermore, we must address the various ways in which the discourse of ecological homecoming continuously reasserts itself even in the present. Any environmental philosophy that deploys the direct or indirect language of *going back (to Nature)* will always retain unrecognized and hence uneasy links with the coloniality of Nature.

So a question arises. If eco-phenomenology is not adequate for thinking contingency, is there another phenomenology that might be?

In concluding, I adopt an admittedly speculative mode to propose an alternative I call *exo-phenomenology*. The chief aim of exo-phenomenology is to trouble all forms of home(l)y rhetoric that continue to influence environmental discourse. The practitioner of this speculative mode is keen to leave behind the idea of ecology as primordial dwelling, and to denaturalize the range of normative ideologies to which that dwelling plays host. In order to accomplish these goals, the exo-phenomenologist understands the need to reeducate their sensorium, attune their bodymind to as-yet alien phenomena, and thereby push the boundaries of ordinary perception. This reeducation does more than just orient the perceiver toward the beauty of Nature, or even toward subtle changes in the biosphere. It also allows them to practice accessing otherwise unsensed affective regimes—both ideological and material—that condition their existence in a given context.

For the settler, whose status *as* settler generally goes unmarked, exo-phenomenology names a perceptual practice aimed at learning what settler coloniality looks and feels like. Important as it may be to know the history of settler occupation in a particular place, the exo-phenomenologist understands that abstract knowledge won't be enough to disrupt the way settler worlds construct themselves from the wreckage of other worlds. Undoing normative modes of settler orientation will require a conscious effort to develop new modes of perception that actively unsettle the comfortable and often unreflective sensation of feeling at home in the world. As one part of this effort, the exo-phenomenologist disavows normative environmental tropes like wholeness, balance, and purity, and instead courts perceptual experiences that honor the realities of fragmentation, dynamic flux, and the impure muckiness of being. Engaged phenomenological curiosity yields a visceral understanding of the body as less of a concretely bound physical entity, and more of an ephemeral resonating-together that persists for a short time before dissolving back into the swirling universe of energetic and material flows.[6] The exo-phenomenologist also accepts the incurably uneven and compromised nature of all relations. Instead of seeking out a feel-good ethics that suppresses discomfort and implicitly disregards responsibility for anyone and anything unilluminated by the hearth's cozy firelight, the exo-phenomenologist cultivates an embodied reckoning with the competing moral obligations from near and far that can only ever be answered partially and imperfectly. No longer sequestered in the *oikos* of the *eco*, the exo-phenomenologist ventures into the great outdoors, not to *face down* but simply to *face* the many known and unknown others that (who) thrive beyond the limits of our conventional phenomenal enclosure.

As the prefix *exo-* indicates, exo-phenomenology recognizes that we only have access to the *outsides* of things, whether that's the physical surfaces of bodies and objects or the figurative surfaces of language and signs.[7] In this sense exo-phenomenology resonates with Emmanuel Levinas's phenomenology of exteriority, which revises Husserl's understanding of transcendence so that it no longer refers to those internal structures that constitute sensible intuitions in and for consciousness, but rather to that which comes from outside the self: the absolute Other (*l'Autrui*). For Levinas, it's in the very nature of the self to allow itself to be alienated—that is, rendered alien—by the Other. The absolute otherness of the absolute Other resists any attempt I make to assimilate it or appropriate it to my own understanding. Coming face-to-face with such a stranger wrenches me from the inwardness of my own dwelling. The encounter necessitates that I reckon with the Other's radical difference; it also forces me to recognize that, though unbroachable, that difference is always already quite near to me. As Levinas puts it: "Neither possession nor the unity of number nor the unity of concepts link me to the Stranger [*l'Etranger*], the Stranger who disturbs the being at home with oneself [*le chez soi*]."[8]

The Other disturbs my being at home with myself not simply by ringing my doorbell and asking to be received, but by always already inhabiting my house. This is what I take Levinas to mean when he writes, "I welcome the Other who presents himself in my home by opening my home to him."[9] Levinasian ethics doesn't simply require me to welcome the Other into my home; it requires me to welcome the paradoxical fact that the Other is already *inside* my house (*dans ma maison*) when I welcome them in. Echoing the ambiguity of the French word *hôte*, which means both "host" and "guest," the Other's perennial proximity renders the terms of hospitality fundamentally unstable. Welcoming the Other into my home as a "guest" entails surrendering my status as "host," effectively becoming a guest in my own home. Far from envisioning a home invasion, Levinas demonstrates how the Other calls me into response-ability by alienating me from the inwardness of my being at home with myself. This alienation disturbs the rootedness of Heideggerian dwelling and, by infusing it with the presence of the absolutely Other, transforms the home into an inn, a waystation for the world's exiles.[10]

Exo-phenomenology attends to the otherness of the Other. But whereas Levinas proved ambiguous on the question of whether his "phenomenology of the face" could extend to other-than-human Others, the exo-phenomenologist answers unambiguously in the affirmative.[11] Exo-phenomenology bypasses the difficulties attendant on determining just what kinds of other-than-human entities have "faces" or can "be faced" in a relationship of ethical accountability. It instead understands *faciality* as a cipher for any manifestation of otherness that calls us out to respond. Openness to the alterity of human and other-than-human "faces" enables a connection to what John Llewelyn has termed "blank ecology," which aims to take any and all "existents" at face value: "defer[ring] consideration of the predicates of things" and "allowing ourselves to be struck by the consideration that for any given existent its existence is a good, at least for that existent."[12] Llewelyn's blank ecology and its emphasis on taking existents at face value in turn resonates with James Gibson's ecological realism and its emphasis on the primacy of information over the mediating apparatus of sensation. Such modes of thinking actively remind us that others aren't just *in* our environment; they *are* our environment.

Though exo-phenomenology shares Levinas's attention to exteriority, it is also necessarily weirder than Levinas in that it reckons with the otherness of interiority. The exo-phenomenologist understands that, however inaccessible to consciousness, it's what's "inside" that counts. As such, they are curious how other beings—both human and other-than-human, organic and artificial—experience the sensation of sentience that Daniel Heller-Roazen calls "the inner touch."[13] Exo-phenomenology's interest in the inner worlds of others links it to what Ian Bogost calls "alien phenomenology," a philosophical project emphasizing the speculative contemplation of how nonhumans experience interiority, or as Bogost says, "what it's like to be a thing." Alien phenomenology describes a critical process for speculating about object perceptions of object worlds—not how human subjects perceive objects, *but how objects perceive other objects*. Bogost's project extends Thomas Nagel's famous 1974 essay "What Is It Like to Be a Bat?," which concludes that, since bat sonar is unlike any mode of perception humans possess, the best we can do is imagine what it *might be like*—a far cry from actually *being* a bat. Nagel's observation highlights the seemingly obvious difference between the internal character of an experience and how an outside observer might imaginatively approximate such an experience. Since alien phenomenology deals with the latter rather than the former, it remains bound by human epistemological limits. And as the phrase "what it's *like* to be a thing" indicates, alien phenomenology operates substitutionally, through simile and metaphor. Regarding this point, Bogost draws on Graham Harman's observation that relations between objects take place "not just *like* metaphor but *as* metaphor."[14] Metaphor thus becomes a kind of strategic essentialism aimed at forging impartial connection with and fragmentary understanding of others. Like Bogost's alien phenomenology, exo-phenomenology sponsors an engaged perceptual curiosity about how our various human and nonhuman others experience their particular worlds.

Crucially, exo-phenomenology is also about exteriorizing one's relation to oneself. It entails recognizing one's own status as an alien and reckoning with the multiplicity of aliens within, but without pursuing a comforting *nostos* to some primordially consolidated self. This is precisely the kind of self-alienating exteriorization Mary Turner experiences in *The Grass Is Singing* when she attempts to gather the farm into a totalizing landscape vision. Whereas for Mary the experience of being exterior to herself transforms her landscape vision into an eco-apocalyptic nightmare, for the exo-phenomenologist such an experience represents a profoundly ecological vision of contingency, one that only initially appears terrifying, and only because it violently displaces an unfounded idealism. Exo-phenomenology must therefore engage with place without needing place to be imaginatively (re)coded as a home(l)y space of cozy self-consolidation. Like D. H. Lawrence in New Mexico, the exo-phenomenologist allows themselves to experience the unsettling vitality of being without belonging. The exo-phenomenologist becomes an *alien* in place, entangled in the vast complexity of human and other-than-human relations, fully alive and yet also refusing to ground that aliveness in a home(l)y metaphysics. This is what uncanny ecological experiences can offer. As Natalie Loveless writes, uncanny experience "invokes not only the ambivalent experience of something simultaneously placing and displacing, but the *conjunction* of these positions—alien, native—in the formation of (psychoanalytic) identity."[15] Advantageously, such uncanny experience can also incite a productive curiosity: "The uncanny instigates a (curious) drive that hovers at the intersection of knowing and not knowing, belonging and not. You can't be curious about something you already know, but you need to know something about it in order to be curious."[16] And where might we find a model for such exo-phenomenological curiosity? *Enter* Magda.

MAGDA THE EXO-PHENOMENOLOGIST

One last, and perhaps unexpected, figure from literature offers an intriguing model for what an exo-phenomenology grounded in contingency thinking might look like. I speak here of Magda, the protagonist of J. M. Coetzee's experimental farm novel *In the Heart of the Country* (1977). Coetzee's novel tells of an Afrikaner father and daughter and their two Black servants, Hendrick and Anna, all of whom live emotionally tortured lives on a farm located "in the heart of nowhere."[17] The bizarre and disturbing events of the novel are narrated by the daughter, Magda, an intelligent but tormented woman who first introduces herself as an absence—"a zero, null, a vacuum towards which all collapses inward" (2). Already in the novel's opening moments, Magda announces herself as the uncannily absent presence at the "heart" of the story, a figure whose black-hole-like singularity renders her a "nothingness" within the "nowhere" of the farm.

Magda's initial self-identification as "a zero, null, a vacuum" may initially strike us as a nihilistic and despairing reflection of an oppressive Afrikaner patriarchy, which reduces a woman to "a hole with a body draped around it" (41). But it may also be read as a sly ecofeminist critique of the dream of a holistic Nature (within which, of course, the Afrikaner *boervrou* "naturally" submits to her husband, the *natuurmens*). Recall that Timothy Morton's notion of the "mesh" of ecology has many holes in it, since "interconnection implies separateness and difference."[18] The image of meshwork further implies that, unlike in the Whole of Nature where every entity "has its proper place within the organic totality and is defined by its relation to all others,"[19] ecological contingency registers that connections can be partial,[20] and that objects can withdraw from relation.[21] Contingency is thus the opposite of holism, a lesson Magda reiterates both early in the novel (e.g., "I am a being with a hole inside me" [10]) and late (e.g., "perhaps in stones there are also holes we have never discovered" [115]). Magda's insistence on her own incompleteness arises most profoundly in her intriguing view of herself as "a straw woman, a scarecrow, not too tightly stuffed, with a scowl painted on my face to scare the crows and in my centre a hollow, a space which the field mice could use if they were very clever" (41). In imagining herself as a hollow woman, Magda offers a strangely ecological revision of T. S. Eliot's poem "The Hollow Men." Eliot's "stuffed men" speak in "dried voices" that "Are quiet and meaningless / As wind in dry grass / or rats' feet over broken glass / In our dry cellar."[22] Whereas Eliot juxtaposes objects in a logic of disconnection, Magda embodies the principle of interconnection. Her "hollowness" also isn't "meaningless," as it is for Eliot's hollow men: her yonic self-description as a "hole" opens a figurative, womblike space for potential inhabitation. Finally, unlike Eliot's rats, afloat in the isolated space of the poetic line, Magda's invites her field mice inside to build a home. This revision of Eliot raises incompleteness to the level of a positive feminine ecological principle. In contrast to an earlier image of herself as imprisoned in her socially constructed and gendered body, here Magda entertains "quite another sense" of herself: a "protectrix of a vacant inner space" that "glimmer[s] tentatively in [her] inner darkness" (41).

The new sense of being Magda seeks to unveil—a sense of being that is shrouded in obscurity, ephemeral as a shimmer, and decidedly feminine—is a profoundly contingent one, and Magda pursues this mode of thinking and perceiving further by learning to notice the holes in her own ways of seeing, knowing, and imagining. Consider the scene where Magda conjures the world-building power of her imagination. After Hendrik returns from a distant village with a new wife, Magda boasts that, even without having visited the village, she can (re)construct it fully in her mind:

> I have never been to Armoede, but with no effort at all . . . I can bring to life the bleak windswept hill, the iron shanties with hessian in the doorways, the chickens, doomed, scratching in the dust, the cold snot-nosed children toiling back from the dam with buckets of water, the same chickens scattering now before the donkey-cart in which Hendrick bears away his child-bride, bashful, kerchiefed, while the six dowry-goats nuzzle the thorns and watch through their yellow eyes a scene in its plenitude forever unknowable to me, the thorn-bushes, the midden, the chickens, the children scampering behind the cart, all held in a unity under the sun, innocent, but to me only names, names, names. There is no doubt about it, what keeps me going (see the tears roll down the slopes of my nose, only metaphysics keeps them from falling on the page, I weep for that lost innocence, mine and mankind's) is my determination, my iron determination, my iron intractable risible determination to burst through the screen of names into the goats-eye view of Armoede and the stone desert, to name only these, in despite of all the philosophers have said. (17–18)

To demonstrate her imaginative bravado, Magda conjures Armoede by listing a number of elements that might populate the village, all of which she envisions to be "held in a unity under the sun." However, she quickly admits to herself that this "unity" is really just a jumble of objects held together by nothing more than her mind and organized in a list ("names, names, names") for her own convenience and amusement. Intriguingly, in the midst of this world-building exercise, Magda's apparently all-encompassing vision discovers a blind spot when it attempts to incorporate the perceptual *Umwelt* of Anna's "dowry-goats."[23] In contrast to the predictable and familiar elements of Armoede's human world (e.g., the buildings, the village, and the hilly landscape that frames them picturesquely), the goats and their unfamiliar goat vision seem to Magda unimaginable, and she acknowledges with remorse that the goat world will remain "forever unknowable to me."

The failure of her imagination leads Magda to abandon her world-building exercise in midsentence. This moment also marks her turn away from the imagination and toward speculation. Whereas imagination seeks to hold objects together within a unity, situating them in a human context of meaning, speculation, as in Bogost's alien phenomenology, seeks out the inner lives of objects, attempting to envision how they perceive their own worlds. This is what Magda suggests when she announces her newfound "determination to burst through the screen of names into the goatseye view of Armoede." Renouncing the metaphysical world-building power of the imaginative faculty, Magda adopts an alternative lens that will help her attend to the strangeness of other entities (albeit obscurely) without reducing their existence to something meaningful within her own particular perceptual world. As she puts it elsewhere in the novel, "This is what I was meant to be: a poetess of interiority, an explorer of the inwardness of stones, the emotions of ants, the consciousness of the thinking parts of the brain" (35). Magda abandons anthropocentric imagination for exo-phenomenological speculation.

Magda's speculative turn was primed in her youth, when she first developed a "love of nature, particularly of insect life, of the scurrying purposeful life that goes on around each ball of dung and under every stone" (6). She describes a childhood spent attending to the worlds of insects—playing with beetles, tumbling ants down the sides of conical ant heaps, and crushing scorpions with a stick. "I have no fear of insects," she proclaims: "I would have no qualm . . . about living in a mud hut, or indeed under a lean-to or branches, out in the veld, eating chickenfeed, talking to the insects" (6). Although her sense of play sounds curiously like torture (not at all the proper comportment for an exo-phenomenologist!), Magda's reflection on the lives of insects leads her to consider more serious matters about ownership and belonging. In particular, she begins to think about insect populations as the most primordial inhabitants of the veld. As for young John Coetzee in *Boyhood*, this poses a problem for any claims to human ownership.

Troublingly, for Magda this applies as much to the African laborers on her farm as it does to the European colonials: "This is not Hendrick's home. No one is ancestral to the stone desert, no one but the insects" (18). In this vein, Magda offers a "speculative history" that reframes the origin of *all* human habitation of the veld—Indigenous pastoralists included—in terms of settler colonialism: "Hendrick's forebears in the olden days crisscrossed the desert with their flocks and their chattels, heading from A to B or from X to Y, sniffing for water, abandoning stragglers, making forced marches. Then one day fences began to go up—I speculate of course—men on horseback rode up and from shadowed faces issued invitations to stop and settle that might also have been orders and might have been threats, one does not know, *and so one became a herdsman,* and one's children after one, and one's women took in washing" (18–19; emphasis added). Magda's speculative history is undoubtedly problematic. Aside from the crude bluntness of her revision in the context of South African racial politics, Magda incorrectly assumes that herding only began among Indigenous communities following the arrival of the Dutch, even though archaeological records show that sheep and cattle herding as well as mixed farming had been practiced in the region prior to the sixteenth century.[24]

Problematic though her account may be, however, Magda offers a bold revision of colonial history. She speculates that roving Indigenous pastoralists widely adopted the identity of the more settled "herder" only after entering the Dutch colonial economy. Before this settlement, pastoralists had "crisscrossed" the veld without laying claim to any particular place. For Magda, what this speculative history means is that, since white and Black *settlement* (though not inhabitation) happened concurrently, neither group has any truly "ancestral" claim to the stony desert. The settlement of the South African veld began only with the advent of the settler colony, which gathered disparate peoples into specific economic relationships that in turn instituted systemic violence:

> There is another great moment in colonial history: the first merino is lifted from shipboard, with block and tackle, in a canvas waistband, bleating with terror, unaware that this is the promised land where it will browse generation after generation on the nutritious scrub and provide the economic

base for the presence of my father and myself in this lonely house where we kick our heels waiting for the wool to grow and gather about ourselves the remnants of the lost tribes of the Hottentots to be hewers of wood and drawers of water and shepherds and body-servants in perpetuity and where we are devoured by boredom and pull the wings off flies. (19)

Although her speculative history initially appears to serve a conservative white agenda that implicitly legitimizes European settlement by qualifying the legitimacy of African claims, Magda in fact makes the rather different point that settlement as such inaugurates an all-encompassing coloniality that affects all entities, both human and nonhuman. The large-scale human settlement of the veld instigated a universal system of suffering: the imported sheep are disoriented and terrified, the Africans suffer lives of servitude (if not slavery), and the Europeans simultaneously bear the weight of responsibility and the burden of boredom. Yet, given her assertion that only the insects are truly ancestral to the veld, it would seem that the flies have fallen furthest from grace. Whereas the sheep provide "the economic base" for the colony, the flies and other insects remain completely outside the colonial economy. Being utterly useless in the reproduction of a settler livelihood, the flies are thus reduced to the victims of the most casual and unconscious cruelty: "we are devoured by boredom and pull the wings off flies." Magda's revisionist history therefore shockingly suggests that the European subjugation of Africans conceals a yet more ethically objectionable violence, one that humans perpetrated against the stony landscape's ur-indigenes: the insects. Furthermore, this violence is truly lost to history. Much like Michael Marder's description of plants, the "absolute silence" of the flies and other insects within human social and political worlds puts them "in the position of the subaltern," and although an absence of voice "does not preclude their spatial, material self-expression[,] . . . it does pose additional hurdles to [their] ethical treatment."[25] It is on this account that Magda identifies herself with the insects: she, too, "fight[s] against becoming one of the forgotten ones of history" (3).

From a postcolonial perspective, Magda's attempt to identify herself with the insects undoubtedly appears to serve as an alibi for her own claim to belonging. And yet, Magda's "speculative entomology" (19) and its implication of insect oppression is linked to her reflection on her own vexed—and historically invisible—status as a woman who grew up squeezed between two cultural imperatives. She describes how she "grew up with the servants' children" and, in both linguistic and cultural senses, "spoke like one of them before I learned to speak like this" (6). With time, however, Magda learned Afrikaans, and with it imbibed the mythic nostalgia of the Afrikaner imaginary: "At the feet of an old man I have drunk in a myth of the past when beast and man and master lived a common life as innocent as the stars in the sky" (6–7). Painfully caught between allegiances to those subjected by colonial history and to those who mythologize a lost origin, Magda asks how it is possible for her to interpret the challenge of her own situation without resorting to a language that aestheticizes or allegorizes it: "How am I to endure the ache of whatever it is that is lost without a dream of a pristine age, tinged perhaps with the violet of melancholy, and myth of expulsion to interpret my ache to me?" (7). A paradigm of alienation becomes crucial here, generating the beautiful myth of an ideal and harmonious past that serves as recompense for the violent expulsion of others that her own people perpetrated. As a woman, however, Magda cannot fully inhabit this mythic history: "My lost world is a world of men, of cold nights, woodfire, gleaming eyes, and a long tale of dead heroes in a language I have not unlearned" (7). Though at once implicated in colonial history and subject to the nostalgia of an ancient past, Magda also cannot lay meaningful claim to it; it represents a white *male* inheritance to which she was never privy in the first place.

Magda's turn to speculative history allows her to articulate the origins of her own double bind, squeezed as she is between her (masculine) identity as "colonizer" and her (feminine) identity as "colonized." It is precisely this in-between status that renders her invisible (a "hole") and threatens her own erasure from history. What makes Magda's case especially interesting is that her speculative turn requires her to consider her own epistemological and ontological limitations. Thus, rather than stopping at the recognition of a double bind, Magda opens herself to something like an ecological double-consciousness.

W. E. B. Du Bois famously defines the double-consciousness of the "American Negro" as the "sense of always looking at one's self through the eyes of others, of measuring one's soul by the tape of a world that looks on in amused contempt and pity. One ever feels his two-ness,—an American, a Negro; two souls, two thoughts, two unreconciled strivings; two warring ideals in one dark body, whose dogged strength alone keeps it from being torn asunder."[26] Du Bois's double-consciousness describes a "two-ness" of being that emerges from the racist bifurcation of American society, which makes African Americans forever aware of their two opposed (and imposed) identities. The double-consciousness that emerges from such a violent socioeconomic institution as slavery is both painful and inescapable, but it also endows its sufferer with the spiritual strength necessary to hold opposites together, even without completely unifying them.

If Magda, as a white woman, can be said to experience anything like a similar condition, such a condition emerges from different historical circumstances that keep her conscious of how her racial and gender identities entrap her in powerful social contradictions. Yet the "tape" against which Magda measures her "soul" is not only sociohistorical; it is also ecological. In addition to the intersectional contradictions of her social being, Magda also ever feels the two-ness of her "social" and "animal" selves. Her social self is characterized by its need to produce meaning—she is, by her own admission, a creature hermeneutically inclined: "I signify something" (9). Social being of this sort requires the subject to be legible within the systems of signification (linguistic, social, political, juridical) in which it participates. Subjects want to *know* and to *be knowable*. In embracing her animal self, however, Magda also recognizes that the logocentric metaphysics of signification are inessential properties of being as such. In this light, she begins to think of herself less as a subject and more as an object, completely alien to herself and only ever partially knowable. In order to hold these two parts of herself together, Magda must curb her desire for knowledge and develop her nascent powers of speculation. This compromise constitutes her ecological double-consciousness.

Magda therefore turns her attention back on herself. As already suggested, this is not the same comforting return to self inspired by traditional landscape vision. For Magda, the return to herself reveals her intrinsic multiplicity, an irreducibly hermeneutic and animal being who is, furthermore, full of "holes." Whereas Mary Turner's encounter with her uncannily split self pushed her to the edge of psychosis, Magda proves much more resilient to the knowledge of her own unconsolidated nature. Often when she meditates on herself, Magda imagines that her physical body is constituted by a number of disparate entities. Unlike Walt Whitman's "I," which contains abstract "multitudes," Magda's "I" contains a multitude of concrete "its." In one example, Magda laments the social expectation for her to develop "the jowls, the bust, the hips of a true country goodwife." She characterizes her failure to meet this expectation as the failure of her willpower to protect her fat molecules from attack: "For alas, the power of my will . . . has not after all been great enough to keep me forever pristine against those molecules of fat: perishing by the millions in their campaigns against the animalcules in my blood, . . . a tide a blind mouths, that is how I imagine it, as I sit year after year across the table from my silent father, listening to the tiny teeth inside me" (21). Moving swiftly among vastly different scales, Magda imagines a connection between her socially inflected body image and a biological battle playing out internally between microscopic protozoa ("animalcules") and lipocytes ("molecules of fat"). This conflict manifests a connection between her social and animal selves that generates an uncanny, defamiliarized image of the self as a strange assemblage of warring molecular soldiers. Although her specific imagery may not be biochemically accurate, it nevertheless conjures the essential strangeness of the human body and the many entities—both organic and inorganic, "native" and "alien"— that make it function. (Indeed, if you have ever peered inside a human cell, you know it looks like an otherworldly circus, full of microscopic weirdos doing high-wire stunts.)[27]

As Magda continues to contemplate the microscopic scales of life within her, she begins to see herself more and more in relation to other forms of life. The churning microbiomes in her body bear an uncanny resemblance to the macroscopic world that surrounds her:

> I lie hour after hour concentrating on the sounds inside my head. In a trance of absorption I hear the pulse in my temples, the explosion and eclipse of cells, the grate of bone, the sifting of skin into dust. I listen to the molecular world inside me with the same attention I bring to the prehistoric world outside. I walk in the riverbed and hear the cascade of thousands of grains of sand, or smell the iron exhalation of rocks in the sun. I bring my understanding to the concerns of insects—the particles of food that must be carried over mountaintops and stored in holes, the eggs that must be arranged in hexagons, the rival tribes that must be annihilated. The habits of birds, too, are stable. It is therefore with reluctance that I confront the gropings of human desire. Clenched beneath a pillow in a dim room, focussed on the kernel of pain, *I am lost in the being of my being*. This is what I was meant to be: a poetess of interiority, an explorer of the inwardness of stones, the emotions of ants, the consciousness of the thinking parts of the brain. (35; emphasis added)

As she draws connections between the "molecular world" within and the "prehistoric world" without, Magda once again acknowledges her own confinement in the prison house of consciousness. And once again, coming up against this limit is painful. This is the first sense intended in her complaint, "I am lost in the being of my being"; her pain keeps her locked inside herself, destined (doomed) to remain "a poetess of interiority" only. But this moment also represents an opening into a mode of thinking that toggles between an exploration of her own interiority and speculation on the "inwardness of stones." The narrative patterning of this passage is important, moving as it does from self to other and back to self. Magda begins by listening to the sounds of her nervous system. She then attempts to translate this experience into an approximate understanding of beings that are external to her. Following this, she returns to her own experience yet again, but this time with a difference. Instead of just thinking about consciousness, she claims to contemplate "the consciousness *of the thinking parts of the brain*." Magda breaks her consciousness down into a constituency of distinct working parts, a complex physiological apparatus that seems less of a mystifying whole and more of a disarticulated, discontinuous assemblage. Curiously, Magda's intuitive observation is not far off the mark from contemporary neuroscience. As the neuroscientist Christof Koch explains, "Conscious perception is synthesized from activity at many essential nodes," and in this sense it represents a "multi-focal activity."[28] The perception of a face, for example, is synthesized from neurons firing in various parts of the brain at different times and at different intervals, ranging from 0.2 to 0.5 seconds. What this means is that perception "may be asynchronous, with different regions generating microconsciousness for color, motion, form, and so on at different times."[29] Perception therefore comes, as Deleuze has said, "only in bits and pieces."[30] The more Magda contemplates the "bits and pieces" of her own inwardness, the more she sees herself as a strangely unconsolidated, contingent assemblage of disparate parts. No longer a Cartesian *res intellectus*, or a *thing* that thinks, Magda reframes herself as a multitude of thinking things. In reducing and fracturing human consciousness in this way, Magda "flattens" it, making it just a thing (or assemblage of things) that can coexist with other things (or assemblages of things) on a shared ontological plane.

Just as ontology becomes flat in Magda's thought, her speculation also further highlights the difference between complicity and contingency. She does this by staging two contrary apocalyptic visions. These two visions are inverted reflections of one another, like a photograph and its negative: in the first vision, the world disappears around Magda; in the second, she disappears from the world. The first vision indulges in a Berkeleyesque subjective idealist fantasy in which Magda imagines that closing her eyes causes the world to disappear. In an anxious moment, she asks: "How can I afford to sleep? If for one moment I were to lose my grip on the world, it would fall apart: Hendrick and his shy bride would dissolve to dust in each other's arms and sift to the floor, the crickets would stop chirping, the house would deliquesce to a pale abstract of lines and angles against a pale sky." She goes on: "All that would remain would be me, lying for that fatal instance in a posture of sleep on an immaterial bed above an immaterial earth before everything vanished" (72–73). Given Magda's recognition of the metaphysical violence implicit in totalizing landscape visions like the one she attempted for Armoede, this fantasy serves to reaffirm both the absurdity and the danger of subjective idealism, which posits (human) perception as the condition of possibility for all existence.[31] She quickly admits that "I make it all up," but this ideational act proves difficult to undo: "I cannot stop now." It is precisely this inescapability that speaks to a sense of complicity, of being bound up in ideological systems that she cannot help but reproduce.

Magda's second vision, however, proves more obscure, more ecological, and somehow more *real* despite being pure fantasy. It is this vision that echoes the world-without-us visions in Schreiner and Lessing and that inches toward a notion of contingency. Magda has this vision while she waits beside her dying father, whose putrefying stomach wound has drawn a cloud of hungry flies. At first frustrated by the insect invasion—"I will not stand for any more flies" (78)—Magda speculates on their motivations: "The flies, which ought to be in transports of joy, sound merely cross. Nothing seems to be good enough for them. For miles around they have forsaken the meagre droppings of the herbivores and flown like arrows to this gory festival. Why are they not singing?" (78). Confused, Magda reassesses her own assumptions: "But perhaps what I take for petulance is the sound of insect ecstasy. Perhaps their lives from cradle to grave, so to speak, are one long ecstasy, which I mistake. . . . Perhaps if I talked less and gave myself more to sensation I would know more of ecstasy. Perhaps, on the other hand, if I stopped talking I would fall into pains, losing my hold on the world I know best. It strikes me that I am faced with a choice that flies do not have to make" (78). By attending to the flies and considering what she calls their "ecstasy," Magda develops an alternative response to her frustration. Contemplation leads her to think of herself in relation to the fly and to reflect, first, on the gap of alterity that opens up between her and them, and second, on the possibility that this gap may not be so unbroachable after all—that she, too, could cultivate an ability to approximate fly-like ecstasy. But again, Magda's attempt to see herself in relation to the fly reasserts her sense of difference: ecstasy would require her to abandon her own perceptual world and privilege only one part of herself, which is not the case for the flies. Magda's attempt to step back from her initial frustration with the flies and engage with the difference that separates them constitutes an instance of ecological ethics—a particularly important moment given Magda's earlier "speculative history" in which she implied that, more than anyone else, it is the insects (and especially the flies) who are truly ancestral to the veld and hence the "real" victims of human settlement.

As the flies' "gory festival" escalates, however, Magda feels an incipient threat. The swarm appears to be on the attack, preparing to conquer the farmhouse. Despite her provisional ethics, this turn puts Magda on the defensive and leads her into an apocalyptic fantasy of insect takeover:

> If I abandon this room, locking the door, stuffing the cracks with rags, I will in time find myself abandoning another room, and then others, until the house is all but lost, its builders all but betrayed, the roof sagging, the shutters clapping, the woodwork cracking, the fabrics rotting, the mice having a field day, only a last room intact, a single room and a dark passage where I wander night and day tapping at the walls, trying for old times' sake to remember the various rooms, the guest-room, the dining-room, the pantry in which the various jams wait patiently, sealed under candle wax, for a day of resurrection that will never come; and then retire, dizzy with sleepiness, for even mad old women, insensible to heat and cold, taking their nourishment from the passing air, from motes of dust and drifting strands of spiderweb and fleas' eggs, must sleep, to the last room, my own room, with the bed against the wall and the mirror and the table in the corner where, chin in hand, I think my mad old woman's thoughts and where I shall die, seated, and rot, and where the flies will suck at me, day after day, to say nothing of the mice and the ants, until I am a clean white skeleton with nothing more to give the world and can be left in peace, with the spiders in my eyesockets spinning traps for the stragglers to the feast. (78–79)

The feverish intensity of this sprawling sentence narrates a drama in which the horde of flies progressively conquers the farmhouse and leads the way for history's other forgotten creatures—mice, spiders, fleas, etc.—to reclaim their ancestral territory. Losing ground as the flies advance, Magda imagines herself retreating through melancholy rooms, feeling as though she's betrayed the human ingenuity that went into designing, building, and furnishing the farmhouse. As she retreats, she makes an inventory of the house's dilapidated architectural and design features, as well as other objects, like the wax-sealed jam jars, whose meaning will be lost without a human presence. Strangely, though, by emphasizing how these home(l)y features reflect her miserable sense of never having felt at home in this house, Magda's nightmare of retreat doubles as a fantasy of escape.

Magda is thus at once a fugitive running from a human mode of habitation (and, even more so, from farmwife domesticity) and a refugee running toward an alternative way of being with others. This is why, despite the fear and the terror that is evident in this unsettling fantasy of insect revolt, Magda's apocalyptic nightmare ultimately moves toward a vision of peace and rest—the only such vision afforded in the entire novel. Instead of simply disappearing from the world, her speculative death allows her to enter a much more intimate relationship of coexistence with her entomological kin. Magda fantasizes about her physical self entering utterly different, nonhuman economies of use. Her flesh will be metabolized by the flies, mice, and ants, leaving behind her skeletal bone-house. Reversing an earlier image of herself as a hermit crab inhabiting the shells of dead creatures (43), here her body becomes another's shelter: a spider takes up residence in her skull, installing stringy webs to make itself more comfortable and secure additional nourishment. Magda's speculative death finally gives her what she has always wanted—to transcend her own consciousness and become an integral part of nonhuman worlds. While she will never know the "inwardness" of these entities, she will go on living inside of them, just as they will go on living inside of her.

THE "NATURAL" ALIEN

Magda is a budding practitioner of a contingent exo-phenomenology—that is, a phenomenology that seeks out relations of worldly engagement which do not bring her home to some primordial dwelling but rather send her out of herself, enabling her to reckon with both the social and ecological problematics of her entrenchment within a settler colonial paradigm. Her exo-phenomenological mode works against her entanglement with the coloniality of Nature in productively unsettling ways. Yet the politics of

Magda's exo-phenomenological practice remain challenging, even controversial in a way that cursory readings might reject as an imperialist apologetics. If Magda considers both "indigenous" Africans and "exogenous" Europeans to be *settlers*, does she not, in one gesture, erase both the historical and ongoing violence of settler colonialism? Not exactly. Magda well understands the violence of her colonialist inheritance, for this violence has contributed greatly to her own suffering. But she also sees that that the social, political, and economic contexts of colonial occupation coexist uneasily with an ecological reality that at once comes under settler control and far exceeds it.

Magda refuses to affirm any claims to primordial belonging, which would always rely on some human metaphysics of land ownership, whether African or European. Instead, she turns to contingency thinking in the exo-phenomenological mode in order to resituate colonial violence within her own, idiosyncratic ecological realism. What results is a profound and unsettling shift in vision. No longer imagined as a human drama playing out in the foreground against a sprawling landscape, settler occupation loses the framing that ensured its uniquely human dimension. The settlers and their various human others find themselves caught up in the thoroughly disorienting, multidimensional reality of ecological enmeshment. As Timothy Morton has put it, the mesh of ecology isn't a background against which individual entities appear: "It is the entanglement of all strangers."[32] This "meshy" understanding of ecology not only means that all entities, both human and other-than-human, constitute each other's environment in a dynamic, contingent, and mutually implicated unfolding. It also means that, however complexly bound up with each other, the entities that collectively constitute an environment are also ontologically strange—that is, they are strangers existing in a context made up of other strangers. Morton terms these doubly estranged entities "strange strangers" in an effort to underscore the bizarre way in which all entities are aliens to one another. Obviously, Magda doesn't borrow from the lexicon of contemporary environment philosophy. Even so, her challenging perspective introduces a paradigm that understands the state of estrangement not as a settler condition, but as the fundamental condition of all entities caught up in tangled knots of multispecies relation. In other words, alienation is natural.

Such a conclusion recalls a key early moment in the emergence of eco-phenomenology that registers alienation as something "natural" to the human animal. In his book *The Natural Alien*, Neil Evernden draws on biological and evolutionary research to make the case that humans are perennial aliens wherever we go. We have no specific niche because we did not evolve with our surroundings, and hence have never developed the kind of relationships with local environments that other lifeforms have. We did not *lose* a primordial sense of deep interconnection, like other ecophenomenologists have argued; *we just never had it.*

Instead of seeking out a home for Evernden's natural alien, I propose to take this figure seriously and suggest that, as Magda comes to recognize in Coetzee's novel, the natural alien is what we humans always already are. Such a recognition recalls the work of the geographer John Wylie, who claims that landscape cannot be a homeland not only because "there are no such homelands for us to inhabit," but even more crucially (and unsettlingly) because "there are no original inhabitants."[33] Wylie's claim may initially sound like it erases Indigenous claims to prior occupation, but it could also be read as implicitly recognizing that indigeneity had to be invented. As Natchee Blu Barnd puts it: "Indigeneity, or what might be loosely defined as the 'quality of being indigenous,' is deeply embedded within and defined by colonial contestations over land and geography. If settler colonialism is defined by its spatial organization and outcomes, then so too must be indigeneity, a term and concept that codes as the supposed precondition of, as well as ongoing foil to, colonial completion."[34] Indigeneity had to arise as a conceptual counterpoint in relation to colonial ventures; for all that it marks prior inhabitance, it is an imposed identity, one marked, through the very designation of the term "native," by the coloniality of being and power.[35] As such, there may be material and spiritual histories of occupation, but there are no original inhabitants in the deep-historical sense that all humans are diasporic citizens, flung far from the evolutionary locus of their emergence in eastern Africa. Wylie's point about how landscape cannot be a homeland must therefore be understood in line with D. H. Lawrence's reformulation of "home" as a place of being without the promise of belonging.

In closing, I would like to note two of many possible avenues for thinking through notions of ecology, phenomenology, and contingency without redeploying the figure of a home(l)y Nature as a place of primordial belonging. Both avenues come from recent work by Indigenous scholars. The first of these is the critical ethnic studies scholar Natchee Blu Barnd, who works productively through the notion of "inhabiting" as defined in the phenomenological anthropology of Tim Ingold. As Barnd describes it, inhabiting is primarily physical, somatic, and relational: "At its base, inhabiting signals the moment(s) when a body is situated in a particular physical location. It is also a verb, implying some sort of spatially defined and relational set of actions."[36] Importantly, inhabiting is not the same thing as homemaking; instead, it "describes a frame *used for establishing* belonging or home."[37] The distinction Barnd makes here is important, since it allows for a recognition of the plurality of modes of inhabitance, and of how these different modes organize different lived spatialities. Perhaps most importantly, inhabiting can serve as a reminder that modes of being-in-relation are not set in stone, but are themselves dynamic: "Inhabiting reminds us how spatial enactments can be practiced and (re)arranged in sometimes unexpected ways toward different kinds of relations to lands, or different geographies."[38] Barnd is certainly also interested in "the process of making meaning in relation to the land where bodies are situated;"[39] yet his distinction between inhabiting and homemaking creates space for a mode of being that is not initially predicated on belonging—understood as an individual's belonging-to the land or the land's belonging-to the individual.

Barnd's attendance to the practice of inhabiting dovetails productively with the Iroquois story of Skywoman and her arrival on the North American landmass, whose original name is Turtle Island. In a recent retelling by the Potawatomi writer and scientist Robin Wall Kimmerer, Skywoman falls from the Skyworld, and as she plunges down, light peeks through a hole in the sky and illuminates the darkness below: "But in that emptiness there were many eyes gazing up at the sudden shaft of light. They saw there a small object, a mere dust mote in the sudden shaft of light. As it grew closer, they could see that it was a woman, arms outstretched, long black hair billowing behind as she spiraled toward them."[40] A flock of geese rushes to break her fall and deposits her on the back of a great turtle. A council of aquatic critters then assembles to determine what to do with this new arrival, and they come to the understanding that Skywoman needs land to thrive. Loon, Beaver, and Otter volunteer to dive into the depths and find mud. Many perish in the dangerous venture, including Muskrat, whose small body resurfaces clutching a small clawful of mud. Turtle offers Skywoman his back as a foundation, and Skywoman spreads the mud on his shell. Then, "moved by the extraordinary gifts of the animals, she sang in thanksgiving and then began to dance, her feet caressing the earth. The land grew and grew as she danced her thanks, from the dab of mud on Turtle's back until the whole earth was made. Not by Skywoman alone, but from the alchemy of all the animals' gifts coupled with her deep gratitude. Together they formed what we know today as Turtle Island, our home."[41] Though Kimmerer concludes this part of Skywoman's story with a reference to Turtle Island as *home*, she is careful not to take this term for granted: "It is good to remember that the original woman *was herself an immigrant*."[42] Following her dance of creation upon Turtle's back, Skywoman, "like any good guest," presents her new home with a gift: a bundle of seeds from Skyworld, which she proceeds to scatter across the land: "Wild grasses, flowers, trees, and medicines spread everywhere."[43] Though not primordially *of* this land, Skywoman becomes native to it through active contribution: "It was through her actions of reciprocity, the give and take with the land, that the original immigrant became indigenous."[44] Kimmerer, who is a biologist invested in environmental restoration, concludes her account by suggesting that the Skywoman story might serve us as a form of "re-story-ation"—that is, "not as an artifact from the past but as instructions for the future."[45] These instructions can be found in all that Skywoman cultivated and left behind, in both the animals and, especially, in the plants: "Plants know how to make food and medicine from light and water, and then they give it away. . . . The plants can tell us [Skywoman's] story; we need to learn to listen."[46]

As a settler/trespasser myself, I recognize that rehearsing Kimmerer's retelling of the Skywoman story at the end of this book may be perceived as a last-ditch gesture at what Eve Tuck (Unangax̂) and K. Wayne Yang have termed a "settler move to innocence."[47] Indeed, such a rehearsal could easily be understood as an attempt to appropriate an Indigenous story in the interest of clearing an alternative path for settler homecoming. However, I take the risk of courting this particular criticism out of a belief that it will only be through widespread, concerted acts of multispecies community building and restor(y)ation that "we," as entities of this planet, will be able to develop a meaningful decolonial perspective and endeavor to achieve mutual flourishing.

In the absence of imminent decolonization, we settlers, along with other guests on the land, urgently need to engage in deeply unsettling decolonial work. This work will entail renewed efforts to learn about each other across our various intersectional differences (species, race, class, sexuality, gender, dis/ability, and so on), and courage to address the many ways global coloniality has produced and leveraged these differences in ways that have instituted astonishing inequities. It will also require fortitude to develop practices inclusive and pluriversal enough to avoid what Mariana Ortega has called the "decolonial woes" that arise when one decolonial perspective makes others invisible.[48] I therefore emphasize that though Indigenous modes of storytelling and worldmaking offer powerful models, I do not want to romanticize them, or fetishize them as the only models. The Hebraic tradition of exile and arrival that animates Levinas's phenomenology of alterity may offer another possible model—though also a cautionary one, as the political Zionism to which perennial homelessness has historically given rise has also tended to underwrite violent settler colonial homemaking.[49] We should also attend to Donna Haraway's lessons on making kin as a deeply collaborative, sympoietic endeavor with companion species.[50] We can learn from powerful examples of multispecies ethnography by the likes of Eduardo Kohn, Juno Salazar Parreñas, Anna Lowenhaupt Tsing, and others.[51] We have much to gain from Eduardo Viveiros de Castro's theory of multinaturalist perspectivism,[52] as well as from anthropologists' explorations of pluriversal worldings.[53] And all of these perspectives need to be articulated with the profound lessons about what Sylvia Wynter has grandiosely (but not *too* grandiosely) termed "the coloniality of being/power/truth/freedom."

A multifocal articulation of these and other important discourses across the social and ecological fields can yield a deeply interdisciplinary framework to help support the continued development of a contingent exophenomenology, one that persistently recognizes that becoming "native" to a place must rely, not on an original claim, but rather on a continuous unfolding through sustained practices of reciprocity and acts of solidarity with both human and other-than-human others. The project of unsettling Nature may well require us ongoingly to reckon with the idea that all entities are natural aliens, guests on contested land.

Notes

PROLOGUE

1. I borrow the term "slow violence" from Nixon, *Slow Violence and the Environmentalism of the Poor.*
2. See Tepler, "Forest Park Guide."
3. Higgins, "Revered, Then Reviled," n.p.
4. I borrow the phrase "my own particular beloved place" from Ranger, *Voices from the Rocks*. See the introduction of this work for further discussion.
5. Rifkin, *Settler Common Sense*, xv.
6. Rifkin, *Settler Common Sense*, xvi.

INTRODUCTION

1. Derrida echoes Levinas most closely in *Of Hospitality*, where he conceives of a home as that which must remain open to receiving the other. This argument hinges on the ambiguity of *hôte*, a French word meaning both "host" and "guest."
2. See Wigley, *The Architecture of Deconstruction*. Incidentally, the paradox of the abyss as foundation comes straight from Heidegger. In *Identity and Difference* Heidegger describes *Er-eignis* (the "event of appropriation") as a "self-suspended structure [*in sich schwingenden Bau*]" (38) built upon a grounding abyss.
3. Marder, "Ecology as Event," 159.
4. See Barad, "Troubling Time/s and Ecologies of Nothingness."
5. Kirby, "Un/Limited Ecologies," 123; emphasis added.
6. Kirby, "Un/Limited Ecologies," 139.
7. The editors of the volume *Eco-Deconstruction* explicitly link eco-deconstruction to eco-hermeneutics and eco-phenomenology (see Fritsch, Lynes, and Wood, Introduction, 2–5). For key essays in eco-hermeneutics, see Clingerman, Treanor, Drenthen, and Utsler, eds., *Interpreting Nature*.
8. Utsler, Clingerman, Drenthen, and Treanor, Introduction, 5.
9. Fritsch, Lynes, and Wood, Introduction, 4–5.
10. C. Wolfe, "Wallace Stevens's Birds," 321–22.
11. See Uexküll, *A Foray into the Worlds of Animals and Human* 17

12. Barad, "Troubling Time/s and Ecologies of Nothingness," 213.
13. I owe this topological metaphor to Cary Wolfe, who in turn borrows it from Michel Serres (see C. Wolfe, "Wallace Stevens's Birds," 331).
14. Donohoe, "The Place of Home," 32.
15. Husserl, *Shorter Works*, 230.
16. Heidegger, *Introduction to Metaphysics*, 39.
17. Tellingly, however, some of Merleau-Ponty's eco-phenomenological interpreters have projected explicit home(l)y rhetoric onto the philosopher's work (see, e.g., Bigwood, "Logos of Our Eco in the Feminine"; and Davis, "*Umwelt* and Nature in Merleau-Ponty's Ontology").
18. Merleau-Ponty, *Phenomenology of Perception*, vii; emphasis added.
19. Moya, "Habit and Embodiment in Merleau-Ponty," n.p.
20. See esp. Bachelard, *Poetics of Space*; see also, e.g., Angelova, "Phenomenology of Home"; and Donohoe, "The Place of Home."
21. See, e.g., Frank, *Literary Architecture*; and Fryer, *Felicitous Space*.
22. See, e.g., Leith, "Home Is Where the Heart Is . . . or Is It?"; and Stefanovic, "The Experience of Place."
23. See, e.g., Dekkers, "Dwelling, House, and Home."
24. Brown and Toadvine, "Eco-Phenomenology: An Introduction," xx.
25. Husserl, *Logical Investigations*, 168.
26. Heidegger, *Basic Writings*, 358.
27. See the ubiquitous references that appear in edited collections such as Brown and Toadvine, eds., *Eco-Phenomenology*; and Smith, Smith, and Verducci, eds., *Eco-Phenomenology*. See also Diehm, "Here I Stand"; and Mattingly, "The Ecology of Being." Work in ecopoetics also frequently leverages etymology to gesture to the natural environment as humanity's home (see, e.g., Bate, *The Song of the Earth*). Bate points to the paradigmatic question: "Ecopoetics asks in what respects a poem may be a making (Greek *poiesis*) of the dwelling-place—the prefix eco- is derived from Greek *oikos*, 'the home or place of dwelling'" (75).
28. Bachelard, *Poetics of Space*, xxxvi.
29. Liddell and Scott, *Greek–English Lexicon*, s.v. "οἶκος."
30. See, e.g., Ricklefs, *The Economy of Nature*.
31. See, e.g., Worster, *Nature's Economy*.
32. Derrida, *Of Grammatology*.
33. Martin Dillon sees this link between ecology and home(coming) as a reason to abandon the term altogether (see Dillon, "Merleau-Ponty and the Ontology of Ecology").
34. See, e.g., White, "The Historical Roots of Our Ecologic Crisis."
35. On the latter point, see Abram, *The Spell of the Sensuous*.
36. See, e.g., Morton, *Dark Ecology*; and Quinn, *Ishmael*.
37. See Sale, *After Eden*.
38. I extrapolate this point from the view that the human species is in itself inherently violent and destructive—a view that is at least as old as Thomas Hobbes's *Leviathan*.
39. See R. Williams, *The Country and the City*, 9–12 and passim.

40. Thomashow, *Bringing the Biosphere Home*, 12.
41. Thomashow, *Bringing the Biosphere Home*, 83.
42. Tom Lynch, for instance, cites Thomashow in support of his argument that a similar perceptual ecology practice in desert spaces like the southwestern United States can lead to a "xerophilia" that no longer privileges lush greenness as a signifier of ecological well-being (see Lynch, *Xerophilia*, 195–97).
43. Abram, *Becoming Animal*, 8.
44. Abram, *Becoming Animal*, 3.
45. Ray, *The Ecological Other*, 25.
46. Ahmed, "A Phenomenology of Whiteness." In a related project Ahmed investigates how sexual orientation phenomenologically (dis)orients the body in (hetero)normative spaces (see Ahmed, *Queer Phenomenology*).
47. See "Rewild.com."
48. Krech, *The Ecological Indian*.
49. The anthropologist Johannes Fabian refers to the latter tactic as a "denial of coevalness" (see Fabian, *Time and the Other*).
50. Bachelard, *The Poetics of Space*, 33.
51. Thomashow, *Bringing the Biosphere Home*, 57; emphasis added.
52. Thomashow, *Bringing the Biosphere Home*, 57 and 83.
53. Veracini, *Settler Colonialism*, 97.
54. Veracini, *Settler Colonialism*, 98–99.
55. Ranger, *Voices from the Rocks*, 11–12.
56. Ranger, *Voices from the Rocks*, 12.
57. Ranger, *Voices from the Rocks*, 13.
58. Veracini, *Settler Colonialism*, 33.
59. Veracini, *Settler Colonialism*, 34.
60. Tuck and Yang, "Decolonization Is Not a Metaphor."
61. P. Wolfe, "Settler Colonialism and the Elimination of the Native," 388.
62. P. Wolfe, "Settler Colonialism and the Elimination of the Native," 388.
63. Cronon, Foreword, 20.
64. Cronon, Foreword, 21.
65. Cronon, "The Trouble with Wilderness," 72.
66. Cronon, "The Trouble with Wilderness," 73.
67. Cronon, "The Trouble with Wilderness," 76.
68. Cronon, "The Trouble with Wilderness," 80.
69. Cronon, "The Trouble with Wilderness," 81.
70. Cronon, "The Trouble with Wilderness," 86–87.
71. Cronon, "The Trouble with Wilderness," 89; emphasis added.
72. Cronon, "The Trouble with Wilderness," 79.
73. Merchant, "Reinventing Eden," 137.
74. Recent years have seen the publication of historical research that more fully centers the alliance between settler homemaking, discourses of Nature, and Indigenous displacement. For two very different examples, see Quammen, *American Zion*; and Taylor, *The Rise of the American Conservation Movement*.

75. See, e.g., Maracle (Stó:lō), *Memory Serves;* Watts (Mohawk [Bear Clan, Six Nations] and Anishinaabe), "Indigenous Place-Thought and Agency amongst Humans and Non-humans"; and Whyte (Potawatomi), "Settler Colonialism, Ecology, and Environmental Injustice." Other noteworthy and allied critiques of settler colonial paradigms, also by prominent Indigenous scholars, include Byrd (Chickasaw), *The Transit of Empire;* Moreton-Robinson (Goenpul tribe of the Australian Aboriginal Quandamooka nation), *The White Possessive;* and Tuck (Unangax̂) and Yang, "Decolonization Is Not a Metaphor"; among many others. For other work on the coimplication of settler colonialism and ecological violence by non-Indigenous scholars, see, e.g., Bacon, "Settler Colonialism as Eco-social Structure and the Production of Colonial Ecological Violence"; and Francis, "The Tyranny of the Coloniality of Nature and the Elusive Question of Justice."

76. Whyte, "Settler Colonialism, Ecology, and Environmental Injustice," 137.
77. Mignolo and Walsh, *On Decoloniality,* 4.
78. Quijano, "Coloniality of Power, Eurocentrism, and Latin America," 533.
79. Wynter, "Unsettling the Coloniality of Being/Power/Truth/Freedom," 280.
80. Wynter, "Unsettling the Coloniality of Being/Power/Truth/Freedom," 296.
81. Quijano, "Coloniality of Power, Eurocentrism, and Latin America," 533.
82. See Foucault, *The Order of Things.* For a decolonial revision of Foucault, see Wynter, "Unsettling the Coloniality of Being/Power/Truth/Freedom," 318.
83. Mignolo and Walsh, *On Decoloniality,* 119.
84. See Dussel, *The Invention of the Americas;* and Quijano, "Coloniality of Power, Eurocentrism, and Latin American."
85. See Mignolo, "The Geopolitics of Knowledge and the Colonial Difference"; and Quijano and Wallerstein, "Americaneity as a Concept."
86. See Maldonado-Torres, "On the Coloniality of Being."
87. See Lugones, "The Coloniality of Gender"; and Perez, *The Decolonial Imaginary.*
88. Mignolo and Walsh, *On Decoloniality,* 114.
89. Mignolo, Introduction, 160. Mignolo uses the phrase in reference to an analysis of the colonization of Indigenous food epistemologies.
90. Alimoda, "The Coloniality of Nature," n.p.
91. Alimoda, "The Coloniality of Nature," n.p.
92. Gómez-Barris, *The Extractive Zone,* xvi.
93. Gómez-Barris, *The Extractive Zone,* 3.
94. Gómez-Barris, *The Extractive Zone,* 11–12.
95. Gómez-Barris, *The Extractive Zone,* 39.
96. Gómez-Barris, *The Extractive Zone,* 40–41.
97. Gómez-Barris, *The Extractive Zone,* 41; emphasis added.
98. Francis, "The Tyranny of the Coloniality of Nature and the Elusive Question of Justice," 39.
99. Mignolo and Walsh, *On Decoloniality,* 153–54.
100. Wynter, "Unsettling the Coloniality of Being/Power/Truth/Freedom."
101. Mignolo and Walsh, *On Decoloniality,* 161–62.

102. For an allied critique of Anthropocene discourse and the foundation violence of its geo-logics, see Yusoff, *A Billion Black Anthropocenes or None.*
103. Mignolo and Walsh, *On Decoloniality*, 158–59.
104. Tuck and Yang, "Decolonization Is Not a Metaphor," 7.
105. Tuck and Yang, "Decolonization Is Not a Metaphor," 7.
106. By contrast, the ecological sciences use the term *landscape* to designate a particular scale of study that encompasses multiple, interacting ecosystems.
107. Nancy, "Uncanny Landscape," 57.
108. Nancy, "Uncanny Landscape," 58.
109. Wylie, "A Landscape Cannot Be a Homeland," 409.
110. Wylie, "A Landscape Cannot Be a Homeland," 413.
111. Wylie, "A Landscape Cannot Be a Homeland," 414.
112. Wylie, "A Landscape Cannot Be a Homeland," 410, 409.
113. Wylie, "A Landscape Cannot Be a Homeland," 411.
114. Freud, "The Uncanny," 226.
115. Freud, "Repression," 147.
116. Freud, "The Uncanny," 241.
117. Freud, "The Uncanny," 236.
118. Freud, "The Uncanny," 247–48.
119. Freud, "The Uncanny," 245.
120. Researchers at the Weizmann Institute in Israel estimate that the average human body contains approximately thirty-nine trillion bacteria cells and thirty trillion human cells (see Sender, Fuchs, and Milo, "Revised Estimates for the Number of Human and Bacteria Cells in the Body").
121. See, e.g., Koch, *The Quest for Consciousness.*
122. On this point, see Weheliye, *Habeas Viscus*, 17–32.
123. Deckard, "Ecogothic," 174.
124. Oloff, "Greening the Zombie," 31–32.
125. Freud, "The Uncanny," 248.
126. Freud, "The Uncanny," 251.
127. Freud, "The Uncanny," 251.
128. Qtd. in Seppänen, "Lost at Sea," 205.
129. Seppänen, "Lost at Sea," 205.
130. Burke, *A Philosophical Inquiry into the Origin of Our Ideas of the Sublime and Beautiful*, 36.
131. Kant, *Critique of the Power of Judgment*, 252–53.
132. Schreiner, *Thoughts on South Africa*, 49–50.
133. Schreiner, *Thoughts on South Africa*, 35; emphasis added.
134. Schreiner, *Thoughts on South Africa*, 41.
135. Kant, *Critique of the Power of Judgment*, 261–62.

1. MARTIN HEIDEGGER AND THE COLONIALITY OF NATURE

1. Heidegger, *Introduction to Metaphysics*, 35.
2. For alternative readings of Heidegger as a philosopher of home(coming), see Gauthier, *Martin Heidegger, Emmanuel Levinas, and the Politics of Dwelling*, 1–102; and O'Donoghue, *A Poetics of Homecoming*.
3. Heidegger and Fink, *Heraclitus Seminar*, 126.
4. Bachelard, *The Poetics of Space*, 213.
5. Heidegger, *Zollikon Seminars*, 120.
6. Adorno, *The Jargon of Authenticity*, 43.
7. The first study to pursue this connection was penned by one of Heidegger's former students, the Chilean historian Victor Farías (see Farías, *Heidegger and Nazism*; see also Faye, *Heidegger*; and Wolin, *The Heidegger Controversy*).
8. See, e.g., Harvey, *The Condition of Postmodernity*, 207–9; and Leach, "The Dark Side of the Domus."
9. Malpas, *Heidegger's Topology*, 20.
10. Maldonado-Torres, "On the Coloniality of Being," 251; emphasis added.
11. I borrow the phrase "colonial matrix of power" from Walter Mignolo, who developed it from Aníbal Quijano (see, e.g., Mignolo, *The Darker Side of Western Modernity*).
12. Zimmerman, *Contesting Earth's Future*, 122–48. See also Antolick, "Deep Ecology and Heideggerian Phenomenology."
13. See, e.g., Brown and Toadvine, eds., *Eco-Phenomenology*; and Foltz, *Inhabiting the Earth*.
14. See, e.g., Rigby, *Topographies of the Sacred*.
15. See Garrard, "Heidegger Nazism Ecocriticism," 262.
16. See, e.g., Glotfelty and Fromm, eds., *The Ecocriticism Reader*; Hiltner, ed., *Ecocriticism*; and Westling, *The Cambridge Companion to Literature and the Environment*.
17. See, e.g., Buell, *The Future of Environmental Criticism*; and Garrard, *Ecocriticism*.
18. Qtd. in Garrard, "Heidegger Nazism Ecocriticism," 252.
19. See, e.g., Bate, *Song of the Earth*; Goodbody, "Heideggerian Ecopoetics and the Nature Poetry Tradition"; Peters and Irwin, "Earthsongs"; and Rigby, "Earth, World, Text."
20. Zimmerman, "Rethinking the Heidegger–Deep Ecology Relationship."
21. Zimmerman, "Heidegger's Phenomenology and Contemporary Environmentalism."
22. Zimmerman, "Heidegger's Phenomenology and Contemporary Environmentalism," 86.
23. Garrard, "Heidegger Nazism Ecocriticism," 254.
24. Claborn, "Toward an Eco-Ontology."
25. In particular, see Marder, *Heidegger*, 69–92.
26. Maldonado-Torres, "The Topology of Being and the Geopolitics of Knowledge," 30.

27. Maldonado-Torres, "On the Coloniality of Being," 242.
28. Maldonado-Torres, "On the Coloniality of Being," 249.
29. See, e.g., Mignolo, *The Darker Side of Western Modernity*, xvi.
30. Maldonado-Torres, "The Topology of Being and the Geopolitics of Knowledge," 32.
31. Maldonado-Torres, "The Topology of Being and the Geopolitics of Knowledge," 32.
32. Maldonado-Torres, "The Topology of Being and the Geopolitics of Knowledge," 32.
33. Mignolo and Walsh, *On Decoloniality*, 4.
34. Maldonado-Torres, "On the Coloniality of Being," 250.
35. Maldonado-Torres, "On the Coloniality of Being," 249. The Cameroonian philosopher Achille Mbembe elaborates on this perennial state of war under the sign of "necropolitics" (see Mbembe, "Necropolitics").
36. Maldonado-Torres, "On the Coloniality of Being," 251; emphasis added.
37. Maldonado-Torres, "On the Coloniality of Being," 253.
38. See, e.g., Ferreira da Silva, *Toward a Global Idea of Race*; and Mbembe, *Critique of Black Reason*.
39. See Sharpe, *In the Wake*.
40. Tansi, "An Open Letter to Africans," 271.
41. Wynter, "Unsettling the Coloniality of Being/Power/Truth/Freedom," 260.
42. See, e.g., Hartman, *Scenes of Subjection*; Jackson, *Becoming Human*; Sexton, "Unbearable Blackness"; Sharpe, *In the Wake*; and Weheliye, *Habeas Viscus*.
43. Fred Moten, for instance, makes frequent reference to Heidegger throughout his "consent not to be a single being" trilogy: *Black and Blur*, *Stolen Lives*, and *Universal Machine*.
44. Warren, *Ontological Terror*, 5.
45. Warren, *Ontological Terror*, 2.
46. Heidegger, *Introduction to Metaphysics*, 1.
47. Warren, *Ontological Terror*, 9.
48. Warren, *Ontological Terror*, 5. Warren clarifies why he places "being" under erasure: "I use the word ~~being~~ in the term *black ~~being~~* simply to articulate the entity of blackness that bears the weight of unbearable nothing[,] . . . to indicate the double bind of communicability and to expose the death of blackness that constitutes the center of being" (179n1).
49. Warren, *Ontological Terror*, 33.
50. Warren, *Ontological Terror*, 9.
51. Though unconventional, from this point on I capitalize "World" and "Earth" to signal the terms' philosophical resonances as *Welt* and *Erde*, respectively.
52. Heidegger, *Basic Writings*, 167–68.
53. Wigley, *The Architecture of Deconstruction*, 61; emphasis added.
54. Fell, *Heidegger and Sartre*, 196–97.
55. Heidegger, *Basic Writings*, 172.
56. Heidegger, *Basic Writings*, 170.

57. In *The Fundamental Concepts of Metaphysics*, Heidegger infamously extends his argument for human exceptionalism, claiming that, whereas humans are "world-forming," animals are "poor in world" (*Weltarm*) and stones are "worldless" (*Weltlos*).
58. Harman, *Quadruple Object*, 39.
59. Heidegger, *Being and Time*, 68.
60. Heidegger, *Being and Time*, 68.
61. Heidegger, *Being and Time*, 85.
62. Heidegger, *Being and Time*, 85.
63. Heidegger, *Being and Time*, 85–86.
64. Heidegger, *Being and Time*, 172; emphasis added.
65. Qtd. in Shaviro, *The Universe of Things*, 17.
66. Heidegger and Fink, *Heraclitus Seminar*, 126.
67. Heidegger, *Basic Writings*, 358.
68. Husserl, *Logical Investigations*, 168.
69. See, e.g., Ingold, *The Perception of the Environment*, 172–88.
70. Foltz, *Inhabiting the Earth*, 171.
71. Heidegger, *The Fundamental Concepts of Metaphysics*, 169–273.
72. Foltz, *Inhabiting the Earth*, 172.
73. See Malpas, *Heidegger and the Thinking of Place*; and Malpas, *Heidegger's Topology*.
74. Malpas, *Heidegger's Topology*, 31.
75. Heidegger, *Basic Writings*, 188.
76. Heidegger, *Basic Writings*, 189.
77. Stambaugh, Introduction, 14.
78. Heidegger, *Basic Writings*, 186.
79. Heidegger, *Introduction to Metaphysics*, 7.
80. Fried and Polt, "Translators' Introduction," ix.
81. Fried and Polt, "Translators' Introduction," xi.
82. Heidegger, *Being and Time*, 70.
83. See Heidegger, *Basic Writings*, 321.
84. Heidegger, *Basic Writings*, 354.
85. Malpas, *Heidegger's Topology*, 198.
86. Heidegger, *Being and Time*, 63.
87. Heidegger, *Poetry, Language, Thought*, 164.
88. Malpas, *Heidegger's Topology*, 76.
89. Heidegger, *Basic Writings*, 360.
90. Warren, *Ontological Terror*, 33.
91. Warren, *Ontological Terror*, 179n3; emphasis added.
92. Zimmerman, "Heidegger's Phenomenology and Contemporary Environmentalism," 86.

2. WILLA CATHER AND THE HOME(L)Y METAPHYSICS OF LANDSCAPE

1. Cather, *The Kingdom of Art*, 422–23.
2. Cather, *The Kingdom of Art*, 422.
3. Cather, *The Kingdom of Art*, 422.
4. Cather, *The Kingdom of Art*, 423.
5. The Prairie Trilogy includes *O Pioneers!* (1913), *The Song of the Lark* (1915), and *My Ántonia* (1918).
6. The most critical readings of Cather's imaginative homemaking include Ammons, "Cather and the New Canon"; Fischer, "Pastoralism and its Discontents"; and Westling, *The Green Breast of the New World*. For other critical readings, see, e.g., Davidson, *Willa Cather and F. J. Turner*; Gorman, "Jim Burden and the White Man's Burden"; Reynolds, *Willa Cather in Context*; Schubnell, "The Decline of America"; and Urgo, "The Cather Thesis."
7. See Fryer, *Felicitous Space*.
8. Bachelard, *The Poetics of Space*, xxxvi.
9. Bachelard, *The Poetics of Space*, 5.
10. See, e.g., Glotfelty, "A Guided Tour of Ecocriticism"; Mezei and Briganti, "Reading the House"; Moseley, "Spatial Structures and Forms in *The Professor's House*"; Mutter, "Godfrey St. Peter's 'Picturesque Shipwreck'"; Oehlschlaeger, "Indisponibilité and the Anxiety of Authorship in *The Professor's House*"; and Russell, *Between the Angle and the Curve*.
11. See Forster, *Aspects of the Novel*, 43–83; and Wood, *How Fiction Works*, 91–128.
12. Cather, *Willa Cather on Writing*, 80.
13. Cather, *Willa Cather in Person*, 79.
14. Quirk, *Bergson and American Culture*, 167.
15. For a fuller analysis of Ruskin's influence on Cather's fiction than I provide here, see Murphy, "Cather's Ruskinian Landscapes." For a similarly in-depth examination of Bergson's influence on Cather, see Quirk, *Bergson and American Culture*.
16. Cather, *The Kingdom of Art*, 400.
17. Cather, *The Kingdom of Art*, 401.
18. See Ruskin, *Modern Painters*, 3:114: "Though we cannot, while we feel deeply, reason shrewdly, yet I doubt if, except when we feel deeply, we can ever comprehend fully."
19. Ruskin, *The Seven Lamps of Architecture*, 151.
20. Tschumi, *Architecture and Disjunction*, 3.
21. Ruskin, *The Seven Lamps of Architecture*, 149.
22. Ruskin, *The Seven Lamps of Architecture*, 150; emphasis added.
23. Ruskin, *Modern Painters*, 3:156.
24. Bergson, *Creative Evolution*, 39.
25. Bergson, *Creative Evolution*, 2.
26. Bergson, *Creative Evolution*, 4.
27. Bergson, *Creative Evolution*, 126.

28. Bergson, *Creative Evolution*, 128.
29. Bergson, *Creative Evolution*, 47–48.
30. Murphy, "Cather's Ruskinian Landscapes," 229–30.
31. Murphy, "Cather's Ruskinian Landscapes," 228–29, 236.
32. Cather, *Willa Cather on Writing*, 79–80.
33. See, e.g., Frank, *Literary Architecture*, which includes chapters on Pater as well as Gerard Manley Hopkins, Henry James, and Marcel Proust.
34. See James's 1908 preface to *The Portrait of a Lady*, reprinted in James, *The Art of the Novel*, 40–58.
35. Cather, *Willa Cather on Writing*, 40.
36. Cather, *Willa Cather on Writing*, 42.
37. Cather, *Willa Cather on Writing*, 39–40; emphasis added.
38. Cather, *Willa Cather on Writing*, 43.
39. Stilgoe, Foreword, x.
40. See, e.g., Apthorp, "Re-visioning Creativity"; and Gelfant, *Women Writing in America*.
41. Cather, *Willa Cather on Writing*, 103.
42. See, e.g., Hemingway, *Death in the Afternoon*. For a sustained meditation on Cather's relationship to other "moderns," see Stout, *Cather among the Moderns*.
43. See Karush, "Bringing Outland Inland in *The Professor's House*."
44. Cather, *Willa Cather on Writing*, 31.
45. Cather, *Willa Cather on Writing*, 31.
46. Cather, *Willa Cather on Writing*, 31–32.
47. For a version of this chapter that more explicitly frames Cather's use of landscape description as a (macro)focalization device, see Eggan, "Landscape Metaphysics."
48. Cather, *The Professor's House*, 3. Further citations from this book appear parenthetically in the main text.
49. Slote, "The Kingdom of Art," 79.
50. Perhaps not incidentally, Tom's admiration of the cliffs' architectural artistry recalls the words of James Simpson, an employee of the U.S. Topographical Corps of Engineers who surveyed Chaco Canyon in 1849 and described the canyon and its ruins in a way that implicitly dismisses contemporary Indigenous claims to both. Simpson describes how his team "discover[ed] in the materials of which [the Chaco Canyon ruins] are composed, as well as in the grandeur of their design and superiority of their workmanship, a condition of architectural excellence beyond the power of the Indians or New Mexicans of the present day to exhibit" (qtd. in Byszewski, "Colonizing Chaco Canyon," 57).
51. Bergson, *Creative Evolution*, 4.
52. Quirk, *Bergson and American Culture*, 146.
53. This reading admittedly makes the novel's ending seem more straightforward than it really is. However, even Godfrey's thoughts about death and his ambiguous suicide attempt can be understood as a form of ontological homecoming. Godfrey seems to envision the grave as itself a kind of house when he quotes from Longfellow: "For thee a house was built / Ere thou was born; / For thee a mould was

made / Ere thou of woman came" (248). In these lines, the threat of the tomb transforms into the welcoming warmth of the home, a womb–tomb that further recalls Heidegger's vision of death as a kind of homecoming: "As the shrine of Nothing, death is the shelter of Being" (*Poetry, Language, Thought*, 176).

54. Wharton and Codman, *The Decoration of Houses*, 65, 67.
55. Eliade, *The Sacred and the Profane*, 32.
56. Veracini, *Settler Colonialism*, 81.
57. Veracini, *Settler Colonialism*, 93.
58. See Vizenor, ed., *Survivance*.
59. Steinhagen, "Dangerous Crossings," 64.
60. Steinhagen, "Dangerous Crossings," 79.
61. Steinhagen, "Dangerous Crossings," 65.
62. Steinhagen, "Dangerous Crossings," 65.
63. This image has been variously contested (see, e.g., Krech, *The Ecological Indian*). Sarah Jacquette Ray has also written incisively of the problem of the "ecological Indian": "Ecological subjects seek to emulate the Indian body's connection to nature, obscuring the ways that those very bodies have been drawn into environmentally exploitative colonial, capitalist, and military projects, and how both Indian bodies and those environments have been sacrificed" (Ray, *The Ecological Other*, 9–10).
64. Steinhagen, "Dangerous Crossings," 77.
65. Steinhagen, "Dangerous Crossings," 78.

3. D. H. LAWRENCE AND THE ECOLOGICAL UNCANNY

1. Luhan, *Lorenzo in Taos*, 4.
2. Luhan, *Lorenzo in Taos*, 4.
3. Franks, Introduction, xiii; emphasis added.
4. Lawrence, *Sea and Sardinia*, 7.
5. Lawrence, *Sea and Sardinia*, 7.
6. Lawrence, *Sea and Sardinia*, 7.
7. Lawrence, *Sea and Sardinia*, 8.
8. Lawrence, *Sea and Sardinia*, 8
9. Luhan, *Lorenzo in Taos*, 3.
10. Qtd. in Merrild, *With D. H. Lawrence in New Mexico*, 28.
11. Luhan, *Lorenzo in Taos*, 5.
12. Newmark, "Sensing Re-placement in New Mexico," 161.
13. See, e.g., Roberts, *D. H. Lawrence, Travel and Cultural Difference*.
14. Merrild, *With D. H. Lawrence in New Mexico*, 105.
15. Lawrence, *Phoenix*, 141.
16. Lawrence, *Phoenix*, 143; emphasis added.
17. Lawrence, *Phoenix*, 146–47.
18. Luhan, "Lawrence of New Mexico," 9.
19. Huxley, *With D. H. Lawrence in New Mexico*, xvii.

20. Huxley, *With D. H. Lawrence in New Mexico*, xvii–xviii.
21. Huxley, *With D. H. Lawrence in New Mexico*, xix; emphasis added.
22. Lawrence, *Studies in Classic American Literature*, 12.
23. See, e.g., Ehlert, "There's a Bad Time Coming"; and Michelucci, *Space and Place in the Works of D. H. Lawrence*.
24. See, e.g., Alcorn, *The Nature Novel from Hardy to Lawrence*; and Ebbatson, *Lawrence and the Nature Tradition*.
25. See, e.g., LaChapelle, *D. H. Lawrence: Future Primitive*.
26. See, e.g., Delany, "D. H. Lawrence and Deep Ecology."
27. Lawrence, *Studies in Classic American Literature*, 12.
28. See Tindall, *D. H. Lawrence and Susan His Cow*.
29. Lawrence, *The Letters of D. H. Lawrence*, 183.
30. Lawrence, *The Letters of D. H. Lawrence*, 183.
31. Lawrence, *The Rainbow*, 111.
32. Lawrence, *A Selection from Phoenix*, 161.
33. Lawrence, *Phoenix*, 528.
34. Lawrence, *Phoenix*, 528.
35. Lawrence, *Phoenix*, 531.
36. Lawrence, *Phoenix*, 517.
37. Lawrence, *Phoenix*, 757.
38. Lawrence, *Phoenix*, 757.
39. Lawrence, *Phoenix*, 756.
40. Lawrence, *Phoenix*, 759.
41. Lawrence, *Studies in Classic American Literature*, 16.
42. Lawrence, *Studies in Classic American Literature*, 13.
43. Lawrence, *Studies in Classic American Literature*, 16.
44. Lawrence, *A Selection from Phoenix*, 454.
45. Lawrence, *A Selection from Phoenix*, 453–54.
46. Lawrence, *A Selection from Phoenix*, 455.
47. Lawrence, *A Selection from Phoenix*, 453.
48. Bogost, *Alien Phenomenology*, 11.
49. Lawrence, *A Selection from Phoenix*, 453.
50. Lawrence, *A Selection from Phoenix*, 458.
51. Lawrence, *A Selection from Phoenix*, 454.
52. See, e.g., Gutierrez, "The Ancient Imagination of D. H. Lawrence"; LaChapelle, *D. H. Lawrence: Future Primitive*; and Tindall, "D. H. Lawrence and the Primitive."
53. Skrbina, *Panpsychism in the West*, 19.
54. Lawrence, *Studies in Classic American Literature*, 119.
55. Lawrence, *Studies in Classic American Literature*, 151–52.
56. Lawrence, *Phoenix*, 580.
57. For a study of how Cézanne and Lawrence shared an "intuitive consciousness" that enabled both artists to represent this flux in their respective media, see Janik, "Toward 'Thingliness.'"
58. Lawrence, *Phoenix*, 580–81.

59. "Weird, adj. 1 and 2a," *OED Online*.
60. "Anima, n. 1," *OED Online*.
61. Lawrence, *Fantasia of the Unconscious*, 42.
62. Lawrence, *Fantasia of the Unconscious*, 43.
63. Lawrence, *Fantasia of the Unconscious*, 43.
64. Lawrence, *Fantasia of the Unconscious*, 46; emphasis added.
65. Lawrence, *Fantasia of the Unconscious*, 22.
66. Lawrence, *The Woman Who Rode Away/St. Mawr/The Princess*, 41. Further citations from this book appear parenthetically in the main text.
67. Lasdun, Introduction, xv–xvi.
68. Lasdun, Introduction, xvi.
69. Gutierrez, "The Ancient Imagination of D. H. Lawrence," 191.
70. Gutierrez, "The Ancient Imagination of D. H. Lawrence," 192.
71. See, e.g., Booth, "Lawrence in Doubt"; and Roberts, *D. H. Lawrence, Travel and Cultural Difference*.
72. See, e.g., Chaudhuri, *D. H. Lawrence and "Difference"*; and Oh, *D. H. Lawrence's Border Crossing*.
73. On this point see Wood, *The Broken Estate*, 122–36.
74. Lawrence, "Certain Americans and an Englishman," 3.
75. Baer, "—your Ghost-Work . . . ," 118.
76. Newmark, "Sensing Re-Placement in New Mexico," 158.
77. Lawrence, *Phoenix*, 92.
78. Snyder, "When the Indian Was in Vogue," 673.
79. See P. Wolfe, "Settler Colonialism and the Elimination of the Native."
80. Baer, "—your Ghost-Work . . . ," 122.
81. Lawrence, *Studies in Classic American Literature*, 57, 41.
82. Lawrence, *Studies in Classic American Literature*, 41.
83. Lawrence, *Studies in Classic American Literature*, 41.
84. Lawrence, *Studies in Classic American Literature*, 56.
85. Baer, "—your Ghost-Work . . . ," 134.
86. Baer, "—your Ghost-Work . . . ," 134–35.
87. On this topic, see the essay "Democracy," reprinted in Lawrence, *Phoenix*, 699–718.
88. Lawrence, *Mornings in Mexico*, 58, 60.
89. Baer, "—your Ghost-Work . . . ," 186.
90. Deloria, *Playing Indian*, 4.
91. Thacker, *In the Dust of This Planet*, 8–9.
92. See Harman, *Weird Realism*.
93. Thacker, *In the Dust of This Planet*, 53.
94. Derrida, *Specters of Marx*, xx.
95. Lawrence, *Mornings in Mexico*, 55.
96. Baer, "—your Ghost-Work . . . ," 122.
97. Lawrence, *Studies in Classic American Literature*, 57.
98. Lawrence, *Studies in Classic American Literature*, 68.

EXCURSUS I. ECOLOGICAL REALISM

1. Morton, *The Ecological Thought*, 92; emphasis added.
2. See Gibson, *Reasons for Realism*.
3. Mace, "James J. Gibson's Strategy for Perceiving," 43.
4. As this veiled reference to Heidegger's concept of "thrownness-into-being" (*Geworfenheit-ins-Dasein*) suggests, Gibson's theory of environmental affordances resonates with the ontology presented in Heidegger's early work (see G. Williams, "Ecological Realism and Affordance Ontology").
5. Reed, "Knowers Talking about the Known," 17; emphasis added.
6. Reed, "Knowers Talking about the Known," 17–18.
7. Lukács, *The Meaning of Contemporary Realism*, 76. Lukács understood "naturalism" and its ideological sympathies as a form of modernism rather than an outgrowth of realism.
8. Lukács, *The Meaning of Contemporary Realism*, 24.
9. Lukács, *The Meaning of Contemporary Realism*, 19.
10. Firestein, *Ignorance*, 7.
11. I lack space to mount a full-scale attack on the long-standing dominance of postmodern cynicism in literary theory. However, recent scholarship has begun to challenge this dominance. For a reassessment of literary realism specifically, see Wood, *How Fiction Works*. For a general essay on the limits of the hermeneutics of suspicion, see Felski, *The Limits of Critique*.
12. Ndebele, "The Rediscovery of the Ordinary," 152.
13. I owe the connection between Lukács and Ndebele to Lazarus, "Realism and Naturalism in African Fiction."
14. Lukács, *The Meaning of Contemporary Realism*, 24.
15. Lukács, *The Meaning of Contemporary Realism*, 39.
16. Lukács, *The Meaning of Contemporary Realism*, 24.
17. Crane, "Surface, Depth, and the Spatial Imaginary," 78–79.
18. Best and Marcus, "Surface Reading," 9.
19. Best and Marcus, "Surface Reading," 13; emphasis added.
20. I owe this distinction to Reilly, "Always Sympathize!"
21. Best and Marcus, "Surface Reading," 9.
22. Morton, *The Ecological Thought*, 47.
23. Iser, *The Act of Reading*, 182.
24. Iser, *The Act of Reading*, 182–83.
25. Evernden, *The Natural Alien*, 148.
26. I owe the black-hole analogy to Bogost, *Alien Phenomenology*, 33.
27. Jameson, *The Antimonies of Realism*, 56.
28. Jameson, *The Antimonies of Realism*, 47.
29. Jameson, *The Antimonies of Realism*, 32.
30. Jameson, *The Antimonies of Realism*, 37.
31. Jameson, *The Antimonies of Realism*, 59, 64.
32. Jameson, *The Antimonies of Realism*, 37.

33. Jameson, *The Antimonies of Realism*, 38.
34. Jameson, *The Antimonies of Realism*, 52.
35. Jameson, *The Antimonies of Realism*, 54–55; emphasis added.
36. See Lukács, *Studies in European Realism*, 97–125.

4. (UN)SETTLING THE SOUTHERN AFRICAN FARM/WORLD

1. See Heidegger, *Being and Time*, 79.
2. See, e.g., Mazoyer and Roudart, *A History of World Agriculture*.
3. See, e.g., Ramachandra Guha, *The Unquiet Woods*; and Ranajit Guha, *A Rule of Property for Bengal*.
4. Stilgoe, *Common Landscape of America*, 137.
5. Stilgoe, *Common Landscape of America*, 137.
6. Crèvecoeur, *Letters from an American Farmer*, 69.
7. Crèvecoeur, *Letters from an American Farmer*, 87.
8. Crèvecoeur, *Letters from an American Farmer*, 54; emphasis added.
9. Lawrence, *Studies in Classic American Literature*, 31.
10. The connections between Indigenous genocide and African transport and enslavement adds a further complicating factor that has only recently received scholarly attention (see, e.g., King, *The Black Shoals*; Rifkin, *Fictions of Land and Flesh*; and Wynter, "Unsettling the Coloniality of Being/Power/Truth/Freedom").
11. Qtd. in Spillman, *British Colonial Realism in Africa*, 175.
12. Spillman, *British Colonial Realism in Africa*, 175; emphasis added.
13. Barnard, *Hunters and Herders in Southern Africa*, 242.
14. Locke, *Two Treatises of Government*, 292.
15. Marx, *Capital*, 1:90.
16. Qtd. in Giliomee, *The Afrikaners*, 2.
17. Giliomee, *The Afrikaners*, 130–60.
18. In his revision of this narrative, the Shona historian Lawrence Vambe rejects the notion that the Shona were under Ndebele rule. He argues that this lie was in fact perpetuated by the Ndebele ruler Lobengula, who was as much a self-interested power seeker as Rhodes (see Vambe, *An Ill-Fated People*).
19. Vambe, *An Ill-Fated People*, 42–43.
20. Ranger, *Peasant Consciousness and Guerrilla War in Zimbabwe*, 44–45.
21. Ranger, *Peasant Consciousness and Guerrilla War in Zimbabwe*, 46.
22. Qtd. in Ranger, *Peasant Consciousness and Guerrilla War in Zimbabwe*, 66.
23. This is a fundamental trope in African environmental history, which has amply documented the centrality of ecological discourse (e.g., preservationism, conservationism) in the vilification of African agricultural practices as well as in relation to the colonial alienation of land. For a touchstone example, see Neumann, *Imposing Wilderness*.
24. Ranger, *Peasant Consciousness and Guerrilla War in Zimbabwe*, 69.
25. See, e.g., Malherbe, *Die meulenaar* (The miller); van Bruggen, *Ampie*; and van den Heever, *Harvest Home*.

26. Coetzee, "Farm Novel and Plaasroman in South Africa," 15.
27. See, e.g., Brink, *A Chain of Voices;* and Leroux, *To a Dubious Salvation.*
28. For useful histories of the farm novel genre, see Coetzee, "Farm Novel and Plaasroman in South Africa"; Devarenne, "Nationalism and the Farm Novel in South Africa"; Olivier, "The Dertigers and the *Plaasroman*"; Marquard, "The Farm"; and Rooney, "Narratives of Southern African Farms."
29. See Eliade, *The Sacred and the Profane.*
30. Coetzee, "Farm Novel and Plaasroman in South Africa," 2.
31. Coetzee, "Farm Novel and Plaasroman in South Africa," 6.
32. Reprinted in Haresnape, "Why and How I Became an Author," 151.
33. Head, *When Rain Clouds Gather,* 28, 16.
34. Head, *A Question of Power,* 19.
35. Coetzee, *White Writing,* 83.
36. Coetzee, *White Writing,* 85.
37. Coetzee, *White Writing,* 85.
38. Coetzee, *White Writing,* 88.
39. Van den Heever, *Harvest Home,* 17, 89. Further citations from this book appear parenthetically in the main text.
40. Kolodny, *The Lay of the Land,* 4; emphasis added.
41. Van Niekerk, "Afrikaner Woman and Her 'Prison,'" 147.
42. Devarenne, "Nationalism and the Farm Novel in South Africa," 636.
43. Devarenne, "Nationalism and the Farm Novel in South Africa," 636.
44. For an alternative reading that interprets Leroux's novel as reinforcing Afrikaner nationalism, see Sheer, "Etienne Leroux's *Sewe Dae by die Silbersteins.*"
45. Brink, *A Chain of Voices,* 125.
46. Brink, *A Chain of Voices,* 24–25.
47. Brink, *A Chain of Voices,* 24; emphasis added.
48. Brink, *A Chain of Voices,* 39.
49. See Jameson, "Third-World Literature in the Era of Multinational Capital."
50. See, e.g., Ahmad, *In Theory,* 95–122.
51. See, e.g., George, *Relocating Agency,* 105–44.
52. See JanMohamed, *Manichean Aesthetics.*
53. On this point, see Coetzee, *White Writing,* 5; and Wenzel, "The Pastoral Promise and the Political Imperative," 94–95.
54. For a recent example of postcolonial allegory that works outside the confines of a Manichean analytic, see DeLoughrey, *Allegories of the Anthropocene.*
55. Veracini, *Settler Colonialism,* 16.
56. Veracini, *Settler Colonialism,* 30.
57. Crosby, *Ecological Imperialism.* For a more recent study of ecological imperialism, see, e.g., Casid, *Sowing Empire.*
58. Coetzee, *White Writing,* 7.
59. Anthony, "Buried Narratives," 12.
60. Gray, *Southern African Literature,* 139.
61. Coetzee, *Boyhood,* 79; emphasis added.

62. Coetzee, *Boyhood*, 91.
63. Coetzee, *Boyhood*, 91.
64. Coetzee, *Boyhood*, 95–96.

5. ALLEGORY, REALISM, AND UNCANNY ECOLOGY ON OLIVE SCHREINER'S AFRICAN FARM

1. Spillman, *British Colonial Realism in Africa*, 191.
2. See, e.g., Horton, *Difficult Women, Artful Lives;* and Lewis, *White Women Writers and Their African Invention*.
3. McClintock, *Imperial Leather*, 265.
4. McClintock, *Imperial Leather*, 265.
5. Schreiner, *The Story of an African Farm*, 1. Further citations from this book appear parenthetically in the main text.
6. McClintock, *Imperial Leather*, 265.
7. McClintock, *Imperial Leather*, 266.
8. Anthony, "Buried Narratives," 12.
9. Horton, *Difficult Women, Artful Lives*, 181.
10. Coetzee, "Farm Novel and Plaasroman in South Africa," 1.
11. Coetzee, "Farm Novel and Plaasroman in South Africa," 4.
12. Coetzee, "Farm Novel and Plaasroman in South Africa," 2.
13. Coetzee, "Farm Novel and Plaasroman in South Africa," 4.
14. Coetzee, "Farm Novel and Plaasroman in South Africa," 2.
15. Fletcher, *Allegory*, 2.
16. Fletcher, *Allegory*, 21.
17. Fletcher, *Allegory*, 23.
18. McClintock, *Imperial Leather*, 279–80.
19. Fletcher, *Allegory*, 102, 105.
20. Fletcher, *Allegory*, 7.
21. Fletcher, *Allegory*, 7.
22. De Man, *Allegories of Reading*, ix.
23. Jameson, *Antinomies of Realism*, 33.
24. I use "broken" intentionally to invoke the theory of reading outlined in Barthes, *S/Z*.
25. Monsman, "Olive Schreiner's Allegorical Vision," 50.
26. Gorak, "Olive Schreiner's Colonial Allegory," 61.
27. Gorak, "Olive Schreiner's Colonial Allegory," 56.
28. Gorak, "Olive Schreiner's Colonial Allegory," 61.
29. Gorak, "Olive Schreiner's Colonial Allegory," 56.
30. Gorak, "Olive Schreiner's Colonial Allegory," 56–57.
31. McClintock, *Imperial Leather*, 280.
32. Spillman, *British Colonial Realism in Africa*, 193.
33. See R. Williams, *The English Novel from Dickens to Lawrence*.

34. Spillman, *British Colonial Realism in Africa*, 177, 180.
35. Schreiner, *Thoughts on South Africa*, 49.
36. Schreiner, *Thoughts on South Africa*, 49–50.
37. Schreiner, *Thoughts on South Africa*, 61.
38. Here is one example of this rhetorical gesture: "Must we for ever remain a vast, inchoate, invertebrate mass of humans, divided horizontally into layers of race, mutually antagonistic, and vertically severed by lines of political state division, which cut up our races without simplifying our problems, and which add to the bitterness of race conflict the irritation of political division? Is national life and organization unattainable by us? We believe that no one can impartially study the condition of South Africa and feel that it is so" (*Thoughts on South Africa*, 60).
39. Schreiner, *Thoughts on South Africa*, 63.
40. Hegel, *Philosophy of Mind*, 57.
41. Qtd. in Horton, *Difficult Women, Artful Lives*, 176–77.
42. Qtd. in Horton, *Difficult Women, Artful Lives*, 177.
43. See, e.g., Azzam, *The Alien Within*.
44. Cohen, *Stone*, 99.
45. See Kolbert, *The Sixth Extinction*, 23–46.
46. Schreiner, *Thoughts on South Africa*, 38; emphasis added.
47. Schreiner, *Thoughts on South Africa*, 39–40.
48. Schreiner, *Thoughts on South Africa*, 40.
49. Schreiner, *Thoughts on South Africa*, 40.
50. I borrow the term *imperial eyes* from Pratt, *Imperial Eyes*.
51. Schreiner, *Thoughts on South Africa*, 38.
52. Schreiner, *Thoughts on South Africa*, 41.
53. Anthony, "Buried Narratives," 12.
54. Morton, *The Ecological Thought*, 107. Morton cites the following passage: "There's a fine young feller aboard of it, Mrs. Dempster wagered, and away and away it went, fast and fading, away and away the aeroplane shot; soaring over Greenwich and all the masts; over the little island of grey churches, St. Paul's and the rest till, on either side of London, fields spread out and dark brown woods where adventurous thrushes hopping boldly, glancing quickly, snatched the small snail and tapped him on a stone, once, twice, thrice" (Woolf, *Mrs. Dalloway*, 28).
55. Coetzee, *Boyhood*, 95.
56. Coetzee, *Boyhood*, 91, 79.
57. Schreiner, *Thoughts on South Africa*, 38.

6. DORIS LESSING'S ECOLOGICAL REALISM

1. Lessing, *Under My Skin*, 50.
2. Lessing, *Under My Skin*, 54.
3. Lessing, *Under My Skin*, 50.
4. Lessing, *Under My Skin*, 52.

5. Lessing, *Under My Skin*, 54–55.
6. Lessing, *African Laughter*, 35.
7. Lessing, *Under My Skin*, 202.
8. Lessing, *A Small Personal Voice*, 98.
9. Lessing, *A Small Personal Voice*, 99.
10. Lessing, *African Laughter*, 35.
11. Lessing, *The Golden Notebook*, 397.
12. Lessing, *The Golden Notebook*, 398.
13. Lessing, *The Golden Notebook*, 397–98.
14. Lessing, *The Golden Notebook*, 398.
15. Lessing, *The Golden Notebook*, 398.
16. Lessing, *Under My Skin*, 124.
17. Lessing, *Under My Skin*, 62.
18. Lessing, *Under My Skin*, 62.
19. Morton, *The Ecological Thought*, 41.
20. Qtd. in Spillman, *British Colonial Realism in Africa*, 181.
21. Lessing, *Under My Skin*, 275.
22. Lessing, *Under My Skin*, 293.
23. Lessing, *Under My Skin*, 275.
24. Lessing, *A Small Personal Voice*, 4–5.
25. Lessing, *A Small Personal Voice*, 4.
26. The Children of Violence series includes *Martha Quest* (1952), *A Proper Marriage* (1954), *A Ripple from the Storm* (1958), *Landlocked* (1965), and *The Four-Gated City* (1969).
27. Lessing, *Time Bites*, 257.
28. Qtd. in Lessing, *Under My Skin*, unnumbered page.
29. Lessing, *Time Bites*, 259.
30. Lessing, *Time Bites*, 259.
31. Lessing, *Canopus in Argos*, xii.
32. Leonard, "The Spacing Out of Doris Lessing."
33. Lessing, *Canopus in Argos*, xii.
34. Lessing, *Under My Skin*, 10.
35. Lessing, *A Small Personal Voice*, 7.
36. Lessing, *A Small Personal Voice*, 7.
37. Lessing, *A Small Personal Voice*, 9.
38. Fishburn, "Wor(l)ds within Words," 187, 186.
39. Fishburn, "Wor(l)ds within Words," 188.
40. Lessing, *The Golden Notebook*, 521.
41. Lessing, *African Stories*, 62.
42. Lessing, *African Stories*, 63.
43. Lessing, *African Stories*, 63.
44. Lessing, *African Stories*, 63.
45. Lessing, *African Stories*, 64.
46. Lessing, *African Stories*, 64.

47. Lessing, *African Stories*, 64–65.
48. Lessing, *African Stories*, 65.
49. See, e.g., Lessing, "The Antheap," reprinted in *African Stories*, 353–403.
50. Lessing, *The Grass Is Singing*, 1. Further citations from this book appear parenthetically in the main text.
51. Marquard, "The Farm," 299.
52. Rooney, "Narratives of Southern African Farms," 431.
53. Fishburn, "Manichean Allegories," 6; emphasis added.
54. Fishburn, "Manichean Allegories," 8.
55. See, e.g., Achebe, "An Image of Africa."
56. Fishburn, "Wor(l)ds within Words," 191.
57. Lessing, *A Small Personal Voice*, 17.

EXCURSUS II. EXO-PHENOMENOLOGY

1. Slemon, "Unsettling the Empire," 39.
2. Bennett, *Influx & Efflux*, xxiv–xxv.
3. Sheldrake, *Entangled Life*, 19.
4. See Kirby, "Un/Limited Ecologies."
5. The phrase "more-than-human world" comes from Abram, *The Spell of the Sensuous*.
6. See, e.g., Deleuze and Guattari, *A Thousand Plateaus*, 3–25; and Evernden, *The Natural Alien*, 35–54.
7. "Language" and "signs" need not be limited to human semiotics. For an illuminating project that uses Peircean semiotics to study multispecies communication, see Kohn, *How Forests Think*.
8. Levinas, *Totality and Infinity*, 39.
9. Levinas, *Totality and Infinity*, 171.
10. For a perceptive study of how Levinas's ethical approach to hospitality functions as a rebuttal to Heidegger's ontological approach to dwelling, see Gauthier, *Martin Heidegger, Emmanuel Levinas, and the Politics of Dwelling*, 103–54.
11. In this I take the lead from many commentators who have demonstrated how Levinas's humanism might be rehabilitated to allow for a more-than-human ethics (see, e.g., Edelglass, Hatley, and Diehm, eds., *Facing Nature*; and Larios, "Emmanuel Levinas and the Meaning of Ecological Responsibility").
12. Llewelyn, "Writing Home," 177.
13. Heller-Roazen, *The Inner Touch*.
14. Bogost, *Alien Phenomenology*, 67.
15. Loveless, *How to Make Art at the End of the World*, 46.
16. Loveless, *How to Make Art at the End of the World*, 47.
17. Coetzee, *In the Heart of the Country*, 4. Further citations from this book appear parenthetically in the main text.
18. Morton, *The Ecological Thought*, 47.

19. Bryant, *The Democracy of Objects*, 276.
20. See, e.g., Haraway, *Staying with the Trouble*, 58 and passim.
21. The concept of object withdrawal plays an important role in object-oriented ontology. For instance, its observation that the "essence" or "virtual proper being" of an object is withdrawn from all access helps account for its larger claim that no object depends on another for its ontological ground (see, e.g., Bryant, *The Democracy of Objects*; and Morton, *Realist Magic*).
22. Eliot, *The Complete Poems and Plays*, 56.
23. Here I adopt Jakob von Uexküll's term *Umwelt* to reference the perceptual world of a species (see Uexküll, *A Foray into the Worlds of Animals and Humans*).
24. See Thompson, *A History of South Africa*, 10–30.
25. Marder, *Plant-Thinking*, 186.
26. Du Bois, *The Souls of Black Folk*, 38.
27. See, e.g., BioVisions Lab, "The Inner Life of the Cell."
28. Koch, *The Quest for Consciousness*, 35.
29. Koch, *The Quest for Consciousness*, 255.
30. Deleuze, "Bartleby; or the Formula," 82.
31. This represents the basic definition of what Quentin Meillassoux has termed "correlationism" (see Meillassoux, *After Finitude*, 5).
32. Morton, *The Ecological Thought*, 47.
33. Wylie, "A Landscape Cannot Be a Homeland," 414.
34. Barnd, *Native Space*, 2–3.
35. Wynter, "Unsettling the Coloniality of Being/Power/Truth/Freedom," 266, 288.
36. Barnd, *Native Space*, 5.
37. Barnd, *Native Space*, 5; emphasis added.
38. Barnd, *Native Space*, 7.
39. Barnd, *Native Space*, 5.
40. Kimmerer, *Braiding Sweetgrass*, 3.
41. Kimmerer, *Braiding Sweetgrass*, 4.
42. Kimmerer, *Braiding Sweetgrass*, 8; emphasis added.
43. Kimmerer, *Braiding Sweetgrass*, 5.
44. Kimmerer, *Braiding Sweetgrass*, 9.
45. Kimmerer, *Braiding Sweetgrass*, 9.
46. Kimmerer, *Braiding Sweetgrass*, 10.
47. Tuck and Yang, "Decolonization Is Not a Metaphor," 9–28.
48. Ortega, "Decolonial Woes and Practices of Un-knowing." For an allied critique emphasizing the dominance of northern academics in decolonial theory, see Rivera Cusicanqui, *Ch'ixinakax utxiwa*, 46–70.
49. I refer only in part to the Jewish political Zionism that helped establish the settler state of Israel and underwrites its ongoing violence against Palestinians. (Indeed, a truly Levinasian approach to exo-phenomenology would have to address Levinas's own ambivalent perspective on the Palestinian Other [see Caro, "Levinas and the Palestinians"].) The Mormon settlement of present-day Utah was likewise

founded on displacement, in this case of the Northern and Southern Paiute peoples (see Quammen, *American Zion*).

50. See Haraway, *Staying with the Trouble*, esp. 58–98.

51. See Kohn, *How Forests Think;* Parreñas, *Decolonizing Extinction;* and Tsing, *The Mushroom at the End of the World*.

52. See, e.g., Viveiros de Castro, "Cosmological Deixis and Amerindian Perspectivism."

53. See, e.g., de la Cadena, *Earth Beings;* and de la Cadena and Blaser, eds., *A World of Many Worlds*.

Bibliography

Abram, David. *Becoming Animal: An Earthly Cosmos*. New York: Pantheon, 2010.
———. *The Spell of the Sensuous: Perception and Language in a More-Than-Human World*. New York: Pantheon, 1996.
Achebe, Chinua. "An Image of Africa: Racism in Conrad's *Heart of Darkness*." In *Hopes and Impediments: Selected Essays*, 1–20. New York: Anchor, 1988.
Adorno, Theodor. *The Jargon of Authenticity*. London: Routledge, 2007.
Ahmad, Aijaz. *In Theory: Classes, Nations, Literatures*. London: Verso, 1992.
Ahmed, Sara. "A Phenomenology of Whiteness." *Feminist Theory* 8, no. 2 (2007): 149–68.
———. *Queer Phenomenology: Orientations, Objects, Others*. Durham, NC: Duke University Press, 2006.
Alcorn, John. *The Nature Novel from Hardy to Lawrence*. New York: Columbia University Press, 1977.
Alimoda, Héctor. "The Coloniality of Nature: An Approach to Latin American Political Ecology." Translated by Alexander D'Aloia. www.alternautas.net/blog/2019/6/10/the-coloniality-of-nature-an-approach-to-latin-american-political-ecology.
Ammons, Elizabeth. "Cather and the New Canon: 'The Old Beauty' and the Issue of Empire." *Cather Studies* 3 (1996): 256–66.
Angelova, Lidiya. "Phenomenology of Home." Ph.D. diss., University of South Florida, 2010.
Anthony, Loren. "Buried Narratives: Masking the Sign of History in *The Story of an African Farm*." *Scrutiny2* 4, no. 2 (1999): 3–13.
Antolick, Matthew. "Deep Ecology and Heideggerian Phenomenology." Master's thesis, University of South Florida, 2003.
Apthorp, Elaine Sargent. "Re-Visioning Creativity: Cather, Chopin, Jewett." *Legacy* 9, no. 1 (1992): 1–22.
Azzam, Julie Hakim. "The Alien Within: Postcolonial Gothic and the Politics of Home." Ph.D. diss., University of Pittsburgh, 2007.
Bachelard, Gaston. *The Poetics of Space*. Translated by Maria Jolas. Boston: Beacon, 1994.
Bacon, J. M. "Settler Colonialism as Eco-Social Structure and the Production of Colonial Ecological Violence." *Environmental Sociology* 5, no. 1 (2019): 59–69.

Baer, Ben Conisbee. "'—your Ghost-Work . . .': Figures of the Peasant and the Autochthon in Literature and Politics, 1880s–1940s." Ph.D. diss., Columbia University, 2006.
Barad, Karen. "Troubling Time/s and Ecologies of Nothingness: Re-turning, Re-membering, and Facing the Incalculable." In Fritsch, Lynes, and Wood, *Eco-Deconstruction*, 206–48.
Barnard, Alan. *Hunters and Herders of Southern Africa: A Comparative Ethnography of the Khoisan Peoples*. Cambridge: Cambridge University Press, 1992.
Barnd, Natchee Blu. *Native Space: Geographic Strategies to Unsettle Settler Colonialism*. Corvallis: Oregon State University Press, 2017.
Barthes, Roland. *S/Z: An Essay*. Translated by Richard Miller. New York: Hill and Wang, 1974.
Bate, Jonathan. *The Song of the Earth*. Cambridge, MA: Harvard University Press, 2000.
Bennett, Jane. *Influx & Efflux: Writing Up with Walt Whitman*. Durham, NC: Duke University Press, 2020.
Bergson, Henri. *Creative Evolution*. 1911. Translated by Arthur Mitchell. New York: Dover, 1998.
———. *Matter and Memory*. 1896. Translated by N. M. Paul and W. S. Palmer. New York: Zone, 1988.
———. *Time and Free Will: An Essay on the Immediate Data of Consciousness*. 1889. Translated by F. L. Pogson. New York: Dover, 2001.
Best, Stephen, and Sharon Marcus. "Surface Reading: An Introduction." *Representations* 108, no. 1 (2009): 1–21.
Bigwood, Carol. "Logos of Our Eco in the Feminine: An Approach through Heidegger, Irigaray, and Merleau-Ponty." In Cataldi and Hamrick, *Merleau-Ponty and Environmental Philosophy*, 93–116.
BioVisions Lab (Harvard University). "The Inner Life of the Cell." www.youtube.com/watch?v=wJyUtbn0O5Y.
Bogost, Ian. *Alien Phenomenology, or What It's Like to Be a Thing*. Minneapolis: University of Minnesota Press, 2012.
Booth, Howard. "Lawrence in Doubt: A Theory of the 'Other' and Its Collapse." In *Modernism and Empire: Writing and British Coloniality, 1890–1940*, edited by Booth and Nigel Rigby, 197–223. Manchester: Manchester University Press, 2000.
Brink, André. *A Chain of Voices*. Naperville, IL: Sourcebooks, 2007.
Brown, Charles S., and Ted Toadvine. "Eco-Phenomenology: An Introduction." In Brown and Toadvine, eds., *Eco-Phenomenology*, ix–xxi.
———, eds. *Eco-Phenomenology: Back to the Earth Itself*. Albany: State University of New York Press, 2003.
Bryant, Levi R. *The Democracy of Objects*. Ann Arbor, MI: Open Humanities Press, 2011.
Buell, Lawrence. *The Future of Environmental Criticism: Environmental Crisis and Literary Imagination*. Malden, MA: Blackwell, 2005.
Burke, Edmund. *A Philosophical Enquiry into the Origin of Our Ideas of the Sublime and Beautiful*. 1757. Oxford: Oxford University Press, 1990.

Byrd, Jodi A. *The Transit of Empire: Indigenous Critiques of Colonialism.* Minneapolis: University of Minnesota Press, 2011.

Byszewski, Berenika. "Colonizing Chaco Canyon: Mapping Antiquity in the Territorial Southwest." In *Formations of United States Colonialism,* edited by Alyosha Goldstein, 57–86. Durham, NC: Duke University Press, 2014.

Caro, Jason. "Levinas and the Palestinians." *Philosophy & Social Criticism* 36, no. 6 (2009): 671–84.

Casid, Jill. *Sowing Empire: Landscape and Colonization.* Minneapolis: University of Minnesota Press, 2005.

Cataldi, Suzanne L., and William S. Hamrick, eds. *Merleau-Ponty and Environmental Philosophy: Dwelling on the Landscapes of Thought.* Albany: State University of New York Press, 2007.

Cather, Willa. *Death Comes for the Archbishop.* 1927. New York: Vintage, 1990.

———. *The Kingdom of Art: Willa Cather's First Principles and Critical Statements 1893–1896.* Lincoln: University of Nebraska Press, 1966.

———. *My Ántonia.* 1918. New York: Vintage, 1994.

———. *O Pioneers!* 1913. New York: Vintage, 1992.

———. *The Professor's House.* 1925. New York: Vintage, 1990.

———. *Shadows on the Rock.* 1931. New York: Vintage, 1995.

———. *The Song of the Lark.* 1915. New York: Vintage, 1999.

———. *Willa Cather in Person: Interviews, Speeches, and Letters.* Lincoln: University of Nebraska Press, 1986.

———. *Willa Cather on Writing: Critical Studies on Writing as an Art.* Lincoln: University of Nebraska Press, 1988.

Chaudhuri, Amit. *D. H. Lawrence and "Difference."* Oxford: Clarendon, 2003.

Claborn, John. "Toward an Eco-Ontology: A Response to Greg Garrard's 'Heidegger Nazism Ecocriticism.'" *ISLE: Interdisciplinary Studies in Literature & Environment* 19, no. 2 (2012): 375–79.

Clingerman, Forrest, Brian Treanor, Martin Drenthen, and David Utsler, eds. *Interpreting Nature: The Emerging Field of Environmental Hermeneutics.* New York: Fordham University Press, 2014.

Coetzee, J. M. *Boyhood.* New York: Penguin, 1997.

———. *Disgrace.* New York: Penguin, 1999.

———. "Farm Novel and Plaasroman in South Africa." *English in Africa* 13, no. 2 (1986): 1–19.

———. *In the Heart of the Country.* New York: Penguin, 1977.

———. *White Writing: On the Culture of Letters in South Africa.* New Haven, CT: Yale University Press, 1988.

Cohen, Jeffrey Jerome. *Stone: An Ecology of the Inhuman.* Minneapolis: University of Minnesota Press, 2016.

Crane, Mary Thomas. "Surface, Depth, and the Spatial Imaginary: A Cognitive Reading of *The Political Unconscious.*" *Representations* 108, no. 1 (2009): 78–97.

Crèvecoeur, Hector St. John de. *Letters from an American Farmer and Sketches of Eighteenth-Century America.* 1782. New York: Penguin, 1986.

Cronon, William. *Changes in the Land: Indians, Colonists, and the Ecology of New England.* New York: Hill and Wang, 1983.
———. Foreword to Cronon, ed., *Uncommon Ground*, 19–22.
———. "The Trouble with Wilderness; or, Getting Back to the Wrong Nature." In Cronon, ed., *Uncommon Ground*, 69–90.
———, ed. *Uncommon Ground: Rethinking the Human Place in Nature.* New York: Norton, 1996.
Crosby, Alfred. *Ecological Imperialism: The Biological Expansion of Europe, 900–1900.* Cambridge: Cambridge University Press, 1986.
Davidson, Marianne. *Willa Cather and F. J. Turner: A Contextualization.* Heidelberg: Universitätsverlag C. Winter, 1999.
Davis, Duane. "*Umwelt* and Nature in Merleau-Ponty's Ontology." In Cataldi and Hamrick, eds., *Merleau-Ponty and Environmental Philosophy*, 117–32.
de la Cadena, Marisol. *Earth Beings: Ecologies of Practice across Andean Worlds.* Durham, NC: Duke University Press, 2015.
de la Cadena, Marisol, and Mario Blaser, eds. *A World of Many Worlds.* Durham, NC: Duke University Press, 2018.
De Man, Paul. *Allegories of Reading: Figural Language in Rousseau, Nietzsche, Rilke, and Proust.* New Haven, CT: Yale University Press, 1979.
Deckard, Sharae. "Ecogothic." In *Twenty-First-Century Gothic: An Edinburgh Companion*, edited by Maisha Wester and Xavier Aldana Reyes, 174–88. Edinburgh: Edinburgh University Press.
Dekkers, Wim. "Dwelling, House, and Home: Towards a Home-Led Perspective on Dementia Care." *Medicine, Health Care, and Philosophy* 14, no. 3 (2011): 291–300.
Delany, Paul. "D. H. Lawrence and Deep Ecology." *CEA Critic* 55 (1993): 27–41.
Deleuze, Gilles. "Bartleby; or the Formula." In *Essays Critical and Clinical*, translated by Daniel W. Smith and Michael A. Greco, 68–90. Minneapolis: University of Minnesota Press, 1997.
Deleuze, Gilles, and Félix Guattari. *A Thousand Plateaus: Capitalism and Schizophrenia.* Translated by Brian Massumi. Minneapolis: University of Minnesota Press, 1987.
Deloria, Philip J. *Playing Indian.* New Haven, CT: Yale University Press, 2007.
DeLoughrey, Elizabeth. *Allegories of the Anthropocene.* Durham, NC: Duke University Press, 2019.
Derrida, Jacques. *Of Grammatology.* Translated by Gayatri Chakravorty Spivak. Baltimore: Johns Hopkins University Press, 1977.
———. *Of Hospitality.* Translated Rachel Bowlby. Stanford, CA: Stanford University Press, 2000.
———. *Specters of Marx: The State of Debt, the Work of Mourning, and the New International.* Translated by Peggy Kamuf. New York: Routledge, 2006.
Devarenne, Nicole. "Nationalism and the Farm Novel in South Africa, 1883–2004." *Journal of Southern African Studies* 35, no. 3 (2009): 627–42.
Diehm, Christian. "'Here I Stand': An Interview with Arne Naess." *Environmental Philosophy* 1, no. 2 (2004): 6–19.

Dillon, Martin C. "Merleau-Ponty and the Ontology of Ecology; or Apocalypse Later." In Cataldi and Hamrick, eds., *Merleau-Ponty and Environmental Philosophy*, 259–72.
Donohoe, Janet. "The Place of Home." *Environmental Philosophy* 8, no. 1 (2011): 25–40.
Du Bois, W. E. B. *The Souls of Black Folk*. 1903. Boston: Bedford, 1997.
Dussel, Enrique. *The Invention of the Americas: Eclipse of the "Other" and the Myth of Modernity*. Translated by Michael D. Barber. New York: Continuum, 1995.
Ebbatson, Roger. *Lawrence and the Nature Tradition: A Theme in English Fiction, 1859–1914*. Brighton, Sussex: Harvester, 1982.
Edelglass, William, James Hatley, and Christian Diehm, eds. *Facing Nature: Levinas and Environmental Thought*. Pittsburgh: Duquesne University Press, 2012.
Eggan, Taylor A. "Landscape Metaphysics: Narrative Architecture and the Focalisation of the Environment." *English Studies* 99, no. 4 (2018): 398–411.
Ehlert, Anna Odenbring. "'There's a Bad Time Coming': Ecological Vision in the Fiction of D. H. Lawrence." Ph.D. diss., Uppsala University, 2001.
Eliade, Mircea. *The Sacred and the Profane: The Nature of Religion*. New York: Harvest, 1987.
Eliot, T. S. *The Complete Poems and Plays: 1909–1950*. New York: Harcourt Brace, 1952.
Emerson, Ralph Waldo. "Nature." In *The Essential Writings of Ralph Waldo Emerson*, 364–77. New York: Modern Library, 2000.
Evernden, Neil. *The Natural Alien: Humankind and Environment*. Toronto: University of Toronto Press, 1993.
Fabian, Johannes. *Time and the Other: How Anthropology Makes Its Object*. New York: Columbia University Press, 1983.
Farías, Victor. *Heidegger and Nazism*. Translated by Paul Burrell, Dominic Di Bernardi, and Gabriel R. Ricci. Philadelphia: Temple University Press, 1989.
Faye, Emmanuel. *Heidegger, l'introduction du nazisme dans la philosophie: Autour des séminaires inédits de 1933–1935*. Paris: Albin Michel, 2005.
Fell, Joseph P. *Heidegger and Sartre: An Essay on Being and Place*. New York: Columbia University Press, 1979.
Felski, Rita. *The Limits of Critique*. Chicago: University of Chicago Press, 2015.
Ferreira da Silva, Denise. *Toward a Global Idea of Race*. Minneapolis: University of Minnesota Press, 2007.
Firestein, Stuart. *Ignorance: How It Drives Science*. New York: Oxford University Press, 2012.
Fischer, Mike. "Pastoralism and Its Discontents: Willa Cather and the Burden of Imperialism." *Mosaic* 23 (1990): 31–44.
Fishburn, Katherine. "The Manichean Allegories of Doris Lessing's *The Grass Is Singing*." *Research in African Literatures* 25, no. 4 (1994): 1–15.
——. "Wor(l)ds within Words: Doris Lessing as Meta-Fictionist and Meta-Physician." *Studies in the Novel* 20, no. 2 (1988): 186–205.
Fletcher, Angus. *Allegory: The Theory of a Symbolic Mode*. Ithaca, NY: Cornell University Press, 1964.
Foltz, Bruce V. *Inhabiting the Earth: Heidegger, Environmental Ethics, and the Metaphysics of Nature*. Amherst, MA: Humanity Books, 1995.

Forster, E. M. *Aspects of the Novel*. New York: Harcourt, Brace and World, 1927.
Foucault, Michel. *The Order of Things: An Archaeology of the Human Sciences*. 1970. Translated by Alan Sheridan. New York: Vintage, 1994.
Francis, Romain. "The Tyranny of the Coloniality of Nature and the Elusive Question of Justice." In *Reimagining Justice, Human Rights and Leadership in Africa: Challenging Discourses and Search for Alternative Paths*, edited by Everisto Benyera, 39–57. Cham, Switzerland: Springer, 2020.
Frank, Ellen Eve. *Literary Architecture: Essays toward a Tradition*. Berkeley: University of California Press, 1979.
Franks, Jill. Introduction to Lawrence, *Sea and Sardinia*, xiii–xxx.
Freud, Sigmund. "Repression." In *The Standard Edition of the Complete Psychological Works of Sigmund Freud*, 17:146–58. London: Hogarth, 1953.
———. "The Uncanny." In *The Standard Edition of the Complete Psychological Works of Sigmund Freud*, 17:217–52. London: Hogarth, 1953.
Fried, Gregory, and Richard Polt. "Translators' Introduction." In Heidegger, *Introduction to Metaphysics*, vii–xix.
Fritsch, Matthias, Philippe Lynes, and David Wood, eds. *Eco-Deconstruction: Derrida and Environmental Philosophy*. New York: Fordham University Press, 2018.
———. Introduction to Fritsch, Lynes, and Wood, eds., *Eco-Deconstruction*, 1–28.
Fryer, Judith. *Felicitous Space: The Imaginative Structures of Edith Wharton and Willa Cather*. Chapel Hill: University of North Carolina Press, 1986.
Garrard, Greg. *Ecocriticism*. New York: Routledge, 2012.
———. "Heidegger Nazism Ecocriticism." *ISLE: Interdisciplinary Studies in Literature & Environment* 17, no. 2 (2010): 251–71.
Gauthier, David J. *Martin Heidegger, Emmanuel Levinas, and the Politics of Dwelling*. Lanham, MD: Lexington, 2011.
Gelfant, Blanche H. *Women Writing in America: Voices in Collage*. Hanover, NH: University Press of New England, 1984.
George, Olakunle. *Relocating Agency: Modernity and African Letters*. Albany: State University of New York Press, 2003.
Gibson, James J. *Reasons for Realism: Selected Essays of James J. Gibson*. Mahwah, NJ: L. Erlbaum, 1982.
Giliomee, Hermann. *The Afrikaners: Biography of a People*. Cape Town, South Africa: Tafelberg, 2003.
Glotfelty, Cheryll. "A Guided Tour of Ecocriticism, with Excursions to Catherland." *Cather Studies* 5 (2003): 28–43.
Glotfelty, Cheryll, and Harold Fromm, eds. *The Ecocriticism Reader: Landmarks in Literary Ecology*. Athens: University of Georgia Press, 1996.
Gómez-Barris, Macarena. *The Extractive Zone: Social Ecologies and Decolonial Perspectives*. Durham, NC: Duke University Press, 2017.
Goodbody, Axel. "Heideggerian Ecopoetics and the Nature Poetry Tradition." In *Nature, Technology and Cultural Change in Twentieth-Century German Literature: The Challenge of Ecocriticism*, edited by Goodbody, 129–67. London: Palgrave Macmillan UK, 2007.

Gorak, Irene. "Olive Schreiner's Colonial Allegory: 'The Story of an African Farm.'" *ARIEL: A Review of International English Literature* 23, no. 4 (1992): 53–72.

Gordimer, Nadine. *The Conservationist*. New York: Penguin, 1974.

Gorman, Michael. "Jim Burden and the White Man's Burden: *My Ántonia* and Empire." *Cather Studies* 6 (2006): 28–57.

Gray, Stephen. *Southern African Literature: An Introduction*. Cape Town, South Africa: David Phillips, 1979.

Guha, Ramachandra. *The Unquiet Woods: Ecological Change and Peasant Resistance in the Himalaya*. Berkeley: University of California Press, 2000.

Guha, Ranajit. *A Rule of Property for Bengal: An Essay on the Idea of Permanent Settlement*. Durham, NC: Duke University Press, 1996.

Gutierrez, Donald. "The Ancient Imagination of D. H. Lawrence." *Twentieth Century Literature* 27, no. 2 (1981): 178–96.

Haraway, Donna. *Staying with the Trouble: Making Kin in the Chthulucene*. Durham, NC: Duke University Press, 2016.

Haresnape, Geoffrey. "Why and How I Became an Author." *English Studies in Africa* 6, no. 2 (1963): 149–53.

Harman, Graham. *The Quadruple Object*. Alresford: Zero, 2011.

———. *Weird Realism: Lovecraft and Philosophy*. Winchester: Zero, 2012.

Hartman, Saidiya V. *Scenes of Subjection: Terror, Slavery, and Self-Making in Nineteenth-Century America*. New York: Oxford University Press, 1997.

Harvey, David. *The Condition of Postmodernity*. Oxford: Basil Blackwell, 1989.

Head, Bessie. *A Question of Power*. Oxford: Heinemann, 1974.

———. *When Rain Clouds Gather*. Oxford: Heinemann, 1987.

Hegel, G. W. F. *Philosophy of Mind*. Translated by W. Wallace and A. V. Miller. Oxford: Clarendon, 2007.

Heidegger, Martin. *Basic Writings*. Translated by Albert Hofstadter. New York: HarperCollins, 1993.

———. *Being and Time*. Translated by Joan Stambaugh. Albany: State University of New York Press, 2010.

———. *The Fundamental Concepts of Metaphysics: World, Finitude, Solitude*. Translated by William McNeill and Nicholas Walker. Bloomington: Indiana University Press, 1995.

———. *Identity and Difference*. Translated by Joan Stambaugh. Chicago: University of Chicago Press, 1969.

———. *Introduction to Metaphysics*. Translated by Gregory Fried and Richard Polt. New Haven, CT: Yale University Press, 2000.

———. *Poetry, Language, Thought*. Translated by Albert Hofstadter. New York: HarperCollins, 1971.

———. *Zollikon Seminars*. Translated by Franz Mayr and Richard Askay. Evanston, IL: Northwestern University Press, 2001.

Heidegger, Martin, and Eugen Fink. *Heraclitus Seminar*. Translated by Charles H. Seibert. Evanston, IL: Northwestern University Press, 1993.

Heller-Roazen, Daniel. *The Inner Touch: Archaeology of a Sensation*. New York: Zone, 2007.

Hemingway, Ernest. 1932. *Death in the Afternoon.* New York: Scribner, 1996.
Higgins, Adrian. "Revered, Then Reviled: Tracking the Rise and Fall of Ivy." www
.washingtonpost.com/lifestyle/home/revered-then-reviled-tracking-the-rise
-and-fall-of-ivy/2017/11/28/6513df5e-cf09-11e7-a1a3-0d1e45a6de3d_story.html.
Hiltner, Ken, ed. *Ecocriticism: The Essential Reader.* London: Routledge, 2015.
Horton, Susan R. *Difficult Women, Artful Lives: Olive Schreiner and Isak Dinesen, In and Out of Africa.* Baltimore: Johns Hopkins University Press, 1995.
Husserl, Edmund. *Crisis of the European Sciences and Transcendental Phenomenology: An Introduction to Phenomenological Philosophy.* Translated by David Carr. Evanston, IL: Northwestern University Press, 1970.
———. *Husserl: Shorter Works.* Edited by Peter McCormick and Frederick Elliston. Notre Dame, IN: University of Notre Dame Press, 1981.
———. *Logical Investigations: Volume 1.* Translated by J. N. Findlay. London: Routledge, 2001.
Huxley, Aldous. *With D. H. Lawrence in New Mexico: A Memoir of D. H. Lawrence.* New York: Barnes & Noble, 1965.
Ingold, Tim. *The Perception of the Environment: Essays on Livelihood, Dwelling and Skill.* London: Routledge, 2000.
Iser, Wolfgang. *The Act of Reading: A Theory of Aesthetic Response.* Baltimore: Johns Hopkins University Press, 1978.
Jackson, Zakkiyah Iman. *Becoming Human: Matter and Meaning in an Antiblack World.* New York: New York University Press, 2020.
James, Henry. *The Art of the Novel: Critical Prefaces.* 1934. Chicago: University of Chicago Press, 2011.
Jameson, Fredric. *The Antinomies of Realism.* London: Verso, 2015.
———. *The Political Unconscious: Narrative as a Socially Symbolic Act.* Ithaca, NY: Cornell University Press, 1981.
———. "Third-World Literature in the Era of Multinational Capital." *Social Text* 15 (1986): 65–88.
Janik, Del Ivan. "Toward 'Thingliness': Cézanne's Painting and Lawrence's Poetry." *Twentieth Century Literature* 19, no. 2 (1973): 119–28.
JanMohamed, Abdul R. *Manichean Aesthetics: The Politics of Literature in Colonial Africa.* Amherst: University of Massachusetts Press, 1983.
Kant, Immanuel. *Critique of the Power of Judgment.* 1790. Translated by Paul Guyer and Eric Matthews. Cambridge: Cambridge University Press, 2001.
Karush, Deborah. "Bringing Outland Inland in the Professor's House." *Cather Studies* 4 (1999): 144–71.
Kimmerer, Robin Wall. *Braiding Sweetgrass: Indigenous Wisdom, Scientific Knowledge, and the Teachings of Plants.* Minneapolis: Milkweed Editions, 2015.
King, Tiffany Lethabo. *The Black Shoals: Offshore Formations of Black and Native Studies.* Durham, NC: Duke University Press, 2019.
Kirby, Vicky. "Un/Limited Ecologies." In Fritsch, Lynes, and Wood, eds., *Eco-Deconstruction,* 121–40.

Koch, Christof. *The Quest for Consciousness: A Neurobiological Approach.* Englewood, CO: Roberts, 2004.

Kohn, Eduardo. *How Forests Think: Toward an Anthropology beyond the Human.* Berkeley: University of California Press, 2013.

Kolbert, Elizabeth. *The Sixth Extinction: An Unnatural History.* New York: Henry Holt, 2014.

Kolodny, Annette. *The Lay of the Land: Metaphor as Experience and History in American Life and Letters.* Chapel Hill: University of North Carolina Press, 1975.

Krech, Shepard, III. *The Ecological Indian: Myth and History.* New York: Norton, 1999.

LaChapelle, Dolores. *D. H. Lawrence: Future Primitive.* Denton: University of North Texas Press, 1996.

Larios, Joe Matthew. "Emmanuel Levinas and the Meaning of Ecological Responsibility." Master's thesis, Louisiana State University, 2018.

Lasdun, James. Introduction to Lawrence, *The Woman Who Rode Away/St. Mawr/The Princess,* xi–xx.

Lawrence, D. H. "Certain Americans and an Englishman." *New York Times,* 24 December 1922.

———. *Fantasia of the Unconscious and Psychoanalysis and the Unconscious.* 1922. New York: Penguin, 1977.

———. *The Letters of D. H. Lawrence.* New York: Viking, 1932.

———. *Mornings in Mexico and Etruscan Places.* 1927. New York: Penguin, 1960.

———. *Phoenix: The Posthumous Papers of D. H. Lawrence.* London: Heinemann, 1936.

———. *Sea and Sardinia.* 1921. New York: Penguin, 1999.

———. *A Selection from Phoenix.* Harmondsworth, Middlesex: Penguin, 1971.

———. *Studies in Classic American Literature.* 1923. New York: Penguin, 1977.

———. *The Rainbow.* 1915. New York: Modern Library, 2002.

———. *The Woman Who Rode Away/St. Mawr/The Princess.* New York: Penguin, 2006.

———. *Women in Love.* 1920. New York: Modern Library, 2002.

Lazarus, Neil. "Realism and Naturalism in African Fiction." In Olaniyan and Quayson, eds., *African Literature,* 340–44.

Leach, Neil. "The Dark Side of the Domus: The Redomestication of Central and Eastern Europe." In *Architecture and Revolution: Contemporary Perspectives on Central and Eastern Europe,* edited by Leach, 150–62. London: Routledge, 1999.

Leith, Katherine H. "'Home Is Where the Heart Is . . . or Is It?': A Phenomenological Exploration of the Meaning of Home for Older Women in Congregate Housing." *Journal of Aging Studies* 20, no. 4 (2006): 317–33.

Leonard, John. "The Spacing Out of Doris Lessing." www.nytimes.com/1982/02/07/books/the-spacing-out-of-doris-lessing.html.

Leroux, Etienne. *To a Dubious Salvation.* Translated by Charles Eglington. Harmondsworth: Penguin, 1972.

Lessing, Doris. *African Laughter: Four Visits to Zimbabwe.* New York: HarperCollins, 1992.

———. *African Stories.* 1964. New York: Simon and Schuster, 2014.

———. *Briefing for a Descent into Hell.* 1971. New York: Vintage, 2009.

———. *Canopus in Argos: Archives.* 1979–83. New York: Vintage, 1992.

———. *The Four-Gated City.* 1969. New York: Harper Perennial, 1995.
———. *The Golden Notebook.* 1962. New York: Harper Perennial, 2008.
———. *The Grass Is Singing.* 1950. New York: Harper Perennial, 1991.
———. *Landlocked.* 1965. New York: Harper Perennial, 1995.
———. *Martha Quest.* 1952. New York: Harper Perennial, 1995.
———. *Memoirs of a Survivor.* 1974. New York: Vintage, 1988.
———. *A Proper Marriage.* 1954. New York: Harper Perennial, 1995.
———. *A Ripple from the Storm.* 1958, New York: Harper Perennial, 1995.
———. *A Small Personal Voice: Essays, Reviews, Interviews.* New York: Vintage, 1975.
———. *Time Bites: Views and Reviews.* New York: Harper Perennial, 2005.
———. *Under My Skin: Volume One of My Autobiography, to 1949.* New York: Harper Perennial, 1995.
Letoit, André. *Somer II: 'n plakboek.* Cape Town, South Africa: Perskor, 1985.
Levinas, Emmanuel. *On Escape.* Translated by Bettina Bergo. Stanford, CA: Stanford University Press, 2003.
———. *Totality and Infinity: An Essay on Exteriority.* Translated by Alphonso Lingis. Pittsburgh: Duquesne University Press, 1969.
Lewis, Simon. *White Women Writers and Their African Invention.* Gainesville: University Press of Florida, 2003.
Lidell, H. G., and R. Scott. *Greek–English Lexicon.* 9th ed. Oxford: Clarendon, 1996.
Llewelyn, John. "Writing Home: Eco-Choro-Spectrography." In Fritsch, Lynes, and Wood, eds., *Eco-Deconstruction,* 165–83.
Locke, John. *Two Treatises of Government.* 1689. Cambridge: Cambridge University Press, 1960.
Loveless, Natalie. *How to Make Art at the End of the World: A Manifesto for Research-Creation.* Durham, NC: Duke University Press, 2019.
Lugones, Maria. "The Coloniality of Gender." In *The Palgrave Handbook of Gender and Development: Critical Engagements in Feminist Theory and Practice,* edited by Wendy Harcourt, 13–33. London: Palgrave Macmillan, 2016.
Luhan, Mabel Dodge. "Lawrence of New Mexico." *New Mexico Magazine,* February 1936.
———. *Lorenzo in Taos.* New York: Knopf, 1932.
Lukács, Georg. *The Meaning of Contemporary Realism.* Translated by John and Necke Mander. London: Merlin, 1977.
———. *Studies in European Realism.* Translated by Edith Bone. New York: Grosset and Dunlap, 1964.
Lynch, Tom. *Xerophilia: Ecocritical Explorations in Southwestern Literature.* Lubbock: Texas Tech University Press, 2008.
Mace, William M. "James J. Gibson's Strategy for Perceiving: Ask Not What's Inside Your Head, but What's Your Head Inside Of." In *Perceiving, Acting, and Knowing: Toward an Ecological Psychology,* 43–66. Hillsdale, NJ: Erlbaum, 1977.
Maldonado-Torres, Nelson. "On the Coloniality of Being: Contributions to the Development of a Concept." *Cultural Studies* 21, no. 2–3 (2007): 240–70.
———. "The Topology of Being and the Geopolitics of Knowledge: Modernity, Empire, Coloniality." *City* 8, no. 1 (2014): 29–56.

Malherbe, D. F. *Die meulenaar.* Cape Town, South Africa: Tafelberg, 1978.
Malpas, Jeff. *Heidegger and the Thinking of Place: Explorations in the Topology of Being.* Cambridge: MIT Press, 2017.
———. *Heidegger's Topology: Being, Place, World.* Cambridge: MIT Press, 2006.
Maracle, Lee. *Memory Serves.* Edmonton, Alberta: NeWest, 2015.
Marder, Michael. "Ecology as Event." In Fritsch, Lynes, and Wood, eds., *Eco-Deconstruction,* 141–64.
———. *Heidegger: Phenomenology, Ecology, Politics.* Minneapolis: University of Minnesota Press, 2018.
———. *Plant-Thinking: A Philosophy of Vegetal Life.* New York: Columbia University Press, 2013.
Marquard, Jean. "The Farm: A Concept in the Writing of Olive Schreiner, Pauline Smith, Doris Lessing, Nadine Gordimer and Bessie Head." *Dalhousie Review* 59, no. 2 (1979): 293–308.
Marx, Karl. *Capital Volume I: A Critique of Political Economy.* 1867. Translated by Ben Fowkes. New York: Penguin, 1990.
Mattingly, Wesley Nolan. "The Ecology of Being." Ph.D. diss., Stony Brook University, 2017.
Mazoyer, Marcel, and Laurence Roudart. *A History of World Agriculture: From the Neolithic Age to the Current Crisis.* Translated by James H. Membrez. New York: Monthly Review Press, 2006.
Mbembe, Achille. *Critique of Black Reason.* Translated by Laurent Dubois. Durham, NC: Duke University Press, 2017.
———. "Necropolitics." *Public Culture* 15, no. 1 (2003): 11–40.
McClintock, Anne. *Imperial Leather: Race, Gender and Sexuality in the Colonial Contest.* New York: Routledge, 1995.
Meillassoux, Quentin. *After Finitude: An Essay on the Necessity of Contingency.* Translated by Ray Brassier. London: Continuum, 2009.
Merchant, Carolyn. "Reinventing Eden: Western Culture as a Recovery Narrative." In Cronon, ed., *Uncommon Ground,* 132–59.
Merleau-Ponty, Maurice. *Phenomenology of Perception.* Translated by Colin Smith. London: Routledge, 2002.
Merrild, Knud. *With D. H. Lawrence in New Mexico: A Memoir of D. H. Lawrence.* New York: Barnes and Noble, 1965.
Mezei, Kathy, and Chiara Briganti. "Reading the House: A Literary Perspective." *Signs* 27, no. 3 (2002): 837–46.
Michelucci, Stefania. *Space and Place in the Works of D. H. Lawrence.* Translated by Jill Franks. Jefferson, NC: McFarland, 2002.
Mignolo, Walter. *The Darker Side of Western Modernity: Global Futures, Decolonial Options.* Durham, NC: Duke University Press, 2011.
———. "The Geopolitics of Knowledge and the Colonial Difference." *South Atlantic Quarterly* 101, no. 1 (2002): 57–96.
———. "Introduction: Coloniality of Power and De-colonial Thinking." *Cultural Studies* 21, no. 2–3 (2007): 155–67.

Mignolo, Walter, and Catherine E. Walsh. *On Decoloniality: Concepts, Analytics, Praxis.* Durham, NC: Duke University Press, 2018.

Monsman, Gerald. "Olive Schreiner's Allegorical Vision." *Victorian Review* 18, no. 2 (1992): 49–62.

Moreton-Robinson, Aileen. *The White Possessive: Property, Power, and Indigenous Sovereignty.* Minneapolis: University of Minnesota Press, 2015.

Morton, Timothy. *Dark Ecology: For a Logic of Future Coexistence.* New York: Columbia University Press, 2016.

———. *The Ecological Thought.* Cambridge: Harvard University Press, 2010.

———. *Realist Magic: Objects, Ontology, Causality.* Open Humanities Press, 2013.

Moseley, Ann. "Spatial Structures and Forms in *The Professor's House*." *Cather Studies* 3 (1996): 197–211.

Moten, Fred. *Black and Blur.* Durham, NC: Duke University Press, 2018.

———. *Stolen Lives.* Durham, NC: Duke University Press, 2018.

———. *Universal Machine.* Durham, NC: Duke University Press, 2018.

Moya, Patricia. "Habit and Embodiment in Merleau-Ponty." *Frontiers in Human Neuroscience* 8 (2014). https://doi.org/10.3389/fnhum.2014.00542.

Murphy, Joseph C. "Cather's Ruskinian Landscapes: Typologies of the New World." *Cather Studies* 8 (2010): 228–45.

Mutter, Sarah Mahurin. "Godfrey St. Peter's 'Picturesque Shipwreck.'" *American Literary Realism* 42, no. 1 (2009): 54–71.

Nagel, Thomas. "What Is It Like to Be a Bat?" *Philosophical Review* 83, no. 4 (1974): 435–50.

Nancy, Jean-Luc. "Uncanny Landscape." In *The Ground of the Image*, translated by Jeff Fort, 51–62. New York: Fordham University Press, 2005.

Ndebele, Njabulo. "The Rediscovery of the Ordinary: Some New Writings in South Africa." *Journal of Southern African Studies* 12 (1986): 143–57.

Neumann, Roderick P. *Imposing Wilderness: Struggles over Livelihood and Nature Preservation in Africa.* Berkeley: University of California Press, 1998.

Newmark, Julianna. "Sensing Re-Placement in New Mexico: Lawrence, John Collier, and (Post)Colonial Textual Geographies." In *"Terra Incognita": D. H. Lawrence at the Frontiers*, edited by Virginia Crosswhite Hyde and Earl G. Ingersoll, 157–82. Madison, NJ: Fairleigh Dickinson University Press, 2010.

Nixon, Rob. *Slow Violence and the Environmentalism of the Poor.* Cambridge, MA: Harvard University Press, 2011.

O'Donoghue, Brendan. *A Poetics of Homecoming: Heidegger, Homelessness, and the Homecoming Venture.* Cambridge: Cambridge University Press, 2011.

Oehlschlaeger, Fritz. "Indisponsibilité and the Anxiety of Authorship in *The Professor's House*." *American Literature* 62, no. 1 (1990): 74–86.

Oh, Eunyoung. *D. H. Lawrence's Border Crossing: Colonialism in His Travel Writings and "Leadership" Novels.* New York: Routledge, 2007.

Olaniyan, Tejumola, and Ato Quayson, eds. *African Literature: An Anthology of Criticism and Theory.* Malden, MA: Blackwell, 2007.

Olivier, Gerrit. "The Dertigers and the *Plaasroman*: Two Brief Perspectives on Afrikaans Literature." In *The Cambridge History of South African Literature*, edited by

David Attwell and Derek Attridge, 308–24. Cambridge: Cambridge University Press, 2012.

Oloff, Kerstin. "Greening the Zombie: Caribbean Gothic, World-Ecology, and Socio-ecological Degradation." *Green Letters* 16 (2012): 31–45.

Ortega, Mariana. "Decolonial Woes and Practices of Un-knowing." *Journal of Speculative Philosophy* 31, no. 3 (2017): 504–16.

Parreñas, Juno Salazar. *Decolonizing Extinction: The Work of Care in Orangutan Rehabilitation.* Durham, NC: Duke University Press, 2018.

Perez, Emma. *The Decolonial Imaginary: Writing Chicanas into History.* Bloomington: Indiana University Press, 1999.

Peters, Michael, and Ruth Irwin. "Earthsongs: Ecopoetics, Heidegger, and Dwelling." *Trumpeter* 18, no. 1 (2002): 1–17.

Pratt, Mary Louise. *Imperial Eyes: Travel Writing and Transculturation.* New York: Routledge, 1992.

Quammen, Betsy Gaines. *American Zion: Cliven Bundy, God, and Public Lands.* Salt Lake City, UT: Torrey House, 2020.

Quijano, Aníbal. "Coloniality of Power, Eurocentrism, and Latin America." *Nepantla: Views from the South* 1, no. 3 (2000): 533–80.

Quijano, Aníbal, and Immanuel Wallerstein. "Americaneity as a Concept, or the Americas in the Modern World-system." *International Social Science Journal* 44, no. 4 (1992): 549–57.

Quinn, Daniel. *Ishmael: An Adventure of the Mind and Spirit.* New York: Bantam, 1992.

Quirk, Tom. *Bergson and American Culture: The Worlds of Willa Cather and Wallace Stevens.* Chapel Hill: University of North Carolina Press, 1990.

Ranger, Terence. *Peasant Consciousness and Guerrilla War in Zimbabwe.* London: James Currey, 1985.

——— . *Voices from the Rocks: Nature, Culture & History in the Matopos Hills of Zimbabwe.* Bloomington: Indiana University Press, 1999.

Ray, Sarah Jaquette. *The Ecological Other: Environmental Exclusion in American Culture.* Tucson: University of Arizona Press, 2013.

Reed, Edward S. "Knowers Talking about the Known: Ecological Realism as a Philosophy of Science." *Synthese* 92, no. 1 (1992): 9–23.

Reilly, Ariana. "Always Sympathize! Surface Reading, Affect, and George Eliot's *Romola*." *Victorian Studies* 55, no. 4 (2013): 629–46.

"Rewild.com." www.rewild.com.

Reynolds, Guy. *Willa Cather in Context: Progress, Race, Empire.* New York: St. Martin's, 1996.

Ricklefs, Robert E. *The Economy of Nature.* New York: W. H. Friedman, 2008.

Rifkin, Mark. *Fictions of Land and Flesh: Blackness, Indigeneity, Speculation.* Durham, NC: Duke University Press, 2019.

——— . *Settler Common Sense: Queerness and Everyday Colonialism in the American Renaissance.* Minneapolis: University of Minnesota Press, 2014.

Rigby, Kate. "Earth, World, Text: On the (Im)Possibility of Ecopoiesis." *New Literary History* 35, no. 3 (2004): 427–42.

———. *Topographies of the Sacred: The Poetics of Place in European Romanticism*. Charlottesville: University of Virginia Press, 2004.
Rivera Cusicanqui, Silvia. *Ch'ixinakax utxiwa: On Practices and Discourses of Decolonization*. Translated by Molly Geidel. Cambridge, UK: Polity, 2020.
Roberts, Neil. *D. H. Lawrence, Travel and Cultural Difference*. Houndmills, Basingstoke: Palgrave Macmillan, 2004.
Rölvaag, O. E. *Giants in the Earth: A Saga of the Prairie*. New York: Harper Perennial, 1999.
Rooney, Caroline. "Narratives of Southern African Farms." *Third World Quarterly* 26, no. 3 (2005): 431–40.
Ruskin, John. *Modern Painters*. 3 vols. New York: John Wiley and Son, 1873.
———. *The Seven Lamps of Architecture*. 1849. New York: Dover, 1989.
Russell, Danielle. *Between the Angle and the Curve: Mapping Gender, Race, Space, and Identity in Willa Cather and Toni Morrison*. New York: Routledge, 2006.
Sale, Kirkpatrick. *After Eden: The Evolution of Human Domination*. Durham, NC: Duke University Press, 2006.
Schreiner, Olive. *Dreams*. 1890. London: Wildwood House, 1982.
———. *The Story of an African Farm*. 1883. Oxford: Oxford University Press, 1998.
———. *Thoughts on South Africa*. London: T. Fisher Unwin, 1923.
———. *Trooper Peter Halket of Mashonaland*. Boston: Roberts Bros., 1897.
Schubnell, Matthias. "The Decline of America: Willa Cather's Spenglerian Vision in 'The Professors' House.'" *Cather Studies* 2 (1993): 92–117.
Sender, Roy, Shai Fuchs, and Ron Milo. "Revised Estimates for the Number of Human and Bacteria Cells in the Body." *PLOS Biology* (2016). doi: 10.1371/journal.pbio.1002533.
Seppänen, Janne. "Lost at Sea: The Freudian Uncanny and Representing Ecological Degradation." *Psychoanalysis Culture & Society* 16, no. 2 (2011): 196–208.
Sexton, Jared. "Unbearable Blackness." *Cultural Critique* 90 (2015): 159–78.
Sharpe, Christina. *In the Wake: On Blackness and Being*. Durham, NC: Duke University Press, 2016.
Shaviro, Stephen. *The Universe of Things: On Speculative Realism*. Minneapolis: University of Minnesota Press, 2014.
Sheer, Vivian. "Etienne Leroux's *Sewe Dae by die Silbersteins*: A Reexamination in the Light of Its Historical Context." *Journal of Southern African Studies* 8, no. 2 (1982): 173–86.
Sheldrake, Merlin. *Entangled Life: How Fungi Make Our Worlds, Change Our Minds, and Shape Our Futures*. New York: Random House, 2020.
Skrbina, David. *Panpsychism in the West*. Cambridge, MA: MIT Press, 2007.
Slemon, Stephen. "Unsettling the Empire: Resistance Theory for the Second World." *World Literature Written in English* 30, no. 2 (1990): 30–41.
Slote, Bernice. "The Kingdom of Art." In Cather, *The Kingdom of Art*, 31–112.
Smith, Pauline. *The Beadle*. London: Jonathan Cape, 1926.
———. *The Little Karoo*. London: Jonathan Cape, 1930.

Smith, William S., Jadwiga S. Smith, and Daniela Verducci, eds. *Eco-Phenomenology: Life, Human Life, Post-Human Life in the Harmony of the Cosmos*. Cham, Switzerland: Springer, 2018.
Snyder, Carey. "'When the Indian Was in Vogue': D. H. Lawrence, Aldous Huxley, and Ethnological Tourism in the Southwest." *Modern Fiction Studies* 53, no. 4 (2007): 662–96.
Spencer, Herbert. *First Principles*. 1860. London: Williams and Norgate, 1870.
Spillman, Deborah Shapple. *British Colonial Realism in Africa: Inalienable Objects, Contested Domains*. New York: Palgrave Macmillan, 2012.
Stambaugh, Joan. Introduction to Heidegger, *Identity and Difference*, 7–18.
Stefanovic, Ingrid Leman. "The Experience of Place: Housing Quality from a Phenomenological Perspective." *Canadian Journal of Urban Research* 1, no. 2 (1992): 145–61.
Steinhagen, Carol. "Dangerous Crossings: Historical Dimensions of Landscape in Willa Cather's *My Ántonia*, *The Professor's House*, and *Death Comes for the Archbishop*." *ISLE: Interdisciplinary Studies in Literature & Environment* 6, no. 2 (1999): 63–82.
Stilgoe, John R. *Common Landscape of America, 1580 to 1845*. New Haven, CT: Yale University Press, 1982.
———. Foreword to Bachelard, *The Poetics of Space*, vii–x.
Stout, Janis P. *Cather among the Moderns*. Tuscaloosa: University of Alabama Press, 2019.
Tansi, Sony Labou. "'An Open Letter to Africans': c/o The Punic One-Party State." In Olaniyan and Quayson, eds., *African Literature*, 272–73.
Taylor, Dorceta E. *The Rise of the American Conservation Movement: Privilege, Power, and Environmental Protection*. Durham, NC: Duke University Press, 2016.
Tepler, Benjamin. "Forest Park Guide: Growing Pains." www.pdxmonthly.com /travel-and-outdoors/2011/06/forest-park-invasive-species-july-2011.
Thacker, Eugene. *In the Dust of This Planet: Horror of Philosophy Vol. 1*. Winchester: Zero, 2010.
Thomashow, Mitchell. *Bringing the Biosphere Home: Learning to Perceive Global Environmental Change*. Cambridge, MA: MIT Press, 2002.
Thompson, Leonard. *A History of South Africa*. 4th ed. New Haven, CT: Yale University Press, 2014.
Tindall, William York. *D. H. Lawrence and Susan His Cow*. New York: Columbia University Press, 1939.
———. "D. H. Lawrence and the Primitive." *Sewanee Review* 45, no. 2 (1937): 198–211.
Tschumi, Bernard. *Architecture and Disjunction*. Cambridge, MA: MIT Press, 1994.
Tsing, Anna Lowenhaupt. *The Mushroom at the End of the World: On the Possibility of Life in Capitalist Ruins*. Princeton, NJ: Princeton University Press, 2017.
Tuck, Eve, and K. Wayne Yang. "Decolonization Is Not a Metaphor." *Decolonization: Indigeneity, Education & Society* 1, no. 1 (2012): 1–40.
Turner, Frederick Jackson. *The Frontier in American History*. New York: Dover, 1996.

Uexküll, Jakob von. *A Foray into the Worlds of Animals and Humans.* Translated by Joseph D. O'Neil. Minneapolis: University of Minnesota Press, 2010.

Urgo, Joseph. "The Cather Thesis: The American Empire of Migration." In *The Cambridge Companion to Willa Cather,* edited by Marilee Lindemann, 35–50. Cambridge: Cambridge University Press, 2005.

Utsler, David, Forrest Clingerman, Martin Drenthen, and Brian Treanor. "Introduction: Environmental Hermeneutics." In Clingerman, Treanor, Drenthen, and Utsler, eds., *Interpreting Nature,* 1–16.

Vambe, Lawrence. *An Ill-Fated People: Zimbabwe before and after Rhodes.* Pittsburgh: University of Pittsburgh Press, 1972.

Van Bruggen, Jochem. *Ampie: Die trilogie.* Johannesburg, South Africa: Afrikaanse Pers, 1965.

Van den Heever, C. M. *Harvest Home: An English Rendering.* Translated by T. J. Haarhoff. Johannesburg: A. P. B. Bookstore, 1945.

Van Heerden, Etienne. *Kikuyu.* Translated by Catherine Knox. Cape Town, South Africa: Kwela Books, 1998.

Van Niekerk, Marlene. "Afrikaner Woman and Her 'Prison': Afrikaner Nationalism and Literature." In *Afrikaans Literature: Recollection, Redefinition, Restitution,* edited by Robert Kriger and Ethel Kriger, 141–54. Amsterdam: Rodopi, 1996.

———. *Agaat.* Translated by Michiel Heyns. Portland, OR: Tin House, 2006.

Veracini, Lorenzo. *Settler Colonialism: A Theoretical Overview.* Houndmills, Basingstoke: Palgrave Macmillan, 2010.

Viveiros de Castro, Eduardo. "Cosmological Deixis and Amerindian Perspectivism." *Journal of the Royal Anthropological Institute* 4, no. 2 (1998): 469–88.

Vizenor, Gerald, ed. *Survivance: Narratives of Indian Presence.* Lincoln: University of Nebraska Press, 2008.

Warren, Calvin L. *Ontological Terror: Blackness, Nihilism, and Emancipation.* Durham, NC: Duke University Press, 2018.

Watts, Vanessa. 2013. "Indigenous Place-Thought and Agency amongst Humans and Non-Humans (First Woman and Sky Woman Go on a European World Tour!)." *Decolonization: Indigeneity Education & Society* 2, no. 1 (2013): 20–34.

Weheliye, Alexander G. *Habeas Viscus: Racializing Assemblages, Biopolitics, and Black Feminist Theories of the Human.* Durham, NC: Duke University Press, 2014.

Weisman, Alan. *The World without Us.* New York: Picador, 2007.

Wenzel, Jennifer. "The Pastoral Promise and the Political Imperative: The Plaasroman Tradition in an Era of Land Reform." *Modern Fiction Studies* 46, no. 1 (2000): 90–113.

Westling, Louise H., ed. *The Cambridge Companion to Literature and the Environment.* New York: Cambridge University Press, 2014.

———. *The Green Breast of the New World: Landscape, Gender, and American Fiction.* Athens: University of Georgia Press, 1996.

Wharton, Edith, and Ogden Codman. *The Decoration of Houses.* 1897. New York: Norton, 1978.

White, Lynn, Jr. "The Historical Roots of Our Ecologic Crisis." In Glotfelty and Fromm, eds., *The Ecocritical Reader*, 3–14.

Whyte, Kyle. "Settler Colonialism, Ecology, and Environmental Injustice." *Environment and Society: Advances in Research* 9 (2018): 125–44.

Wigley, Mark. *The Architecture of Deconstruction: Derrida's Haunt*. Cambridge, MA: MIT Press, 1993.

Williams, Gary. "Ecological Realism and Affordance Ontology." www.philosophy andpsychology.wordpress.com/2010/08/31/ecological-realism-and-affordance -ontology/.

Williams, Raymond. *The Country and the City*. Oxford: Oxford University Press, 1973.

———. *The English Novel from Dickens to Lawrence*. Oxford: Oxford University Press, 1970.

Wolfe, Cary. "Wallace Stevens's Birds, or, Derrida and Ecological Poetics." In Fritsch, Lynes, and Wood, eds., *Eco-Deconstruction*, 317–38.

Wolfe, Patrick. "Settler Colonialism and the Elimination of the Native." *Journal of Genocide Research* 8, no. 4 (2006): 387–409.

Wolin, Richard, ed. *The Heidegger Controversy: A Critical Reader*. Cambridge, MA: MIT Press, 1993.

Wood, James. *The Broken Estate: Essays on Literature and Belief*. London: Jonathan Cape, 1999.

———. *How Fiction Works*. 2008. New York: Picador, 2018.

Woolf, Virginia. *Mrs. Dalloway*. 1925. New York: Harcourt Brace, 1999.

Worster, Donald. *Nature's Economy: A History of Ecological Ideas*. Cambridge: Cambridge University Press, 1985.

Wylie, John. "A Landscape Cannot Be a Homeland." *Landscape Research* 41, no. 4 (2016): 408–16.

Wynter, Sylvia. "Unsettling the Coloniality of Being/Power/Truth/Freedom: Towards the Human, After Man, Its Overrepresentation—An Argument." *CR: The New Centennial Review* 3, no. 3 (2003): 257–337.

Yusoff, Kathryn. *A Billion Black Anthropocenes or None*. Minneapolis: University of Minnesota Press, 2018.

Zimmerman, Michael E. *Contesting Earth's Future: Radical Ecology and Postmodernity*. Berkeley: University of California Press, 1994.

———. "Heidegger's Phenomenology and Contemporary Environmentalism." In Brown and Toadvine, eds., *Eco-Phenomenology*, 73–101.

———. "Rethinking the Heidegger–Deep Ecology Relationship." *Environmental Ethics* 15, no. 3 (1993): 195–224.

Index

Abram, David, 18–19, 21, 262n5
abyss, 67; Heidegger's concept of, 243n2
Acoma, 94
Adorno, Theodor, 50
Aeneas, 21
Aeneid (Virgil), 21
affect, 138–40
affordance theory (Gibson), 131–32, 256n4. *See also* ecological realism
African Reserves (Southern Rhodesia), 151–52
Afrikaner nationalism, 153, 258n44. *See also* Apartheid
Afrikaners, 147–50
Afro-pessimism, 57
agriculture, 144; vilification of African, 152, 257n23
Ahmed, Sara, 19, 245n46
alien: as human condition, 237–38; as ontological condition, 106
alienation: and eco-phenomenology, 13; and landscape, 33–34; and repression, 36; right of, 148. *See also* estrangement; exile
alien phenomenology (Bogost), 223–24, 227. *See also* exo-phenomenology; speculative realism
Alimoda, Héctor, 29
allegory: and affect, 139; colonial, 162–63, 164–65; and the ecogothic, 38; Manichean, 162, 212–13; meta-Manichean, 213–14; as metaphysical, 174–75; and Nature, 179; and postcolonialism, 163,

214; and realism, 162, 177; in Schreiner, 176–77; settler colonial, 164–65
allotropy, 109
alterity, 8–9, 37, 45; and exo-phenomenology, 222–23; in Lawrence, 124, 127, 130, 132–33
Anglo-Boer War, 171
animal worlds, 132, 137, 188–89, 190–91, 227, 250n57. *See also Umwelt*
animism, 114–15
Anthony, Loren, 166, 172, 189
Anthropocene, 13, 37, 247n102
anthropocentrism, 117–19
anti-Blackness, 57–58: colonial origins of, 28; and Indigenous genocide, 257n10
antipastoralism, 153, 160
Apartheid, 153, 154
apocalypticism: in Coetzee, 233–36; in Lawrence, 126, 128, 130; in Lessing, 202–3, 208–10
appearance, 129. *See also* World
architecture. *See* dwelling; home
Arendt, Hannah, 50
arrivants. *See* exogenous others
attention: and biospheric change, 18; shifting of, 3; and the uncanny, 39. *See also* inattention
autochthony, 55–56, 74
Autrui. See Other

Bachelard, Gaston, 16, 20–21, 50, 80–81
Baer, Ben Conisbee, 126–30
Barad, Karen, 11, 12, 219

INDEX

Barnd, Natchee Blu, 238–39
Barthes, Roland, 259n24
Bate, Jonathan, 52, 244n27
being: without belonging, 43, 106, 123–24, 166, 217, 224, 237–38; and Blackness, 58, 75–75, 249n48; and dwelling, 50; equipmental, 62; at home in landscape, 79; at home in the world, 9, 10, 222; materiality of, 112–13; topology of, 54, 64–72, 74
being-toward-death (Heidegger), 56, 253n53
Bennett, Jane, 219
Bergson, Henri, 82, 84–85, 251n15
Best, Stephen, 136
Blackfeet, 26
Black Forest (Germany), 117–18
Black studies, 51, 57–58
blank (Iser), 137–38
blank ecology (Llewelyn), 223
Blut und Boden. See Nazism
body: as primordial home, 13. See also embodiment
Boers. See Afrikaners
Bogost, Ian, 113, 223–24, 227
Botswana, 154
Brink, André, 153, 160–62
British Empire Exhibition, 150, 192
British South Africa Company, 150
Brown, Charles, 15, 19, 53
Burke, Edmund, 40
Bursum Bill (New Mexico), 125–26

Carlyle, Thomas, 77–79
Carson, Rachel, 51
Cather, Willa, 43, 166; and Bergson, 82, 84–86; "Concerning Thomas Carlyle," 77–79; critiques of, 251n6; *Death Comes for the Archbishop*, 80, 82, 100–101; homemaking trope in, 80; and narrative description, 87–88; and narrative form, 81–82, 88–90; "The Novel Démeublé," 86–88; "On the Art of Fiction," 87–88; *O Pioneers!*, 146; and Ruskin, 82–84, 85–86; *Shadows on the Rock*, 82; theory of fiction,
81–82, 86–90. See also *Professor's House, The*
Cézanne, Paul, 115–16, 254n57
Chaco Canyon (New Mexico), 252n50
character: and setting, 87; and vitality, 81
chez soi, le (Levinas), 9, 222
Claborn, John, 53
Codman, Ogden, 97
Coetzee, J. M., 45, 153; *Boyhood*, 166–68, 191, 228; *Disgrace*, 153; dream topography, 166, 167–68, 216; on farm novels, 154, 155, 166; lineal consciousness, 157; on Schreiner, 173, 179; and unbelonging, 166–68. See also *In the Heart of the Country*
cognition: and the unconscious, 135
Cohen, Jeffrey Jerome, 185
colonial allegory, 162–63, 164–65; in Lessing, 212–13. See also settler colonial allegory
colonialism: Dutch, 147–49; as narrative structure, 21–22; Spanish, 28; as worlding act, 99. See also coloniality; imperialism; settler colonialism
coloniality, 27; of being, 28, 51, 54–58, 72–73; of gender, 28, 29; of knowledge, 28, 55; of power, 28–29, 55. See also coloniality of Nature
coloniality of Nature, 27–33, 75–76, 99, 220, 245n74; history of, 29–32; and home(l)y metaphysics, 101. See also Nature
colonial matrix of power. See coloniality
colonial narrative, 21–22. See also settler colonial narrative
complicity, 4, 5, 45, 204, 213, 218–19, 220, 233–34. See also contingency
concealment. See Earth
Conrad, Joseph, 214
Constable, John, 85
contingency, 38, 45, 218–20, 224, 225, 233, 234, 237, 238. See also complicity
Cooper, James Fenimore, 126–27
correlationism (Meillassoux), 263n31
Coubert, Gustave, 177
Crane, Mary, 135

Crèvecoeur, Hector St. John de, 144–45
Cronon, William, 24–26
Crosby, Alfred, 164
curiosity: and the uncanny, 224
Cuvier, Georges, 185

damnés de la terre. *See* wretched of the earth
Dasein, 8, 14; and the coloniality of being, 51, 56–58, 72–73; dwelling of, 50, 59–64, 71–72; and landscape, 74–76; and World, 62–63
Deckard, Sharae, 38
decolonial theory, 27–33; critique of Heidegger, 51, 54–57; critiques of, 241, 254n48
decolonization, 24, 33: as metaphor, 23
deconstruction, 8, 10, 11. *See also* eco-deconstruction
deep ecology, 52, 107, 136
dehumanization. *See* human
Deleuze, Gilles, 233
Deloria, Philip J., 128
De Man, Paul, 175
denial of coevalness (Fabian), 245n49
Derrida, Jacques, 8–9, 11, 16–17, 127, 128, 129, 243n1. *See also* deconstruction; eco-deconstruction
Descartes, René, 233
Destruktion (Heidegger), 8, 49
Dickens, Charles, 178
Dillon, Martin, 244n33
Dinge. *See* things
double-consciousness (Du Bois), 230–31
dream topography (Coetzee), 166, 167–68, 216. *See also* empty land, myth of; world-without-us visions
Du Bois, W. E. B., 230–31
durée (Bergson), 84–85, 96
Dutch East India Company, 149
dwelling: in Cather, 84; and Dasein, 50, 59–64, 70–71; and exo-phenomenology, 222; and landscape, 70–71, 80; in Levinas, 262n10; location of, 49–51; materiality of, 64; in Nature, 5; and Nazism, 50–51; as primordial state, 14

Earth (Heidegger), 59–64, 66–67; capitalization of, 249n51; as homeland, 65–66; and World, 61, 63–64, 71
eco-apocalypse: in Coetzee, 234–36; in Lessing, 208–10
eco-deconstruction, 10–12, 220, 243n7
ecogothic, 38
eco-hermeneutics, 11–12, 220, 243n7
ecological homecoming narrative, 11; critique of, 26; in eco-phenomenology, 13–19; and repression, 37; and settler colonialism, 22, 24, 26–27; and the uncanny, 36–42
ecological imperialism. *See* imperialism
ecological Indian (Krech), 20, 253n63
ecological realism, 43–44, 162; and affect, 138–40; and animal focalization, 190–91; and the ecological uncanny, 133, 138; in environmental psychology, 131–33, 223; in Lessing, 200–205, 214–16; in literature, 133–40; as reading practice, 135–38; in Schreiner, 186–91; and settler colonial allegory, 164; and surface reading, 136; and symptomatic reading, 135. *See also* realism
ecological salvation narrative, 151–52
ecological uncanny, 2, 39–42; in Brink, 161–62; and ecological realism, 133, 138; in Lawrence, 107, 117–19, 121; in Lessing, 204–5. *See also* uncanny
ecology: and economy, 9; etymology of, 9, 16–17; and home(l)y discourse, 244n33; and representation, 131
economy: and ecology, 9
eco-phenomenology: and contingency, 200; critique of, 19; and eco-deconstruction, 11–12, 243n7; and the ecological homecoming narrative, 13–19, 220; history of, 11–12; and the restoration of meaning, 15; spiritual valences of, 21; and the uncanny, 36–42
ecopoiesis, 53, 244n27
Edwards, J. A., 151
élan vital (Bergson), 82, 84–85. *See also* vitality
Eliade, Mircea, 97, 98, 99, 153

Eliot, T. S., 213, 225–26
Ellis, Havelock, 196
embodiment: and inhabitation, 14–15; as practice, 19. *See also* body
Emerson, Ralph Waldo, 169, 181
empty land, myth of, 22, 26–27, 97, 98, 147, 184, 187. *See also* dream topography
enframing (Heidegger), 67
environment. *See Umwelt*
environmental ethics, 15, 136; and perception, 18
environmental historiography, 144, 257n23
Erde. See Earth
Erichthonius, 55, 79
estrangement: and ecology, 9–10; and exo-phenomenology, 222; and home(l)y rhetoric, 9; in Lawrence, 106–7, 120–21. *See also* alienation; exile
ethics. *See* environmental ethics
être-là. See Dasein
Evernden, Neil, 237–38
exile: Christian paradigm of, 17, 27; and the ecological homecoming narrative, 32; as precondition for home, 34–35. *See also* alienation; estrangement
exogenous others, 23, 163–64, 237. *See also* Other
exo-phenomenology, 45, 221–24; and being without belonging, 237–38; in Coetzee, 236–37; and curiosity, 224; and estrangement, 222; and the self, 224; as speculative mode, 223–24
extractivism, 29–30

Fabian, Johannes, 245n49
Faerie Queene, The (Edmund Spenser), 175
false life (Ruskin), 83–84, 96. *See also* true life
Fanon, Frantz, 57, 73
Farías, Victor, 248n7
farm: and home(l)y metaphysics, 149; in Rhodesia, 150–52; in South Africa, 147–50, 153–54; in southern African history, 146–47; in U.S. history, 146. *See also* farmworld

farmer: in Heidegger, 143, 159; and natural right, 148, 156; as *natuurmens*, 155; versus husbandman, 144
farm novel, 44, 152–53; Coetzee's commentary on, 154, 155, 156; in Germany (*Bauernroman*), 153; and the land question, 171; Schreiner's contribution to, 172–73
farmwife (*boervrou*), 159
farmworld, 144; in Afrikaans-language literature, 155–59; in Afrikaner history, 149–50; in anglophone South African literature, 154–55; in Lessing, 208; as microcosm of the nation, 145, 158; in Schreiner, 179; in Van den Heever, 159; and world-without-us visions, 166–68
felicitous space (Bachelard), 80–81, 84, 93
Fell, Joseph, 61
Fishburne, Katherine, 202, 213–14
flat ontology, 113, 233
Fletcher, Angus, 174–75
focalization, 90, 252n47; of animals, 190–91
Foltz, Bruce, 65–66, 70, 75
Forster, E. M., 81
Foucault, Michel, 28, 136
fourfold (Heidegger), 64
Franklin, Benjamin, 111–12
Freud, Sigmund, 9, 35–36, 38–39, 133; "The Uncanny," 35
Friedrich, Caspar David: *Wanderer above the Sea of Fog*, 40, 42
frontier, 25, 27
Frontier Wars, 149
Fryer, Judith, 80–81

Garnett, Edward, 109
Garrard, Greg, 53–54
Genesis (Bible), 144, 161
genius loci. See spirit of place
geological time, 184, 189–90
Ge-stell. See enframing
Geviert. See fourfold
Ghost Dance (Hopi), 128
ghosts: in Lawrence, 112, 113, 118, 129, 136; as sign of Indigenous erasure, 127–28; in surface reading, 136

Gibson, James, 131–32, 133, 223, 256n4
Giliomee, Hermann, 149
globalization: and coloniality, 27–28, 32
Gogol, Nikolai, 86
Gómez-Barris, Macarena, 29–32
Gorak, Irene, 176–77
Gordimer, Nadine, 153
Grass Is Singing, The (Lessing), 44, 165, 194–95, 198–99, 200, 202–3, 217–18, 224; allegory and realism in, 214–15; and eco-apocalypse, 208–10; as a farm novel, 207–8; and female hysteria, 209, 215; generic indeterminacy of, 205–10; as Manichean allegory, 212–13; as meta-Manichean allegory, 213–14; and Nature, 214; and social realism, 206; and trope of unconsolidated self, 211–12; world-without-us visions in, 216
Gray, Stephen, 166
Great Karoo. *See* Karoo
Great Trek (*Groot Trek*), 149, 158
Gregg, Lyndall, 184
Gutierrez, Donald, 121–22

Haarhoff, T. J., 156
Haraway, Donna, 241
Hardenberg, Georg Philipp Friedrich Freiherr von. *See* Novalis
Harman, Graham, 62, 129, 224
Harrison, Robert Pogue, 52
Harvey, David, 50
hauntology (Derrida), 129
Haus des Seins. See house of being
Head, Bessie, 154–55
Hegel, G. W. F., 182
Heidegger, Martin, 7, 42: *Being and Time*, 56, 59, 62–64, 69, 143; *Black Notebooks*, 8; and Black studies, 51, 57–58; "Building Dwelling Thinking," 65, 68, 69–70; and the coloniality of Nature, 75–76; concept of landscape, 51, 69–72, 143; and concern (*Fürsorge*), 143; *Contributions to Philosophy*, 59; critique of metaphysics, 8, 14, 49; critiques of, 8, 53–54, 76; and decolonial theory, 51, 54–58; and the environmental humanities, 52–54; and the farmer, 143, 159; *The Fundamental Concepts of Metaphysics*, 7, 250n57; and the German homeland myth, 55–56; influence on eco-phenomenology, 13; and meaningfulness (*Bedeutsamkeit*), 143; and Nazism, 8, 50–51, 53, 73, 248n7; "The Origin of the Work of Art," 60–62, 70; as philosopher of home, 49; "The Question Concerning Technology," 69; thrownness-into-being (*Geworfenheit-ins-Dasein*), 256n4; topology of being, 64–72, 74. *See also* Dasein; dwelling; Earth; World
Heller-Roazen, Daniel, 223
Hemingway, Ernest, 88
Hobbes, Thomas, 244n38
Hölderlin, Friedrich, 53
Holy Ghost (Lawrence), 112, 113, 118, 129
home: as actively made, 26; comfort of, 3; and ecological discourse, 244n33; as ideological space, 16; ontology of, 80–81, 84; topology of, 9, 11, 59–64
homecoming, 17: in Cather, 92, 93, 252n53; philosophy as a form of, 7–8; and the senses, 14; settler narrative of, 10, 20–27. *See also* ecological homecoming narrative
homeland (*Heimat*), 8, 34, 238; German myth of, 55–56, 73–74; Nazi politics of, 8, 50, 153
home(l)y metaphysics, 43, 74, 99, 106; and the coloniality of Nature, 101; in Schreiner, 181; in Van den Heever, 156
home(l)y rhetoric, 1–2, 5: avoidance of, 12; in Cather, 80–81; and dwelling, 50–51; in environmental philosophy, 10–12; and Nazism, 73; in phenomenology, 13–15; and the southern African farm, 148–49
homemaking: as domestic activity, 1–2; as narrative trope, 80; as settler activity, 4, 79, 97, 124, 245n74; versus inhabitation, 239
homesickness (*Heimweh*): in Lawrence, 107, 119; of philosophy, 7–10

Hopi, 94, 107, 128
Hopkins, Gerard Manley, 252n33
horror, 36, 128–29
Horton, Susan, 172, 184
hospitality, 222, 243n1, 262n10
house of being (Heidegger), 8, 14, 49
human: cellular composition of, 247n120; dehumanization of, 57; in Enlightenment philosophy, 78; in Lawrence, 110–11; as natural alien, 237–38; reification of, 18–19, 32, 37; as uncanny boundary object, 38. *See also* Man
hurudza, 151. *See also* farmer
husbandry, 144, 158
Husserl, Edmund, 13–14, 15, 65, 222
Huxley, Aldous, 106–7
hybridity, 164

imperialism: in ancient Britain, 78–79; ecological, 3–4, 164; and landscape, 79, 100. *See also* colonialism; coloniality; settler colonialism
inattention, 2, 38–39. *See also* attention
indigeneity, 74, 238
Indigenous peoples: and architecture, 94, 252n50; dehumanization of, 28; erasure of, 20, 23–24, 98–99, 124, 126–28, 163; expropriation of land, 125, 151, 245n74, 252n50; and metaphysics, 101; romantic stereotypes of, 20, 100
individual: and exo-phenomenology, 224; trope of unconsolidated self, 211–12, 231–33; uncanny nature of, 36, 37
Ingold, Tim, 238
inhabitation: and embodiment, 14–15; versus homemaking, 239
In the Heart of the Country (Coetzee), 45, 153, 165; animals worlds in, 227; and apocalypse, 233–34; and eco-apocalypse, 234–36; and ecological double-consciousness, 231; and exo-phenomenology, 236–37; as a farm novel, 225; and flat ontology, 233; and gender, 225–26, 229–30; and imaginative world building, 226–27; and settler positionality, 229–30; and speculative history, 228–29; and speculative realism, 227; and trope of unconsolidated self, 231–33; world-without-us visions in, 234–36
invasion: as structure, 24
Iser, Wolfgang, 137–38
ivy, English (*Hedera helix*), 1–4

Jacobsen, Dan, 176
James, Henry, 86, 252n33
Jameson, Fredric, 135, 138–40, 162, 175
JanMohamed, Abdul R., 162, 213

Kafka, Franz, 134
Kant, Immanuel, 40, 41
Karoo (South Africa), 184, 186–88
Khoisan, 147–48, 149
Kimmerer, Robin Wall, 239–40
Kiowa Ranch, 104, 106–7, 119
Kirby, Vicky, 10–11, 219
Koch, Christof, 233
Kohn, Eduardo, 241
Kolodny, Annette, 158
Krech, Shepard, III, 20

Lacan, Jacques, 213
Lacoue-Labarthe, Philippe, 50
Land Apportionment Act (Southern Rhodesia), 151
land question (South Africa), 170, 171
landscape: and alienation, 33–34; and Cézanne, 116; and Dasein, 71–72, 75–76; and the domestic interior, 89, 91–92, 97; and dwelling, 51, 69–72; in ecological science, 247n106; as exclusionary mechanism, 75–76; as feminine, 158–59; as home, 34–35, 78–79, 96; and home(l)y metaphysics, 43, 74, 98–99; and identity, 34, 41; and imperialism, 79, 100; and narrative description, 80; and narrative form, 93; as psychic possession, 99–101; as representational mode, 33–34; and settler colonialism, 25; and the sublime, 25; as worlding figure, 69, 70–72, 99
Landschaft. *See* landscape

Lasdun, James, 119
Lawrence, D. H., 43, 199, 224, 238; on allotropy, 109; and alterity, 124, 127, 130, 132–33; and animism, 114–15; and apocalypse, 126, 128, 130; on being without belonging, 43, 106, 123–34, 217, 224, 238; on character, 109, 111; concept of home, 107, 123–24, 238; critique of Crèvecoeur, 145; critique of Franklin, 111–12; and the ecological uncanny, 107, 117–18; and estrangement, 106–7; *Fantasia of the Unconscious*, 113, 116–18; "Indians and an Englishman," 126; and Indigenous erasure, 124, 126–28; "Introduction to Pictures," 115–16; and the materiality of being, 112–13; and modernity, 105, 126; "Morality and the Novel," 110; *Mornings in Mexico*, 127–28; and the New Mexican landscape, 104–7; "New Mexico," 104–5, 130; "The Novel," 110; "The Novel and the Feelings," 110–11, 112; and the occult, 103–4, 105–6, 107–19, 129–30; *The Rainbow*, 109; "Reflections on the Death of a Porcupine," 112–13; and representation, 127–28, 128–30, 131, 133, 134; *Sea and Sardinia*, 102–3, 106; and the spirit of place, 102, 103, 104, 107–8, 112–14, 122–24, 131; *Studies in Classic American Literature*, 107–8, 111–12, 126–27; theory of the novel, 110–11; and unbelonging, 106–7, 123–24; and vitality, 104–5, 113–14; weird anima, 107, 115–16; *Women in Love*, 109. See also *St. Mawr*
Lawrence, Frieda, 102, 103
Leach, Neil, 50
Leonard, John, 200
Leroux, Etienne, 153, 160, 258n44
Lessing, Doris, 44, 165, 217–18, 234; *African Laughter*, 193; and apocalypse, 202–3; *Briefing for a Descent into Hell*, 199; *Canopus in Argos*, 200; childhood of, 192–93; *Children of Violence* (series), 199, 261n26; and Communism, 197–98; and ecological realism, 200–205, 214–16; and the ecological uncanny, 204–5; *Memoir of a Survivor*, 199; and ontology, 203; and psychological fiction, 198–99; and reality's uncanny nature, 195–96; and the Rhodesian landscape, 193–95; and Schreiner, 193, 196; and science fiction, 200; "The Small Personal Voice," 198, 201–2; and social realism, 197–200; and Sufism, 199–200; "A Sunrise on the Veld," 203–5; *Under My Skin*, 195–96, 197–98, 199. See also *Grass Is Singing, The*
Lessing, Gottfried, 197, 198
Letoit, André, 160
Levinas, Emmanuel, 8–9, 222–23, 241, 243n1, 262nn10–11, 263n49
life force. *See* vitality
lineal consciousness (Coetzee), 157
Llewelyn, John, 223
Lobengula (Ndebele king), 257n18
Locke, John, 148
logocentrism, 16–17
logos, 9, 16
Longfellow, Henry Wadsworth, 252n53
Lovecraft, H. P., 129
Loveless, Natalie, 224
Lugones, Maria, 29
Luhan, Mabel Dodge, 102–4, 124–25
Luhan, Tony, 125
Lukács, Georg, 133–34, 140, 200, 256n7
Lummis, Charles, 103
Lynch, Tom, 245n42

Maize Control acts (Southern Rhodesia), 151
Maldonado-Torres, Nelson, 55–57, 73
Malherbe, Daniel François, 152
Malpas, Jeff, 66, 68, 70–71
Man: invention of, 32, 57. *See also* human
Manichean aesthetics (JanMohamed), 162–63, 213–14, 219
Marcus, Sharon, 136
Marcuse, Herbert, 50
Marder, Michael, 9–10, 54, 229
Marx, Karl, 148

INDEX

Maxwell, James Clark, 134
Mbembe, Achille, 249n35
McClintock, Anne, 172, 174
Meillassoux, Quentin, 263n31
Melville, Herman, 99–100
Merchant, Carolyn, 27
Merleau-Ponty, Maurice, 14–15, 244n17
Merrild, Knud, 104, 106
mesh (Morton), 136–37, 225, 237
metaphysics: and allegory, 174–75; Heidegger's critique of, 8, 14; of home, 74; homesickness of, 7–8; of Indigenous peoples, 101; of landscape, 98–99; of presence, 8, 17
Mignolo, Walter, 28, 32–33, 55, 56, 248n11
Mill, John Stewart, 169
modernity/coloniality. *See* coloniality
Monsman, Gerald, 176
mood (*Stimmung*), 134–35, 138, 139–40
more-than-human world (Abram), 19, 39, 220, 262n5
Morton, Timothy, 131, 136–37, 191, 196, 225, 237, 260n54
Mostert, Noël, 184
Moten, Fred, 249n43
mother of the nation (*volksmoeder*), 158
Mount Etna (Italy), 103–4, 106
multinaturalist perspectivism (Viveiros de Castro), 241
Mutapa Empire, 150. *See also* Shona

Nagel, Thomas, 223
Nähe. See nearness
Nancy, Jean-Luc, 33–34
national parks (U.S.), 25
National Socialism. *See* Nazism
native question (South Africa), 171
naturalism, 133, 256n7; and realism, 134
natural right, 148, 156
Nature: and allegory, 179; colonial domestication of, 145; as destructive force, 156; as feminized, 158–59; idealization of, 25–27, 115, 118–19, 181, 182–83; as ideology, 24–26; as mirror, 117; as ontological fiction, 32–33; overdetermination of, 214; as transcendental signified, 17. *See also* coloniality of Nature
Navajo, 100–101
Nazism: *Blut und Boden* policy, 8, 153; and Heidegger, 8, 50–51, 53, 73, 248n7
Ndebele, 23, 150, 257n18
Ndebele, Njabulo, 134, 140
nearness (Heidegger), 71–72
necropolitics (Mbembe), 249n35
negative capability, 129
Neolithic Revolution, 144
New Man, 169
Newmark, Julianna, 104
New Woman, 169
nomos, 9. *See also* homecoming
nonownership, 10; in Coetzee, 167–68. *See also* property ownership; unbelonging
Novalis, 7–8

object-oriented ontology, 263n21
Odysseus, 17, 21
Odyssey (Homer), 21
oikos, 9, 16, 244n27
Oloff, Kerstin, 38
ordinary (N. Ndebele), 134
Ortega, Mariana, 241
Other: ecological, 19; and exo-phenomenology, 222; Levinas's concept of, 9, 222–23. *See also* alterity; exogenous others
Ouzman, Sven, 147
ownership. *See* property ownership

Parreñas, Juno Salazar, 241
Pater, Walter, 86, 252n33
Peirce, Charles S., 262n7
perception: attunement of, 18–19; and ecological realism, 131–32, 133; neuroscience of, 233; selectivity of, 132
phenomenology: decolonial critique of, 29–32; and home(l)y rhetoric, 13–15; of reading, 137–38; and settler perception, 5, 22–23. *See also* eco-phenomenology
Pilgrim's Progress (John Bunyan), 175
pioneer mythos, 80, 98
plaasroman. See farm novel

place: Heidegger's lexicon for, 66–72. *See also* spirit of place
Plato, 98
pluriversality, 241
postcolonialism, 162; and allegory, 163, 214; and environmental history, 144
postmodernism, 134, 256n11
preservationism, 25, 94
Professor's House, The (Cather), 43, 79, 90–99, 100, 101, 124; Cliff City, 93–95; and coloniality, 97–101; and exile, 90, 93; homecoming in, 92, 93, 252n53; and home(l)y metaphysics, 97–99; and narrative form, 88–90, 97; and nostalgia, 92; orphanhood in, 95; preservationism in, 94; property ownership in, 94–96; "Tom Outland's Story," 43, 88, 92–96, 97–99; and vitality, 90–92, 95
property ownership, 4–5, 94–96, 148, 155. *See also* nonownership; unbelonging
Proust, Marcel, 134, 252n33

quantum realism, 132
quantum theory, 11, 219
Quijano, Aníbal, 27–28, 54, 248n11

race: as global paradigm, 32; as uncanny boundary object, 37–38. *See also* anti-Blackness; coloniality
Ranger, Terence, 22–23, 151, 152
Ray, Sarah Jacquette, 19, 253n63
real allegory (Coubert), 177
realism: and affect, 138–40; and allegory, 162, 177; in Lessing, 197–205; in literature, 133–35, 138–39, 170, 177–79; and naturalism, 134; in painting, 115; quantum, 132; in Schreiner, 177–79, 186; speculative, 132. *See also* ecological realism
reality effect, 115, 134
Reed, Edward, 132
repression, 35–36, 119, 135; and alienation, 36; and the ecological homecoming narrative, 37; in Schreiner, 184–85
restor(y)ation (Kimmerer), 240
rewilding, 20, 22
Rhodes, Cecil, 150, 152, 176

Rhodesia. *See* Southern Rhodesia
Rifkin, Mark, 4–5
Rilke, Rainer Maria, 53
Rölvaag, O. E., 146
Romanticism, 77–78; in Schreiner, 196
Rousseau, Jean-Jacques, 25
Ruskin, John, 82–84, 85, 86, 251n15, 251n18. *See also* false life; true life

Sachen. *See* things
Sapir, Edward, 126
Schopenhauer, Arthur, 129
Schreiner, Olive, 40–42, 44, 152, 166, 193, 195, 196, 202, 205, 214, 234; biography of, 170–71; *Dreams*, 176; and ecological realism, 186–91; and the ecological uncanny, 186–91; and the landscape of South Africa, 180–81, 184; and Romanticism, 196; theory of realism, 177–79; *Thoughts on South Africa*, 40–42, 180–81, 186–88, 260n38; *Trooper Peter Halket of Mashonaland*, 176. *See also Story of an African Farm, The*
selfhood. *See* individual
semiotics, 262n7
Serres, Michel, 244n13
setting: and character, 87; and vitality, 81
settler colonial allegory, 164–65. *See also* colonial allegory
settler colonial imagination, 23–24; and exo-phenomenology, 221–22
settler colonialism: as form of ecological imperialism, 4; genocidal logic of, 23–24; and the homecoming narrative, 20–27; as narrative structure, 21–22; as ongoing project, 4; population economy of, 163–64; as spatial practice, 238; as world-making activity, 5; and Zionism, 263n49
settler colonial narrative, 21–22; in Cather, 98; and the ecological homecoming narrative, 22, 24. *See also* colonial narrative
settler common sense (Rifkin), 4–5
settler moves to innocence (Tuck and Yang), 23, 240

Shah, Idries, 199
Sheldrake, Merlin, 219
Shelley, Percy Bysshe, 196
Shona, 23, 150–52, 257n18
Simpson, James, 252n50
Sixth Extinction, 13
Skrbina, David, 114–15
Skywoman, 239–40
slow violence, 3
Smith, Pauline, 154, 173
Snyder, Carey, 126
social Darwinism (Spencer), 114
socioecological impasse, 217–18; in Coetzee, 231
Somer (Van den Heever), 44, 155–59, 163, 171; and the Afrikaner farmer, 156; and the farmworld, 159; and home(l)y metaphysics, 156; and landscape, 157–59; and Nature's destructiveness, 156; and pastoralism, 158–59; shrinking farm trope in, 156–57
Southern Rhodesia, 22–23, 150–52
spacetimemattering (Barad), 11, 219
spectacular (N. Ndebele), 134
spectral rhetoric: in Lawrence, 103–4, 105–6, 107–19, 129–30
speculative realism, 132, 223–24; and Coetzee, 227
Spencer, Herbert, 114, 169, 181
Spillman, Deborah, 179
spirit of place, 79: in Lawrence, 43, 102, 103, 104, 107–8, 112–14, 122–24, 131
Stein, Gertrude, 125
Steinhagen, Carol, 99–101
St. Mawr (Lawrence), 43, 107, 119–24, 137; concept of home in, 123–24; and the ecological uncanny, 121; and egotism, 120; and estrangement, 120–21; and existential exhaustion, 122; homesickness in, 119; and the idealization of Nature, 120; and the spirit of place, 122–24; and unbelonging, 123–24; and vitality, 122–23
Story of an African Farm, The (Schreiner), 44, 152, 165, 193, 196, 202, 214, 215–16, 218; and allegory, 176–77, 179, 186; animal worlds in, 188–89, 190–91; and dream topography, 166; and existential crisis, 182, 188; as farm novel, 171–73; generic hybridity of, 169–70; and geological time, 185, 189–90; and home(l)y metaphysics, 181, 188; and the idealization of Nature, 181, 182–83; and landscape representation, 172–73; in literary history, 170; postcolonial critiques of, 172, 180, 189; and realism, 177–79, 186; and repression, 184–85; world-without-us visions in, 189–91
strange strangers (Morton), 196, 237
Sturm und Drang. *See* Romanticism
sublime, 25, 41–42; versus the ecological uncanny, 39–40
surface reading, 136
surroundings. *See Umgebung*
survivance (Vizenor), 98
symptomatic reading, 135, 175

Table Bay (South Africa), 149
Tansi, Sony Labou, 57
Taos, New Mexico, 102, 103–4, 106, 124–26
Taos Pueblo, 107, 125–26, 130
teleology, 84, 139
terra nullius. *See* empty land, myth of
Thacker, Eugene, 128–29
things (*Sachen, Dinge*), 15, 65, 75
Thomashow, Mitchell, 18, 19, 21, 245n42
Toadvine, Ted, 15, 19, 53
Tolstoy, Leo, 87
topology: of being, 64–72, 74; of home, 9, 11, 59–64
Transvaal (South Africa), 150
true life (Ruskin), 83–84. *See also* false life
Tsing, Anna Lowenhaupt, 241
Tuck, Eve, 23, 33, 240
Turner, J. M. W., 85

Uexküll, Jakob von, 12, 137, 263n23
Umgebung, 65, 70
Umwelt, 138, 227; in Heidegger, 65–66, 70; in Uexküll, 12, 137, 263n27. *See also* animal worlds

unbelonging: in Coetzee, 166–68; in Lawrence, 106–7, 123–24. *See also* nonownership; property ownership
uncanny (*das Unheimliche*), 9, 35; and affect, 139–40; and curiosity, 224; and the ecological homecoming narrative, 36–42; and literature, 39, 133. *See also* ecological uncanny; Freud, Sigmund
unconcealment. *See* World
Unheimliche, das. See uncanny
U.S. Topographical Corps of Engineers, 252n50

Vambe, Lawrence, 150, 257n18
Van Bruggen, Jochem, 152
Van den Heever, Christiaan Maurits, 44, 152, 155–56, 160, 171, 173. See also *Somer*
Van Heerden, Etienne, 153, 160
Van Niekerk, Marlene, 153
Van Riebeeck, Jan, 149
Veracini, Lorenzo, 22, 23, 97–98, 163–64
Victorian aesthetics, 78, 170, 179
violence: ecological, 2; foundational, 74, 247n102; slow, 3
vitality: in Cather, 78, 81–86, 89–90, 90–92, 95; in Lawrence, 104–5, 113–14, 122–23. *See also élan vital*
Viveiros de Castro, Eduardo, 241
Vizenor, Gerald, 98
Voortrekkers, 149. *See also* Afrikaners
Vorhandenheit (Heidegger), 62

Warren, Calvin, 58, 73, 74–75
weird anima (Lawrence), 43, 107, 115–16, 118, 130, 137
Weisman, Alan, 165, 190
Welt. See World
Weltlosigkeit. See worldlessness
Wharton, Edith, 97
Whitehead, Alfred North, 64

Whitman, Walt, 109, 219, 231
Whyte, Kyle, 27
Wigley, Mark, 60–61
wilderness: ideological foundations of, 25–26; as primordial home, 26. *See also* Nature
Williams, Raymond, 17
withdrawal, 129; in object-oriented ontology, 263n21. *See also* Earth
Wohnen. See dwelling
Wolfe, Cary, 12, 244n13
Wolfe, Patrick, 23–24, 126
Wood, James, 81
Woolf, Virginia, 191, 260n54
World (Heidegger), 59–64, 66–67, 138, 143, 250n57; capitalization of, 249n51; and Dasein, 62–63; and Earth, 61, 63–64, 71; as home, 72; and landscape, 69, 70–72
worldlessness (Heidegger), 65, 75
world-without-us visions, 165–68; in Coetzee, 234–36; in Lessing, 216; in Schreiner, 189–91. *See also* dream topography
wretched of the earth (Fanon), 57, 59, 73, 75–76
Wylie, John, 34–35, 238
Wynter, Sylvia, 28, 32, 57, 241

xerophilia (Lynch), 245n42
Xhosa, 149

Yang, K. Wayne, 23, 33, 240

Zambezi River, 150
Zimbabwe. *See* Southern Rhodesia
Zimmerman, Michael, 52–53, 76
Zionism, 241, 263n49
Zola, Émile, 139–40
Zuhandenheit (Heidegger), 62
Zuurveld (South Africa), 149

Recent books in the series
UNDER THE SIGN OF NATURE: EXPLORATIONS IN ECOCRITICISM

Samuel Amago • *Basura: Cultures of Waste in Contemporary Spain*

Marco Caracciolo • *Narrating the Mesh: Form and Story in the Anthropocene*

Tom Nurmi • *Magnificent Decay: Melville and Ecology*

Elizabeth Callaway • *Eden's Endemics: Narratives of Biodiversity on Earth and Beyond*

Alicia Carroll • *New Woman Ecologies: From Arts and Crafts to the Great War and Beyond*

Emily McGiffin • *Of Land, Bones, and Money: Toward a South African Ecopoetics*

Elizabeth Hope Chang • *Novel Cultivations: Plants in British Literature of the Global Nineteenth Century*

Christopher Abram • *Evergreen Ash: Ecology and Catastrophe in Old Norse Myth and Literature*

Serenella Iovino, Enrico Cesaretti, and Elena Past, editors • *Italy and the Environmental Humanities: Landscapes, Natures, Ecologies*

Julia E. Daniel • *Building Natures: Modern American Poetry, Landscape Architecture, and City Planning*

Lynn Keller • *Recomposing Ecopoetics: North American Poetry of the Self-Conscious Anthropocene*

Michael P. Branch and Clinton Mohs, editors • *"The Best Read Naturalist": Nature Writings of Ralph Waldo Emerson*

Jesse Oak Taylor • *The Sky of Our Manufacture: The London Fog in British Fiction from Dickens to Woolf*

Eric Gidal • *Ossianic Unconformities: Bardic Poetry in the Industrial Age*

Adam Trexler • *Anthropocene Fictions: The Novel in a Time of Climate Change*

Kate Rigby • *Dancing with Disaster: Environmental Histories, Narratives, and Ethics for Perilous Times*

Byron Caminero-Santangelo • *Different Shades of Green: African Literature, Environmental Justice, and Political Ecology*

Jennifer K. Ladino • *Reclaiming Nostalgia: Longing for Nature in American Literature*

Dan Brayton • *Shakespeare's Ocean: An Ecocritical Exploration*

Scott Hess • *William Wordsworth and the Ecology of Authorship: The Roots of Environmentalism in Nineteenth-Century Culture*

Axel Goodbody and Kate Rigby, editors • *Ecocritical Theory: New European Approaches*

Deborah Bird Rose • *Wild Dog Dreaming: Love and Extinction*

Paula Willoquet-Maricondi, editor • *Framing the World: Explorations in Ecocriticism and Film*

Bonnie Roos and Alex Hunt, editors • *Postcolonial Green: Environmental Politics and World Narratives*

Rinda West • *Out of the Shadow: Ecopsychology, Story, and Encounters with the Land*

Mary Ellen Bellanca • *Daybooks of Discovery: Nature Diaries in Britain, 1770–1870*

John Elder • *Pilgrimage to Vallombrosa: From Vermont to Italy in the Footsteps of George Perkins Marsh*

Alan Williamson • *Westerness: A Meditation*

Kate Rigby • *Topographies of the Sacred: The Poetics of Place in European Romanticism*

Mark Allister, editor • *Eco-Man: New Perspectives on Masculinity and Nature*

Heike Schaefer • *Mary Austin's Regionalism: Reflections on Gender, Genre, and Geography*

Scott Herring • *Lines on the Land: Writers, Art, and the National Parks*

Glen A. Love • *Practical Ecocriticism: Literature, Biology, and the Environment*

Ian Marshall • *Peak Experiences: Walking Meditations on Literature, Nature, and Need*

Robert Bernard Hass • *Going by Contraries: Robert Frost's Conflict with Science*

Michael A. Bryson • *Visions of the Land: Science, Literature, and the American Environment from the Era of Exploration to the Age of Ecology*

Ralph H. Lutts • *The Nature Fakers: Wildlife, Science, and Sentiment*

Mark Allister • *Refiguring the Map of Sorrow: Nature Writing and Autobiography*

Stephen Adams • *The Best and Worst Country in the World: Perspectives on the Early Virginia Landscape*

Karla Armbruster and Kathleen R. Wallace, editors • *Beyond Nature Writing: Expanding the Boundaries of Ecocriticism*

www.ingramcontent.com/pod-product-compliance
Lightning Source LLC
Chambersburg PA
CBHW021651230426
43668CB00008B/592